SOPWITH AIRCRAFT
1912–1920

Salamanders (row in foreground) and Snipes beyond: Sopwith caption, 'S.705—Sopwith Aviation Co Ltd Kingston. Ham Works. Dec. 1918'.

SOPWITH AIRCRAFT

1912–1920

H. F. KING

M.B.E.

PUTNAM

LONDON

BY THE SAME AUTHOR

Armament of British Aircraft 1909–1939

Aeromarine Origins

The World's Bombers

The World's Fighters

The World's Strike Aircraft

ISBN 0 370 30050 5

Printed in Great Britain for
Putnam & Company Ltd
9 Bow Street, London WC2E 7AL
by Thomson Litho Ltd, East Kilbride,
Scotland
set in Monophoto Times
by Willmer Brothers Ltd, Birkenhead, Merseyside
First Published 1981

CONTENTS

Foreword

Although this volume continues *en suite* with its Putnam predecessors in recording the achievements—and in particular the products—of the great aircraft-builders it has certain points of difference which call for explanation.

In the first place (the bulk notwithstanding, for the number of types and variants described may prove quite startling) the book deals with a company which had a corporate existence of some eight years only (1912—1920)—though its founder, T. O. M. Sopwith, had, as an individual, an enduring association with the Hawker and Hawker Siddeley concerns, as chronicled in the companion volume *Hawker Aircraft since 1920*, by Francis K. Mason.

The second point to be made is that not only was Tom Sopwith's personal involvement an enduring one in the annals of the British aircraft and aerospace industries, but it had its beginnings in times and circumstances of great technical significance (hence the chapter on 'Other Men's Aeroplanes') as well as in the call of high adventure.

Thirdly, one must stress the contributions to Sopwith's success that were made by men in his employ—especially by Harry Hawker and Fred Sigrist, but also men of lesser fame in the persons of Herbert Smith, Harry Kauper, W. G. Carter, R. J. Ashfield and Jack Pollard.

Yet even this readily extensible list does not include such 'pre-Sopwith' men as Howard Wright—men, that is, who in the very dawn of flying helped to light the beckoning and glittering path that stretched ahead for young Tom Sopwith, beyond the glaring years of 1914—18, through flickering fortune on to undiminished fame.

The present book being one mainly about aeroplanes, however, it would be unedifying to play upon personal claims and controversies, though for such a course there would be the classic precedent of Anthony Fokker, Sopwiths' arch-opponent (and here the apostrophe is not misplaced) in the world's first great 'war in the air'.

The fourth point that the present writer aims to stress is T. O. M. Sopwith's extremely close involvements with aero-engines of astonishing variety. To engines, then, this volume will give close attention—closer than hitherto accorded them even in the scrupulously detailed Putnam Aeronautical Series—the author being strengthened in this resolve by the editor's ungrudging support.

Should any further justification for this departure be required it must be one that has become increasingly apparent during the recent years of unprecedented research into the history of British aircraft—a period initiated, one believes, by J. M. Bruce's articles in *Flight* during the 1950s (respecting which one served both as midwife and as fosterparent) and which marked the beginning of the now long-established Putnam series just mentioned.

Be these matters as they may: on one aspect of aircraft history we can be clear—that some historians have been too enthusiastically (and thus quite understandably) borne aloft and away by the trappings and the heraldry of war—markings, numbers and the like—thus neglecting unduly the weightier panoply. In that regard the powerplants concerned must be considered eminent among the neglected. In the particular case of Sopwith, for instance, it has sometimes been construed that this great name was one that soared to the heights under the urge of Le Rhône, Clerget and Bentley rotaries—only to be hammered into extinction by ghastly happenings in A.B.C. radials. And yet, as we shall see, Tom Sopwith had a close alliance with A.B.C. from 1912 until his company's end—an end that arrived after orders had come 'pouring in' (as it was declared) for Sopwith-built A.B.C. motor-cycles!

One fact more to press home this particular point about engines: Whatever feelings one may entertain towards these objects, let it not be overlooked that whereas a Sopwith Camel airframe cost less than £900 a 130 hp Clerget rotary cost rather more.

By way of emphasis and explanation the foregoing paragraphs should suffice, leaving the successive chapters to relate their ever-changing tale.

London, 1980 H. F. K.

Origins and Company History

This first chapter (like the very last one) has a curious start, for this present start was made well after much of the main content had been put on paper. The occasion for this state of affairs was a happy one indeed, for not only did it arise from a meeting between Sir Thomas Sopwith and Mike Ramsden, who succeeded the present writer as editor of *Flight International* (and who printed his wholly delightful account in the issue of 6 January, 1979—to which the reader of this Sopwith book is quite impartially commended), but it served to verify and stress much of what one had already written, both in the preceding Foreword and under this present chapter-heading. Especially, it confirmed one's view of the men, other than Sir Thomas himself, who served in establishing the fast-and-furious firm of Sopwith (1912—1920), which made so many types of aeroplanes in so short a time, and the *significance* of which aeroplanes—just as much as their technical characteristics—one aims to set down in this book.

So, without any permission or personal motives whatsoever—though encouraged by Sir Thomas' own pronouncement years ago that 'I remember buying those first copies of *Flight* (priced, I think, at a penny) in 1909, the year before I flew for the first time'—one gives this excerpt. Thus Sir Thomas:

'I was pleased with our First World War aeroplanes. Mind you, the Camel wasn't everybody's cup of tea. It was rather tricky to fly. The Pup that preceded the Camel was a delightful little aeroplane. It had no tricks at all, no vices. You couldn't say that for the Camel. I don't know whether I had expected it to be more successful than the other 80 or 90 or 100 aircraft we produced. It's always very largely a matter of timing. You must remember that up to the beginning of the First World War all our flying was done entirely by feel . . . All our aeroplanes were built entirely by eye. They weren't stressed at all. The Camel was the product of a more scientific approach. We were just learning how to stress at the time of the Camel.'

Then, of men rather than machines: 'It is very hard to say what personal design decisions were mine on the design of the Camel. We worked as a very close team with Freddie Sigrist, Bill Eyre, Harry Hawker and a fellow called Bennett . . . Herbert Smith was an important contributor to Sopwith aircraft design. He gradually worked up from the drawing office. I wouldn't like to say that he designed the Camel or the Pup. I give a lot of that to Freddie Sigrist and to Hawker. Smith really put their ideas together onto a piece of paper. Sigrist was an engineer in the schooner I owned with Bill Eyre. Hawker came a little later when we were running the school at Brooklands . . . Harry Hawker I got on with very well indeed. He was a beautiful pilot. He used his head. Incidentally, I taught him to fly. He was a very competent designer. He didn't work in a drawing office but he brought all his ideas in his head . . . That's why I hesitate to give too much credit to Herbert Smith in those days because an awful lot of it was Hawker himself. Smith edited the ideas, put them on paper. Sigrist was all practical. No theory. I don't think he ever went to school . . . '

Now, while considering 'men rather than machines', it is natural to enquire 'And what about Tom Sopwith himself?' Would-be biographers have always found their path bewildering and daunting; but this present book being one of machines rather than men its purpose will be served by recording that Thomas Octave Murdoch Sopwith (born 18 January, 1888)—unlike his colleague Sigrist—certainly did go to school: at Seafield Engineering College, Fife, and at Cottesmore, Rutland.

The surname Sopwith, though uncommon, is not unknown in Britain's aircraft industry (present associations aside) and nor, for that matter, in the Church—a Thomas Sopwith having been Archdeacon of Canterbury during the Second World War; but though the 'Tom' Sopwith now receiving our attention had a father who was a civil engineer, even this engineer can hardly be equated with the Thomas Sopwith, mining engineer and specialist in measurement, who died in 1879. This last-named Thomas Sopwith nevertheless has passing claim on our attention because he knew the Duke of Argyll, who wrote *The Reign of Law,* deeming therein balloons to be mere toys, and was the first President of the (now Royal) Aeronautical Society. The Thomas Sopwith who died in 1879 has, in fact, a more direct link with our present subject, for he said he had scarcely ever met a man, however poor and simple, or great or intelligent, who would not stop to look at the working of a piece of mechanism that presented some novelty. 'A balloon', he said, is a piece of mechanism, a rude mode of flight, still a novelty because incomplete for practical purposes; therefore everybody runs to see a balloon, and some in their excitement would tear the thing up if they could get at it as if they wanted to see what was inside it.'

Balloons and the young Tom Sopwith of our present story were associated as follows by J. H. Ledeboer (well-known in RAeS affairs) in 1913: 'His first appearance on the stage of aerial navigation dates back some six years when, in partnership with Philip Paddon, he owned and piloted a balloon of somewhat disreputable appearance known as the 'Padsop'. But aviation drew him on with irresistible force . . . ' How those last words echo those just quoted—'as if they wanted to see what was inside it.'

One man who helped to put the insides in balloons and other contrivances was the Hon C. S. Rolls, and though Charles Rolls was among Tom Sopwith's ballooning friends (in 1906, according to Sopwith's recollection) one seemingly curious fact that this book elicits is that Rolls-Royce engines were virtually absent from the run of Sopwith aeroplanes. This is not significant, however, if only for the reasons that Rolls was killed in July 1910 and that Tom Sopwith was destined to build aeroplanes mostly to demand rather than personal choice—and, in any case, loved boats as he loved most things that moved (aeroplanes included). It was, in fact—following a tedious crossing of the English Channel in a yacht with his friend Bill Eyre in September 1910—that the 'aviation bug' (see beginning of next chapter) was transmitted to Sopwith. With Fred Sigrist aboard, looking after the Thornycroft paraffin engine (for anything up to £2.14s a week!) the yacht reached Dover, where Pups were later stationed. Only a few miles distant from the Sopwith/Eyre seaborne outfit was the American John B. Moisant, with his Blériot monoplane. The 'monoplane' one stresses, for the reason that late in 1919 it was declared—and the substance of the declaration has often been repeated, though this book largely refutes it: 'Sopwith was the first to see where the monoplane failed and where the biplane must score in speed-range . . . '

The facts concerning the development of Sopwith aircraft from the 'bug' time forth will probably remain forever hidden (this fact itself demanding unflinching recognition); and while one applauds the research that has distinguished much relevant writing on the theme one is by no means disposed to over-value such minutiae as names and dates on drawings, while the contribution of men so great in stature that their names seldom appeared on a drawing at all passes almost ignored. The decades-old question 'Who *really* designed the Sopwith aircraft?' will continue in contention, Sir Thomas' reassurances, already recorded, notwithstanding. One can only re-express the hope that the smoke of disputation will not obscure the monolithic figures round whom so many lesser ones evolved and revolved.

Here (as will have been gathered) one has especially in mind the quietly great Fred Sigrist, to whom so much of the success of both the Sopwith and the Hawker lines of aircraft is unquestionably due. It is no overstatement, one submits by way of example, that the Sigrist (or Sigrist/Camm) so-called 'Meccano' structures were the true foundation for the incomparable record, in the 1920s and '30s, of the Hart and Fury families; for except in construction and engine there was precious little to choose between these aeroplanes and (for instance) the Curtiss Falcons and Hawks which antedated them. Camm was, in fact, sometimes too conservative for the liking of senior colleagues. Though a Martinsyde man by training Camm had long been close to Sopwith affairs, and concerning aircraft design he once advised the present writer 'I think I only met Mr Herbert Smith twice'—and clearly he was no champion of Smith, though the latter's Sopwith contribution after 1914 was increasingly apparent.

Thus one makes no apology for attempting to fuse the reputations of two great British aircraft companies by the repetition throughout this book of one perpetuated name—Hawker, though if continuity of work and example be the criterion the name must still be that of Sigrist—'Fred, who had a large corner in all our hearts', as Sir Thomas Sopwith remembered him in 1957. Yet still one wonders if the extent of that 'corner' is truly appreciated. So let us bring Sigrist up from that yacht's engine-room and allow C. G. Grey to have his say, as he did in 1945. Thus:

'His many friends will be glad to hear that Mr. F. Sigrist, who, along with Mr. T. O. M. Sopwith, Mr. Bill Ayres [clearly 'Eyre'] and Mr. Harry Hawker (since dead) founded the Hawker Mfg. Co. Ltd. [*sic*] which became Hawker Aircraft Ltd., is going strong in the Bahamas, whither he was ordered by the doctors some years ago . . . He was Chief Engineer for all Mr. Sopwith's speed-boats and fast cars before there was any flying, and he built the first Sopwith aeroplanes at Brooklands himself in 1911. Thereafter, he was primarily responsible for the production of the thousands of Sopwith fighters, which did so much towards winning the 1914—18 War, just as their successors, the Hawker fighters, have done in 1939—45.

'Mr. Sigrist was a very sick man when he went to the Bahamas, but the climate did him so much good that in the early part of this War he was able to go to the West Coast of the U.S.A. as an official of the British Air Commission, to do there a job similar to that which Sir Richard Fairey was doing on the East Coast. His job, as a Commission official put it, was to be "a trouble-shooter, more particularly charged with keeping the heavy-weight contractors in a good mood." Apart from that, Mr. Sigrist's wide experience as an engineer was of great value to our contractors in the U.S.A. In addition, Mr. Sigrist has done much for the welfare of the R.A.F.'

All this (the present writer would observe) was a long time after the yacht's engine-room, the 'Sigrist Bus', as mentioned early under '1½ Strutter (land versions)'—and the precursors and successors thereof. So now, from men and machines to events—and Tom Sopwith the *airman*, an attribute sometimes forgotten.

Towards the end of 1910 one of several significant contests was in prospect, not for T. O. M. Sopwith alone, on his Howard Wright biplane (see chapter 'Other

Men's Aeroplanes'), but likewise Claude Grahame-White (Bristol Boxkite); C. H. Greswell (Farman); Robert Lorraine (Bristol Boxkite); Lieut H. E. Watkins (in Capt Maitland's Howard Wright biplane); and Frank McClean (Short S.28). Geoffrey de Havilland was a prospective entrant but couldn't raise the necessary cash.

The particular contest concerned was for the Baron de Forest prize of £4,000 for the British pilot who could fly the greatest straightline distance nonstop into Europe from the United Kingdom. Thus attracted and challenged, and with 20 gal of petrol and a thermos of meat-extract aboard, Sopwith took-off from Eastchurch on the morning of 18 December, 1910, and headed for Canterbury and Dover. The wireless station at this last-named port announced his speed as 50 mph at a height of 1,200 ft—this speed by courtesy of a following wind of 20 mph. Though visibility was poor, and his compass stuck, Sopwith, who was really aiming for Chalons, fairly near Paris, was able to report for *The Times* on his trip as follows:

'I had nothing to direct myself by, so I just kept flying on. Towns and villages passed below; I knew none of their names. Then the wind began to get more gusty. The machine swayed and lurched and the arm with which I moved the controlling lever began to ache.

'Just as I was flying over a village at about 800 feet a very ugly gust caught my machine on one side and tilted it partly over. To my consternation the aeroplane refused to regain its normal position even when I exerted the full pressure of the small balancing planes [ailerons to present readers]. It was a moment I am not likely to forget. Changing hands quickly on my steering lever I leaned over as far as I could from my driving seat [*sic*] so as to be able to throw the weight of my body against the rising wing. Just when I thought I should slide hopelessly down through the air the machine slowly righted itself, but another gust assailed me and I had to look out for a landing place, although I had 11 gallons of petrol left and the engine had not misfired once. A field near the village presented itself. I planed down and sat still quite exhausted.'

Sopwith was, in fact, in Belgium, some nine miles from the French frontier. The day was a Sunday, and the local telegraph office was closed; but, his exhaustion evidently overcome, this excited young Briton trudged to a railway station and got a cable through to his sister May, by then waiting anxiously at the Lord Warden Hotel, Dover. Fred Sigrist was asked to recover the machine, because, if any contestant did better, then Sopwith proposed to have another try. But no contestant did do better; for a gale had put paid to many of their chances, and though Grahame-White (who is named again in this book, in the chapter on the Gnu) did manage to start from Dover, he was forced back, crashed, and fainted. Hero and victor Tom Sopwith—for so he was, having flown 177 miles (285 km) in 3 hr 40 min—himself drove the redoubtable 'G–W' to London.

Tom Sopwith had now taken on some of the best of Britain's airmen—and had beaten them, just as he was later to beat—and yet sustain in business by sub-contracts—some of his rivals in industry. Even so, Fred Sigrist had been his advisor (as well as sister May); and it was, in fact, Sigrist who advised him to start from Eastchurch, because, being an engine expert, Sigrist reckoned that any powerplant failure would most likely occur in the few minutes after take-off, at full power, and that the Isle of Sheppey was a better place to descend than the sea beyond.

That the memory of Sopwith's early achievements in airmanship (which, after all, helped to set him up in the aircraft business) lived on in Britain's changing aerospace industry, even after the Second World War, was attested by a man of his own stature, in the person of Sir Geoffrey de Havilland, when the Hawker Siddeley

Group began to loom as a mighty edifice in the 1950s and '60s. With his own splendid company especially in mind, Sir Geoffrey nevertheless cast a sporting sidelight on the broader and heavier canvases of aeronautical/financial history by declaring: 'From the handful of people who started at Stag Lane forty years ago has grown a great enterprise which has lately amalgamated with another great enterprise—The Hawker Siddeley Group. But this is not a coming together of strangers. In 1910 Tom Sopwith, having learnt to fly, was preparing his Howard Wright biplane to try for the Baron de Forest prize. I was also going to try for it if I could raise a few hundred pounds for expenses, but had still not succeeded when Sopwith set out and won the prize. Now, as Sir Thomas Sopwith, he is Chairman of the Hawker Siddeley Group . . . '

By 1911, then, Tom Sopwith was shaping-up very nicely in the air world—still a dominantly sporting one. Sporting in more than one sense, as this little item of April 1911 discloses: 'The authorities at Sandown Park Racecourse have sent a polite request to aviators at Brooklands asking them not to fly over Sandown Park on race days. In spite of the enthusiastic reception accorded to Mr. Sopwith, when he flew over Hurst Park, the request is obviously quite reasonable.'

As for the instructional and looming military aspects, could there be any finer testimonal to the Sopwith School of Flying, of Brooklands, Weybridge, Surrey (established early 1912), than the following note by Harald Penrose—which, one has always felt, should have carried the heading 'The Loom of Boom'. Thus: 'Behind the trinity heading the R.F.C. began to loom another figure—a very tall, dark and sombre man, with parchment-coloured face, bristling moustache, and visionary eyes. He was Major Hugh Trenchard, whose active military days had seemed numbered through ill-health after service in Africa. Informed by the War Office that if he learned to fly before he was 40 he would be sent to C.F.S. as assistant to Godfrey Paine, he lost no time before seeking the advice of his acquaintance Tommy Sopwith, who offered to teach him at his new Brooklands flying school on the Burgess-Wright obtained at the conclusion of his very successful and financially rewarding American tour. With Copeland [*sic*—Copland?] Perry, a Sopwith trainee, as his instructor, Trenchard managed to qualify for his *brevet* on July 31st, shortly before his birthday.' (According to Sir Thomas Sopwith's more recent recollections, 'Boom' knocked on the door of his cottage just outside the Brooklands track and asked 'You Sopwith? . . . Can you teach me to fly in ten days?').

In any case, remembering the post-war service of the Sopwith Snipe (as later outlined) and matters intervening, how fitting does it seem to add that, introducing the Air Force Estimates for 1920—21 on 11 March, 1920, Maj H. C. Tryon, the new Under-Secretary of State for Air, remarked how the R.A.F. post-war rebuilding process had been going on under 'that distinguished and able officer, Air Marshal Sir Hugh Trenchard' and that, despite economies, the R.A.F. had been engaged 'against the Bolsheviks in north and south Russia, the Afghans, the Pathans on the Indian frontier, and the Mullah' [in Somaliland]; adding 'and in Mesopotamia the Civil Commissioner has referred appreciatively to co-operation of the R.A.F. in maintaining order and communications, making maps, and even controlling revenue.'

The progress of the Sopwith aeroplanes was chronicled by the 1913 British aviation journals with evident enthusiasm, though this enthusiasm was not invariably matched by commensurate precision or explicitness. The present author is nevertheless moved to reproduce verbatim (with only 'Sopwith' having his own italics) the following excerpt from *Flight*'s commentatorial record 'From the British

Flying Grounds', in the issue of 7 June, 1913, for it gives a close-up view of workaday civil and military activities and record-breaking at Brooklands—all in a few paragraphs (though well-nigh inevitably, the name of that illustrious and long-suffering armament specialist Clark-Hall is misrendered in the opening line).

'Lieut. R. B. Davies (carrying Lieut. Clarke Hall as a passenger) flew, on Wednesday, over from Eastchurch on the *Sopwith* tractor biplane, which was delivered to the Navy some time ago, and after a short stay resumed his journey to Bradfield, Berks.

'Thursday, Lieut. Knight, of the 3rd Royal Munster Fusiliers, went through his *brevet* tests in excellent style on the Vickers biplane. Lieut. Duncan passed the first part of his *brevet* tests on the Bristol biplane. Capt. F. S. Wilson, of the Royal Marines, passed his *brevet* tests in excellent style on a Bristol biplane, after only a week's instruction.

'Mr. Hawker made the first test of another *Sopwith* tractor biplane on Saturday, which started right away, and proved to be as quick a climber as its predecessor. Lieut. Spencer Grey [another misrendering, for the first name was Spenser, and will re-appear quite early in this book] afterwards made some good tests on the same machine. Advantage was taken of the fine exhibition flights of the new Martin-Handasyde monoplane by a representative of a kinematograph company, who was able to secure some excellent films. Mr. Gordon Bell made several flights with passengers, amongst whom were the four winners of the ballots for the free passenger flights. The event of the afternoon was Mr. H. Hawker's attempt on the British Altitude Record on the new *Sopwith* tractor biplane fitted with an 80-h.p. Gnome engine. The wind having dropped and the sky cleared, the weather conditions were perfect . . . The machine used was the one which climbed to 7,500 ft. in 15 min. recently at Hendon . . . The machine climbed steadily upwards [*sic*] for about 45 min., at the end of which time it had reached an altitude of 11,450 ft., beating the previous British test by 950 ft., and as the pilot was experiencing some little difficulty with the mixture, he decided to come down, and shutting his engine off he made a beautiful 9 minutes' glide to terra firma again . . . Mr. *Sopwith* is certainly to be congratulated on having such a first-class pilot as Mr. Hawker to demonstrate the wonderful capabilities of the new *Sopwith* tractor biplanes . . . '

How vividly this tiny episodic record epitomises the skills and daring of Sopwith/Hawker pilots, beginning with Sopwith himself, with the incomparable 'Mr Hawker' assuming ever-growing prominence, aided by Howard Pixton and Victor Mahl, to be followed by Raynham, Bulman, Sayer, Lucas, Wade, Duke, Bedford and the rest.

Yet, as we have already had the word of his chief 'Mr Sopwith', (Sir Thomas after 1953) Harry George Hawker was more than just a 'first-class pilot', as a 1913 commentator has just appraised him—he became, in his way, a Sopwith designer also. This is not as surprising as it might appear, for while he was very young (he was born at Moorabin, Australia, in 1889 and was killed in 1921) he was working for a company of bicycle and motor-car agents. Before coming to England in 1911 he had his own workshop, and having joined Mr Sopwith in June 1912 (and Mr Sopwith, as we have been assured, 'getting along with him very well') contributed to the fame of Sopwith aircraft as just exemplified and as further instanced.

Thus much, thus far, for men, aircraft and aerial skills; but what of the works that were established at Kingston-on-Thames, Surrey, by the newly registered Sopwith Aviation Company (founded 1912)? Happily we have this early-1913 report:

'The new Sopwith works are in the building which at one time was the Kingston Skating-Rink, an establishment which will always be connected by certain inhabitants of Brooklands with exceedingly amusing episodes which occurred there. Though they would probably be unacquainted with the fact, the building has a floor surface of over 30,000 sq. ft., and this should be sufficient accommodation for the manufacture of quite a considerable number of machines. Installation has been made, of course, of hand-saws, rotary saws, thicknessers, planers, and other necessary plant, while the office is of a most palatial description. The lorry employed is a 40-h.p. Daimler shod with pneumatic tyres, and is capable of a speed which is distinctly unusual for vehicles of this description.

'A point with which one cannot fail to be struck is the keenness and real personal interest shown by the workmen, who are under the able supervision of Mr. Sigrist. He, of course, has been with Mr. Sopwith since the early days, and has accompanied him constantly on his successful flying trips in the United States. He has, in consequence, a greater knowledge of the up-keep, and, incidentally, of the repair, of various types of machines than is possessed by the average works manager, while the quality of the work turned out under his care is really beyond all criticism.'

How curious (it strikes one) was this quoted writer's employment of the phrase 'since the early days'—especially so as it was almost duplicated in quite a different context ten years later, in a *Flight* report headed 'Aircraft Activity at Kingston—The H. G. Hawker Engineering Co. Busy', and embodying the following:

'Mr. Sigrist was never easy to please in the matter of workmanship. We remember him in the old days ruthlessly scrapping anything which was not "just so" . . . the work now being turned out in the old Sopwith shops is of the very highest quality.'

This, as remarked, after the corporate name 'Hawker' had superseded that of 'Sopwith'; though I feel bound to introduce at this point a matter which has puzzled me since I joined *Flight* in 1931, and which as far as I know has never been explained. The matter on my mind is this: For the preparation of *Flight*'s three-view drawings the aircraft companies concerned would supply original blueprints, which we usually re-arranged, inked over, cut up, blanked off, whitened out or otherwise 'processed' to suit our particular purpose. So it came about that I saw, at some time in the early 1930s, the 'originals' (as we called them) associated with the 1929 Olympia Aero Show, where the Hawker Hornet (later called Fury) was one of the stars. The 'processed' maker's drawing I remember seeing bore a legend which so struck me that I noted it down, and which, for the first time, I copy out now (for it was never reproduced). This is exactly how the maker's title read on that drawing: 'Sopwith-Hawker Hornet R.R. F11.S. Scale $\frac{1}{2}$in = 1 ft.'

Why—I wondered, and still do—the 'Sopwith' prefix. In the early 1920s such a rendering (say for the Duiker) could well be comprehended. But hardly in 1929.

Much wise recording has been done, and comment made, on the voluntary liquidation of the Sopwith Aviation & Engineering Co. Ltd. in the slump of 1920 (September saw the end of the 'Sopwith Company', as its founder, Sir Thomas, has called it)—and for a reasoned run-down Harald Penrose's Putnam book *British Aviation—The Adventuring Years* sets out facts, and even figures. Yet sadder still for present readership, perhaps, would have been an announcement made late in 1919, when the Interallied Aircraft Corporation, of 185 Madison Avenue, New York City, advertised war-surplus British aeroplanes in these heartbreaking, if breathtaking, terms: 'For undiluted pleasure a red blooded man will always pick an Avro or Sopwith "Camel". Their reliable, up-to-date, easily accessible rotary motors make flying simple—stunting comes as second nature—and repairs and overhauls are like play.'

Hopefully, this present book—only now in its first chapter—offers Camels and their Sopwith kin in a less glaring, if no less illuminating, light.

The first aeroplane to be owned—and flown—by T. O. M. Sopwith was this 1910 Howard Wright Monoplane. (Sopwith caption: 'S.355—Howard Wright Monoplane 40 hp E.N.V. Mr. Sopwith pilot.'). This monoplane was still flying in 1912, with a special A.B.C. engine.

Other Men's Aeroplanes

The aeroplane which first transmitted the 'aviation bug' (as the recipient himself once described the affliction) to Tom Sopwith was a Blériot monoplane belonging to the American John B. Moisant, one of whose distinctions was that of having flown his mechanic, and also his kitten *Mademoiselle Paree,* from Paris to London in under three weeks. The circumstances of this bug-transmission to Sopwith have already been related, and the occurrence was an indication of just how international the sport and business of flying was becoming; so much so, in fact, that during the following year (1911) Sopwith himself was showing-off his own Blériot—in America!

Yet this Blériot of Sopwith's was not the first monoplane he had owned; indeed, his very earliest heavier-than-air craft (apart from his skimming boats) was a monoplane of British design and construction. This machine was a product of Howard T. Wright, an Englishman who had assisted the American-born Hiram Maxim in various experiments and enterprises—notably respecting what Maxim called his 'show apparatus', or 'Captive Flying Machines'. Jointly with his brother Warwick Wright, Howard T. Wright had started a coachbuilding and aircraft business in 1907, under a railway arch at Battersea.

In the autumn of 1910 Sopwith bought a Howard Wright Monoplane (a development of the same designer's 'Avis' series of 1909/10) on which he taught himself to taxi at Brooklands, flew—on 22 October—for something like 300 yards, stalled by reason of inexperience, and crashed. This Sopwith-owned-and-flown Howard Wright Monoplane had a 40 hp E.N.V. engine (these initials signifying that this British powerplant had its cylinders arranged '*en V*') and a four-wheel landing gear, with a skid between each pair of wheels. Above the fuselage were the petrol tank and a kingpost; below it was the radiator. The 'tailskid' was immense, and, as the photograph on the previous page proclaims, had only a remote association with the tail.

Early in November 1910 Sopwith was continuing to put in time on this aircraft (now repaired) with vastly greater success, until the engine burst a cylinder-head; but, even so, he soon had the monoplane airborne once again, and was clearly making progress as a serious-minded aviator.

One later reference to the Howard Wright Monoplane will be made in the context of an A.B.C. engine; but Sopwith's next aeroplane was a biplane—another product of Howard T. Wright—and it was on this sturdy machine, in which he incorporated some of his own modifications, that he really made his name as an airman. As the matter was put in a report of a lecture by Sir Thomas during 1960 (in the lecture theatre of the Institution of Mechanical Engineers—though his audience was dominantly of the RAeS persuasion): 'Crashing the Avis [*sic*] and buying the biplane was expensive, so he decided to try and get money back by going for the Baron de Forest £4,000 prize for the longest flight from England to the Continent.'

The biplane concerned was built in 1910 and was a typical Farman-type pusher 'box-kite' of its period, with two interconnected elevators—one forward, one aft—the latter on the boom-borne monoplane tail. Four ailerons gave lateral control, and the engine was a more powerful (60 hp) E.N.V., of the F series, instead of the original 50 hp Gnome. On 21 November, 1910 (according to Sopwith's own testimony), he spent the morning 'rolling', or taxying, this biplane, and in the afternoon made a few circuits. These led to his qualification for the Aero Club's Aviator's Certificate No.31 on the same day; and on this day also he took up his first passenger—identified by him merely as 'some trusting person'.

To mix our metaphors with the bug that had entered Sopwith's bloodstream, he now had the bit between his teeth; he was flat-out for flying, and could even contemplate beating the great Samuel Franklin Cody at his own game. 'I seized every opportunity to get into the air', he once recalled, 'and by the time I had ten hours' flying behind me I began to feel that I was a really experienced pilot. Col. S. F. Cody had just set up British distance and duration records of 94½ miles in 2 hours 24 minutes, and I thought that something should be done about it and made all preparations (which were not many!). On the first attempt I was fortunate enough to cover a distance of 107 miles in 3 hours and 12 minutes.'

For the sake of historical precision, the 'preparations' (in the way of a good breakfast and extra clothes) appear to have been largely due not to Fred Sigrist or any other of Tom's male counsellors and helpers, but to his sister May, who was not merely proud of her brother, but solicitous for his welfare, and useful in such matters as lap-counting and timekeeping.

This is how *Flight* recorded the feat at the time (issue of 3 December, 1910): 'To Mr. Sopwith, the aviator, and to Messrs. Howard Wright, the builders, we have to extend our hearty congratulations on having put up on Saturday last a new all-British distance record of 107 miles, and at the same time established a British duration record of 3 hrs. 12 mins. for any type of machine, British or foreign, flown in this country.

'Mr. Sopwith has also by the same flight achieved the best performance to date for the British Empire Michelin Cup. The Howard Wright machine on which these records were made is a biplane fitted with a 60-h.p. E.N.V. engine and Spiral tube radiator. It has a Farman type wheel-base [*sic*], monoplane tail and elevator with a

T. O. M. Sopwith in the Howard Wright biplane wherein he began to feel that he was 'a really experienced pilot.'

central rudder above and below the tail plane. Mr. Sopwith first flew a Howard Wright monoplane—this was only some few weeks ago—and we drew attention in a previous issue to the rapid progress he made. He has only had delivery of the biplane a few days, which speaks well for the ease of control of this make of machine.'

Later in its career Sopwith's Howard Wright 1910 biplane was modified to have upper wing extensions, with extra bracing-wires; but the most interesting alterations were those associated with the flight that gained for Sopwith the Baron de Forest prize—a performance that greatly enhanced not only the pilot's personal reputation but the prestige of the nation which later types of aeroplane, then bearing Sopwith's own name, were to defend and even symbolize. Chief among the alterations mentioned were increased petrol tankage, and a windshield (more explicitly a foot-scuttle) which, had it been given three-quarters of a chance, would have grown into a nacelle.

The radiators for the E.N.V. engine were mounted fore-and-aft, one on each side between the wings; and as this flight for the de Forest prize was to be an all-British affair a special word must be said for the powerplant. The '*en V*' connotation already mentioned was no mere whim; for, contrary to the supposition that all the best aero-engines of those times were French (and, indeed, the E.N.V. concern operated a factory in the Paris suburbs) this engine had very close British associations, though it was used successfully by several eminent French pilots.

Thus, when describing a new (100 hp) model early in 1914 *Flight* saw fit to remind its readers: 'The E.N.V. Motor is by no means new to the aeronautical world, for as far back as 1908—09 Mr. Moore-Brabazon had one fitted upon his machine, but the E.N.V. Motor Co., of Willesden, N.W., has now been formed to design and manufacture an entirely new engine . . . ' (The 60 hp E.N.V. presented long ago to the Science Museum was catalogued as having been made in 1910 by the E.N.V. Motor Syndicate Ltd, of London).

True, S. F. Cody once declared publicly: 'I have had a little experience of foreign engines—the E.N.V., for instance. I had an E.N.V. engine with which I failed to fly in Manchester. I tried to get the makers to put it right but they did not . . . we entered into a law suit. I sent the engine on to them and they kept it for four months. They did get it right themselves after breaking a crank shaft and one or two cylinders . . . I then took up the Green.'

Sopwith, too, 'took up the Green', as we shall see; but respecting the sometimes obscure 'origins' of aero-engines generally it could hardly be improper here to observe that the Rolls-Royce range from Eagle to Merlin had its 'origins' in a German Mercedes racing car stored, during 1914, in a Shaftesbury Avenue showroom, and that the 'origins' of the Rolls-Royce Bristol Siddeley Pegasus vectored-thrust turbofan are traceable to a 1956 submission by the Frenchman Marcel Wibault.

A general account of Sopwith's 'de Forest prize' adventure having already been rendered in the opening chapter it remains here to note that, in the development and application of aircraft of those times, second only to the powerplant were the instruments installed; and thus it came to be recorded in those times: 'Mr. Sopwith had fitted a compass to his machine, but as this persisted in sticking at N.W., in whichever direction the machine was steered, Mr. Sopwith backed his own judgment in preference and steered by the sun.'

To continue our perusal of the various aeroplanes that helped Tom Sopwith, in one way or another, to 'originate' his own (and never forgetting the men whose help was ready to his hand nor the powerplants that made their efforts possible) we now transfer attention to the most international of all his early aeronautical ventures—that is, to his American tour of 1911, made within a few months of his having flown the Howard Wright biplane to Windsor for the King's inspection. By the time of this Royal occasion—1 February, 1911—the biplane had acquired not only the wing-extensions already mentioned, but also a new fuel system, of Fred Sigrist's creation.

Although the Howard Wright did not actually accompany its owner to the United States, it was sent on after him; and having been assembled, and thereafter somewhat disarticulated by gale-damage, took part in competitions and displays. Sad to relate further, some time after a package that had been intended by young Sopwith for the deck of the White Star liner *Olympic* had missed its mark and fallen into the water, the Howard Wright itself came in for a similar ducking.

Small wonder that its owner-pilot—himself with salt as well as the 'aviation bug' in his blood—would soon be giving his attention to the building of seaplanes and flying-boats.

The Sopwith-owned aeroplane which had been shipped to America to coincide with Tom's arrival was neither British nor a biplane, but a 70 hp Gnome-engined Blériot, which had been bought in France. This particular monoplane was quickly wrecked by a downdraught; so a cable went off to France for another specimen. Thereafter, both in the USA and Britain, Tom Sopwith in a Blériot became an attractive combination, and as late as 8 June, 1912, it was, in fact, to win Britain's

Aerial Derby at an average speed of 58.5 mph. 'If you want speed the monoplane has it', Sopwith had said in America; and notwithstanding the monumental fact that his company's greatest technical success before the 1914 war (the winning of the Schneider Trophy Contest) was achieved with a biplane of superlative quality—the Tabloid—the dainty Swallow and Scooter monoplanes of later years need follow as no surprise.

Sopwith returned to England in October 1911; and although during his absence he had bought a Wright biplane (not of the British Howard T. or Warwick variety but an authentic American Wilbur/Orville type, licence-built by the Burgess Company, and called in consequence a Burgess-Wright) our next concern is—refreshingly—with a British monoplane. Of the Burgess-Wright there will be more to say later.

Having regard to the neighbourly Sopwith/Martinsyde relationships that were later strengthened—especially in the contexts of the postwar Atlantic flight attempts and in the parallel search for peacetime succour in the building of motor-cycles—particular interest attaches to that beauteous creation the Martin-Handasyde monoplane which was first flown by its new owner Tom Sopwith in November 1911. This machine was a development of the two-seat Dragonfly—otherwise known as Type B.4 or 4B—which, with a 50 hp Gnome engine installed, had been shown at Olympia during the previous March, the price-tag reading £1,100. Strength was reconciled with beauty—for example, in the landing gear, which had 'bungee' shock-absorbers, vertically disposed on a central pylon, attached to the ash-and-plywood fuselage, though even this substantial chassis sustained damage at Sopwith's hands within a few days of the initial flight in his new acquisition.

George Handasyde had a special interest in the French Antoinette engines, and seems to have persuaded from the particular specimen fitted in Sopwith's machine a trifle more than the 65 hp nominally attributed to it. He himself was among Tom's passengers in this splendidly designed, constructed and engined monoplane (others included two of Sopwith's sisters); and, although the airframe, as well as the engine, proclaimed Antoinette influence, it nevertheless incorporated, for the first time, the characteristic 'Martinsyde tail' (long-chord fin-and-rudder assembly, curved at front and bottom, pointed at rear end).

Here it must be remarked that, in regard to the Antoinette engine, Sopwith may well have found himself reminiscing somewhat ruefully during the 1930s, when the Hawker company was experimenting with those snag-plagued steam-cooled Rolls-Royce engines called special Kestrels and Goshawks—for the eight-cylinder Antoinette he had himself flown behind in 1911 and early 1912 had been itself steam-cooled; and of yet higher significance, its fuel system was of direct-injection type. This latter attribute—utilising a variable-stroke pump—is probably better known in connection with the Antoinette engine than is the steam cooling; so to emphasise that this cooling system was indeed a feature it can be recorded that, during February 1912, Graham Gilmour landed this same machine (on which he was soon to be killed) with frozen condensers—not, be it noted, radiators.

Thus did Tom Sopwith's experience with various airframes, and with different kinds and makes of engine, proceed—and having now alluded to the steam-cooled Rolls-Royce engines of the 1930s, would it be unduly hurtful to add that the direct-injection Daimler-Benz engines with which the Merlin-powered Hawker Hurricanes contended *circa* 1940 bestowed on the opposing Messerschmitts a literally positive advantage?

Certainly, that a clean-lined monoplane with an evaporatively cooled, inline, direct-injection engine of excellent power/weight ratio should have been flying—and flying well—in England during 1911 must be accounted remarkable; nor is it irrelevant to observe that the Frenchman initially and dominantly responsible (Leon Levavasseur) had in large degree made his name with racing motor-boats, as had young Tom Sopwith.

Such French connections and associations will recur throughout this book, though it will emerge that these became very much a two-way affair. Their culmination, it appears, was an especially significant one, if only in a technical —and certainly not an industrial—sense. This was the installation of a *supercharged* Hispano-Suiza engine in a *Sopwith* Dolphin airframe (author's italics in both cases). So, although the Armstrong Whitworth Siskin was the first British fighter to enter service with a 'blown' engine, it was not the first to have an experimental installation of such a powerplant—with the urgently required objective of sustaining power at the higher levels for combat superiority.

From steam-cooling, direct fuel-injection and supercharging we must now pursue our essentially chronological narration; and in a technical context this is conveniently achieved by recalling that, in the present writer's earlier Putnam volume *Armament of British Aircraft 1909-1939* it was shown that the Coventry Ordnance Works Ltd was best-known in the development of military aeronautics

For the Military Trials of 1912 Tom Sopwith acted as test pilot for the Coventry Ordnance Works respecting No. 10 (the *Wombus*) shown in these two views, and No. 11, later depicted. Note the names on the hangars in the lower of this pair of pictures, showing No. 10 in rebuilt form.

The second of the two C.O.W. Military Trials biplanes which had Sopwith associations—No. 11, with Chenu engine and distinctive tail.

for its 37 mm guns. Yet this same company had an even earlier association with the design and construction of airframes—an association which had its beginnings in 1911, when the 'C.O.W.' absorbed the Warwick Wright concern, acquiring also the services of Howard T. Wright and W. O. Manning. The talents of these two men—Manning in particular—were evident in two biplanes that were built (under the Battersea railway-arches) for the Military Trials of August 1912; and in this Sopwith story these aeroplanes have a special place because T.O.M. Sopwith was the test-pilot for them both.

Though entered in the Military Trials as No. 10, the first of the pair was otherwise known as the *Wombus,* the first three letters connoting Manning's involvement. This involvement was, in fact, more than a purely technical one, for it was Manning who occupied one of the side-by-side seats when Sopwith made the first ground tests and short flights; but the technical feature calling for special attention here (and certainly for study on Sopwith's part) was the 14-cylinder 100 hp Gnome rotary engine which—after steam-cooling and direct fuel-injection—afforded him experience with a *geared* powerplant. True, the Gnome itself lacked any built-in gearing; but a 2:1 reduction ratio was afforded by a chain drive. The high thrust line thus resulting allowed a propeller diameter of no less than 11 ft 6 in—a figure appreciably greater than the 10-ft diameter that characterised the Sopwith B.1 bomber of 1917. On one occasion the big chain-driven propeller of 1912 enabled Sopwith to carry not one, but two, passengers—though these had to sit on the bottom wing.

Sopwith's second C.O.W. responsibility was the testing and demonstration of another biplane—No.11—which had tandem seating and a Chenu water-cooled engine. There were frequent troubles with this aeroplane, as with its predecessor, and during August 1912 Sopwith set out for America on a second visit, to race not aeroplanes but motor-boats. He returned triumphantly in September.

At this point we may reconsider the American-built Burgess-Wright biplane already briefly mentioned, for in 1912 this was quite extensively rebuilt by Sopwith—to such a degree, indeed, that the present writer was at one time led to contemplate a separate study of the machine. Such treatment was, in fact, quite understandably accorded it by Mr Peter Lewis in his Putnam book *British Aircraft 1809-1914*, under the heading 'Sopwith-Wright Biplane'.

While in the USA during 1911 Sopwith bought a Burgess-Wright biplane which he rebuilt extensively in 1912, and which did service both as a 'school machine' and record-breaker. The close-up picture here shows the offset installation of the A.B.C. engine, while the flying study was said to show the machine at 'extreme angle'.

Fred Sigrist, it seems, was largely, if not primarily, responsible for the reconstruction, which was undertaken in the interests of 'school', or instructional, work, in which Sopwith became quite heavily involved at Brooklands during 1912. (To the credit of the American biplane, in its more-or-less original form, it must be recorded that among its passengers had been a Capt F. H. Sykes, later an eminent figure in the development of British military and civil aeronautics, and better known perhaps to certain readers as Sir Frederick Sykes).

One especially notable modification made to the American aeroplane was to give it side-by-side seating, in a sizable nacelle, with the pilot to starboard. In side elevation the nacelle drooped like a Concorde's nose—though permanently. Of no less interest was the fitting (after a 35 hp Green) of an A.B.C. engine instead of the original 50 hp Gnome. Together with its petrol tank, this A.B.C. engine was offset to port; it was nominally of 40 hp, though was sometimes credited with 45, and it drove two pusher propellers by means of crossed chains, housed in tubes.

Thus, by virtue of this last arrangement, Sopwith could now add contra-rotating propellers to his repertoire of exotic powerplant installations.

On the aeroplane just mentioned (which was described contemporaneously, if somewhat dubiously, as a 'Sopwith British-built biplane' or as a 'Sopwith-Wright') Harry Hawker secured the 1912 British Empire Michelin Cup No.1 (and £500) by staying airborne for 8 hr 23 min. The date of this performance, which constituted a new British record for duration, was 24 October, 1912.

A particularly fine view of the rebuilt Burgess-Wright (sometimes called Sopwith-Wright) on which Harry Hawker stayed airborne for 8 hr 23 min.

In the Sopwith 'zoo'—for many later members had menagerial names—the tractor biplane shown in these two views, with fuselage both naked and draped, was strictly a mongrel, and has, in fact, come to be known as the 'hybrid'. The rudder was distinctive, but its form was not perpetuated in later Sopwith tractors.

That the foregoing was not Sopwith's first association with A.B.C. engines is affirmed by this report, published as early as March 1912: 'The 40–50 h.p. vertical four-cylinder A.B.C. engine, which earlier in the year was put through some severe tests by its makers, has recently been put into one of the earlier Deperdussin monoplanes, and without any tuning up of the machine it flew at the first attempt, Lieutenant Porte, R.N., who piloted the machine, said that he had never flown at such a speed.

'The same engine has now been refitted into Sopwith's Howard Wright monoplane and is provided with a new water-heated White and Poppe carburetter which has been specially tuned up by the makers, with the result that the engine is giving about twenty per cent. more power than ever.'

Thus here we find yet one more seemingly exotic powerplant—an A.B.C.

As was noted a few paragraphs earlier—immediately prior to our just-concluded appraisal of the rebuilt Burgess-Wright biplane—Sopwith left for some motor-boat racing in America during August 1912 and returned in the following month. Early in the previous July he was making the first flights of a yet more drastically revised machine, in the form of a tractor, though using American Wright-type wings ('Wright planes pure and simple', as one account stated—adding, to make things perfectly clear, that they were 'built roughly on the Wright model')!

For this newly created tractor the name 'Sopwith Three-Seater Tractor Biplane' was used at the time of its introduction, though the description 'Hybrid' (with or without initial capital) has now gained currency, and, apart from being descriptive, serves to differentiate it from the much-improved 'Three-Seater Tractor' shown at Olympia in 1913.

These matters being so, it will be well to consider the 1912 'hybrid' as the true precursor of the Sopwith aeroplanes that form the subject-matter of this book, and to regard it not so much as the ending of the present chapter but as the beginning of the next.

'Three-seaters' and Derivatives

The Sopwith-developed 'hybrid' tractor biplane that T. O. M. Sopwith tested in July 1912 had a wing cellule which closely resembled in plan form, section and bracing that of the Burgess-Wright pusher which he had bought in the USA during 1911, but the span was increased from 38 ft 9 in (11.9 m) to 44 ft (13.4 m). This, at least, seems to have been true of the form in which it was first publicised, though by then it had already been repaired after a crash when Gordon Bell and J. Charteris were setting out, on 12 July, 1912, to fly it down to Cowes, where Sopwith was practising in a Saunders-built craft for the motor-boat racing scheduled for his second American trip. As was so often the case, the Brooklands sewage farm had received the fragmented structure after a sideslip. 'Reconstruction', it was reported, proceeded 'rapidly'.

In any case, the engine was a 70 hp Gnome, driving a Chauvière propeller of 9 ft 6 in (2.9 m) diameter, and the fuselage was set between the wings, in the style of the later Bristol Fighter—though with a prominent chordwise 'gap-filler' fairing between it and the bottom wing. Here, it might reasonably be suggested, was C.O.W. influence discernible. At first the fuselage was left uncovered aft of the rear (pilot's) cockpit, though later it was completely fabric-covered. The two passengers sat side by side, well forward, in a separate open cockpit. Silver spruce was largely

Three seats and twin skids (the latter in respect of main landing gear and tail-protection likewise) characterised the finest of the 'new line' of Sopwith tractor biplanes, represented here tail-up and tail-down—in the latter instance with modified rudder and other alterations (especially affecting the windows).

used for the wing structure, and even for the four-wheeled, twin-skid landing gear, which was subsequently exchanged for a twin-wheel type.

A very noticeable feature of this 'hybrid' was the vertically-divided rudder, of which it was observed at the time: 'The rudder constitutes an important variation from Wright practice, being situated above and below the elevator, which can be given a warping angle of 6 in (150 mm) in either direction.'

When Sopwith returned from his American motor-boat races he tested the rebuilt machine, and on 8 October, 1912, flew it to Farnborough. There, with one passenger (in the form that 'the military' would be likely to use it) the biplane climbed to 3,000 ft (915 m) in just under 3 min, though the maximum speed—about 55 mph (88.5 km/h)—was poor enough to evoke the comment from a Brooklands observer: 'The Sopwith tractor biplane made its first flight since its repair, piloted by Sopwith. The machine carried a passenger, and left the ground after a very short run, but it is certainly slow.'

Such slowness (it occurs to the present writer) may be a sensation by this time being experienced by any reader who may have had sufficient of 'antediluvian' Sopwith types and is becoming impatient to get along towards the 'real wartime Sopwiths' (as he may well regard the rightfully dominant subjects of this book). In some degree, at least, such readers may now be given satisfaction; for there existed *two* tractor biplanes of the general form just described. These—Nos.27 and 33, with 80 hp Gnome engine—were flying early in the war from Eastchurch, and are said to have been used for armament practice. In this regard, clearly, they might have made better targets than 'gun machines', by reason of their low performance. On reflection, indeed, they might have made perfect 'sitting ducks' as ground targets—easier to hit than the airborne feathered duck that the Eastchurch armament pioneer Lieut (later Air Marshal) R. H. Clark-Hall brought down to the Swale Marshes—if not to the cooking-pot—from a Short pusher some time before the war.

Further concerning the two early Naval Sopwith tractors, it seems worth recording that a demi-official drawing once existed showing just such a machine, though with top-plane extensions, revised engine installation and other alterations. This drawing may well have been a mere pastiche; but the tell-tale tail certainly obtruded.

With a tractor aeroplane that was only a little faster than one of his motor-boats—and bearing in mind that the contemporary 43,000-ton *Titanic* was good for over 24 knots (26.5 mph = 42.6 km/h)—Sopwith could hardly be content; nor was the just-described crude derivation—for 'conception' or 'innovation' would be terms far too enobling—much to the liking of the tiny but talented team now assembling round him.

So it came about that on the occasion of the 1913 Olympia Aero Show, held in February of that year, the designation 'Sopwith Three-seater', jointly with that of 'Bat Boat', borne by Britain's first successful flying-boat, as later shown, removed the name and fame of 'Sopwith'—the image as it would be termed today—from the sporting and promotional scene to that of original design and construction. It would, in fact, be hardly overstating the matter to declare that the 1913 Three-seater marked the true inauguration of the 'all-Sopwith' range of tractor landplanes and floatplanes, the lineage whereof will be traced through many pages to follow. More explicitly, the 'hybrid' represented an archetype, whereas the design now under scrutiny was a prototype in the accepted sense.

A particularly pleasing aspect of the original Three-seater—bespeaking its excellent performance and carrying capacity using relatively low power.

Of even greater significance in a broader sense, the present writer would go so far as to proclaim the early-1913 Sopwith 'land tractor' (as it was sometimes called) as the true begetter of a line of British transport biplanes built—with notable success—to a characteristic formula: that is, with the passengers in a forward fenestrated compartment, with the pilot behind them, and having no more engine-power than was strictly necessary to perform (jointly with generous wing-area) a rigorous operation with exemplary economy. The line of aeroplanes one has in mind came to full fruition in the well-nigh incomparable de Havilland series of the inter-war years, culminating in the little Fox Moth.

So evident, in any case, were the merits of Sopwith's new tractor that they were acclaimed even in their time—for pioneers and prophets pre-1914 went frequently unhonoured, even though meticulous records were supposedly maintained to chart aeronautical progress. So true was this last assertion that publication (for instance) of Octave Chanute's *Progress in Flying Machines* stood as a beacon on the very course it charted.

Less renowned than such a bible—and less well-known even than *Flight* or *The Aeroplane*—was the English journal *Aeronautics* which, in August 1913, made this declaration: 'If one were asked to give the name of the most successful machine in existence, at least one of British design and construction would be among the claimants for place among those for final selection. How many constructors would undertake to build a machine which, when fitted with an 80-h.p. Gnome, is satisfactorily capable of carrying pilot and two passengers in addition to four hours' fuel? The Sopwith biplane can accomplish this and, at the same time, fly within a wide range of speed. This range, naturally, is not so great as the range obtained when the standard load of three hours' fuel and one passenger is carried, but reasonably large nevertheless.' Having mentioned also the Bat Boat, the journal continued: 'The Sopwith Aviation Co. have undertaken an enormous amount of work, and though this fact might have tended to cause a lack of thoroughness in the firm's productions, no trace of this is visible.'

As exhibited at Olympia in February 1913 the Three-seater differed very noticeably from its hybrid precursor—bearing, in fact, much the same relationship to it as did the Avro 504 of 1913 to the Avro Type D of 1911. Staggered wings were only one of its identifying features. The tail and landing gear were altogether new, and—most striking of all the novelties—three big non-inflammable celluloid panels were let in to each side of the forward fuselage, to improve the downward-and-sideways (and in some degree forward) field of view for the three occupants.

One contemporary assessment tantalisingly invited elucidation. This was to the effect that the new machine was 'similar to the one recently supplied to the Admiralty' (meaning, presumably, the hybrid)—excepting that it had staggered wings and an 80 hp Gnome engine (instead, it was implicit, of a 70 hp unit of the same type). Clearly, though, here was a machine calling for detailed study even in its day, and even more so on this present occasion. Thus may history be honoured, and today's investigations be satisfied, by the following first-hand appraisal:

'Staggered planes are used, and these appear to be of such a camber as would permit the machine to be of an exceedingly fast description. An 80-h.p. Gnome is fitted in front, driving a tractor propeller.' (Present writer's note: Clearly, the observer under quotation wished to impress on his readers that this aeroplane was of *tractor* form, though in order to acquit him from naivety or tautology it must be recalled that certain types of the period had the engine in front, though driving a pusher propeller. The Mersey Military Trials Monoplane of 1912 was a case in point, likewise the Grahame-White Type 6 Military Biplane—exhibited at Olympia in 1913, as was the Sopwith now reviewed). To continue the contemporary appraisal under quotation:

'The main chassis consists of a pair of wheels mounted on what may best be described as a V inclined forward, while a pair of smaller wheels are mounted on another pair of struts in front. The whole strikes one as being amply qualified for resisting any shock, within limits, to which it may be subjected, and the fitting of an additional and non-weight-carrying pair of wheels forward has found favour in certain quarters, on account of its tendency to prevent any possibility of the machine's complete inversion should a somewhat steep landing be made. The inter-plane struts situated at the point where the planes abut upon the fuselage, and the fuselage uprights, are one and the same member—this naturally reducing weight and head resistance to a considerable extent.

'The reduction of these two bugbears, in fact, has obviously been the end and aim of the designers. Their efforts to minimise the former can be seen everywhere and are distinctly worthy of emulation. In almost every instance, not only in fuselage longitudinals and uprights and in the ordinary struts, but even in the thin members which support the fabric in the various members of the tail, has weight been brought down by channelling to what in some cases almost approaches an I section. In other cases, where this method of procedure has been impossible, the spars and struts are built up hollow, the centre being flat and composed of ash, and surrounded on each side by a curved section of spruce. They are bound at intervals with oil-silk ribbon. The lightness of the spars is astonishing, while their rigidity and strength should leave nothing to be desired. It is interesting to note that the ash centres of the main spars are of approximately the same-sized section as that of the wing-spars on the standard 70-h.p. Blériot, though in the latter case they are not strengthened by spruce coverings! This is another instance of the fact that, though the reduction of unnecessary weight has been one of the chief aims of the designers, this has not been accomplished at the expense of the strength of parts. One may at first question the necessity of drilling eighteen to twenty holes in a steel clip which is only five inches by about two or three inches in dimensions, for the weight of a single example cannot be greatly reduced thereby, but one realises eventually that, if this is done to every clip upon the machine, the desired effect is produced to a considerable extent. Some of the clips are really the most remarkable that the writer has yet seen, representing as they do the outcome of most thoughtful design, while the labour and expense that must necessarily have been involved is by no means inconsiderable. The most remarkable of these clips is a socket which holds no fewer than four members, viz, the main-plane spar, the inter-plane strut, the fuselage upright, and one of the undercarriage struts! Even the small clips for attaching the fuselage uprights and longitudinals together are carefully drilled, and are of the pattern introduced by the Hanriot Company, *i.e.*, are actually clipped to the wood by compressing it, no bolts in consequence piercing it.

'A novel and extremely laudable point is the employment of steel tube for the trailing edge and extreme tips of the planes. This, in the latter case, permits the camber to extend right up the whole length of each wing, which fact must obviously give greater efficiency. The struts for the undercarriage are also constructed of steel, though in this case they are of 18-gauge steel plate bent into a tube of marvellous sections.

'One point which will undoubtedly pass unnoticed by the majority of those who examine the tractor-biplane is not really of any great structural importance, though it shows the care which has been devoted to the small details of design, and will be appreciated by those who have to make any adjustments or alterations to that part of the machine referred to. The tail-plane, as mentioned, has its outer framework constructed of steel tube, the latter being attached to the longitudinals by U-bolts. The minor, though all-important feature of this attachment, however, is that the bolt is kept from falling out when the nuts on its top extremities are loosened.'

The general conclusion was that the two new Sopwiths—the Three-seater landplane and the Bat Boat pusher flying-boat—represented 'the last word' in aeroplane construction. Yet conservatism lingered, and lateral control on both machines was by wing-warping, the warping pulleys being ingeniously combined with the rear-spar hinges, and actuated by a warp-wheel on a vertical 'elevator column'. Ailerons came later, on developed versions; and certainly by August 1913 there were at Brooklands 'two Sopwith tractor biplanes—one fitted with warping wings and the other with ailerons.'

'Last word' or not, it was the 1913 Three-seater that not only put Sopwith in business as an aircraft constructor in general, but, in particular, as a supplier to the Services; for more or less coinciding with the Olympia exhibition of February 1913 (the displayed example having already been tested by Sopwith himself, and shown to be capable of over 70 mph) came orders for two Naval specimens of the same general type. The show-machine itself, in fact, was actually the first to be delivered to the Navy, the pilots to whom it was handed over—on 1 March, 1913—being that breezy pair Lieuts Spenser Grey and L'Estrange Malone. To this same aeroplane the Service number 103 was allocated, and Spenser Grey, with Lieut Bigsworth as passenger, was reported as having looked in at Hendon with the 'Naval 80 hp Sopwith Tractor' on 24 May.

During May also Harry Hawker flew the Navy's No.104 to Farnborough, to demonstrate for the War Office such useful attributes as a speed range of 35–75 mph. As *Aeronautics* had already noted, the standard load (for Service use, certainly) was one passenger only, with fuel for three hours; and it was assuredly not as a three-seater that No. 104 was to be found at Eastchurch in January 1915 on 'Zeppelin stand-by'.

Brooklands track—familiar in so many views of Sopwith and Hawker aeroplanes—stretches ribbon-like across this study of a Three-seater.

Viewed close at hand, this particular Three-seater displays the nose-bearing mounting for the Gnome engine, and was used by the Navy, generally, one may suppose, as a two-seater, though Sir Thomas Sopwith once said that the first '3-seater tractor' he supplied to the Navy (possibly meaning the 'hybrid') was used to collect oysters for the officers' mess at Eastchurch.

Nor was the prewar public in ignorance of the (nominally) Three-seater's capabilities; for with Hawker as pilot, an aircraft of the type won the Cross-country Race at Brooklands on Whit-Monday 1913, then climbed to 7,500 ft in 15 min. Whereafter things got better still, for on 31 May the 'great Harry' set a (solo) British height record of 11,450 ft.

Matters then progressed from excellent to better still, for on 16 June Hawker took a single passenger to a breath-taking 12,900 ft, and on that same day (half an hour later, in fact) carried two passengers to 10,600 ft. Even with three men, in addition to his slender self, he managed 8,420 ft—on 27 July, 1913.

Nevertheless the Sopwith Tractor Biplane (potential seating capacity going unacknowledged) was chiefly used not for passenger transport, but as a general-duties two-seater, by both the RNAS and the RFC.

Cannibalisation—as the term was to become current in air-service jargon—was rife, as was modification; and certainly distinctive in this regard was the second example constructed (or reconstructed) for this had a deeper fuselage and windows with rounded corners.

The transformation from three-seater to two-seater must not be passed over without noting a clear intention that a military application was foreseen for the former layout. Thus a contemporary declaration: 'The machine is arranged to seat three, two observers side by side at the centre of gravity, and the pilot some little distance behind them. The bottom plane being staggered back by 1 ft, the observers are able to obtain a good view of all that is happening below them. The pilot can get a good view below him too, for the trailing edges of the planes on either side of him are cut away for that purpose.'

For the shapely Olympia-pattern rudder a more angular form was substituted, reminiscent of the earlier hybrid in that the area was distributed above and below the fuselage. But still there was no fin. For competition work, and Service testing also, a 100 hp Green water-cooled engine was fitted experimentally, and in this instance an Integral propeller, instead of a Levasseur or Chauvière, was tried. The petrol tank (on a typical Three-seater, at least) was under the passengers' seat.

Although passing references have already been made to the landing gear, it must now be stressed that the agglomeration of struts, wheels and skids tended to obscure one particular feature that was eventually incorporated—that is, the so-called 'split axle', which was to become almost a Sopwith trademark on the later and better-known war machines (a design with each wheel mounted on a half-axle, the inner ends whereof were pivoted at the mid-points of the spreader-bar, or bars). There were, in fact several variations in landing gear, notably with two massive tailskids; but the following description appears to have applied early in 1913:

'The landing chassis is of the combined wheel and skid type. The body of the machine is supported from two long hickory skids by six spruce struts. The two pairs of rear struts are assembled to the skid by a welded steel fitting which is also slotted to take the axle of the two landing wheels. The latter are strapped with rubber cords to the skids. Miniature skid-tip wheels are fitted. They are 13 ins. in diameter, and are each supported by a pair of beaten steel fittings. The main skids, the sides of which are hollowed out for lightness, are continued back 3 ft. or so behind the rear chassis struts, in such a manner that there is no necessity to provide a rear tail skid. These continuations of the rear skids have the advantage that they act as most efficient land brakes when it is required to pull the machine up quickly . . .'

Betrayed—or proclaimed—by its windows, though distant in this Farman/B.E./Short/*Astra-Torres* gathering, is a Sopwith tractor biplane of the 'Three-seater' family.

Apart from Nos.103 and 104, already mentioned as going to the Navy, that same Service had No.906 (after it had served as Sopwith's hack and demonstrator) and other recorded numbers for these Sopwith Tractors were 248, 315, 319, 324 and 325—all these for the RFC. That cannibalism and modification was rife in respect of these biplanes has already been observed, and even the number built remains open to question, though in September 1913 it was reported, apparently with good authority, that: 'The Sopwith Co at Kingston continues to test its many machines at Brooklands before their delivery to the two Services. Nine 80-h.p. tractor biplanes, standard type [*sic*] were recently ordered for the Army, while delivery of two was requested by the Admiralty.' The inference was rightly drawn that 'these two sets of orders are only the beginning of a long period of prosperity for Messrs Sopwith'—rightly, that is, if one reckons the 1914–18 war and its immediate aftermath as 'long'.

Before the coming of that war the great event for the Royal Flying Corps (Military Wing) was, of course, the 'Concentration Camp' at Netheravon—the description having no sinister implication, but being analogous with the Naval Review, or mustering, of the same period. Respecting the 'M.W.' camp, I have examined a duplicated typed sheet once possessed by the Camp Commandant himself. Pencilled-in, and detailing one particular 'task for aircraft', could be perceived—after the listing of Blériots, Farmans and B.E.s—the addition '& Sopwiths'.

Two or three years later a typical RFC Order of Battle might have listed 'Sopwiths (& other types)'.

Need more be said?

'Three-seater' (as at early 1913)
(80 hp Gnome)

Span 40 ft (12.2 m); length 29 ft 6 in (9 m); wing area 365 sq ft (33.9 sq m). Empty weight 1,100 lb (500 kg); maximum weight 1,750 lb (790 kg). Maximum speed 70 mph (112 km/h). Climb 500 ft/min (152 m/min); ceiling (with one passenger) 12,900 ft (3,930 m).

Anzani Tractor Seaplane

This otherwise undistinguished aircraft is given one distinction, in nomenclature at least, by prefacing its name by the *marque* of engine installed—thus conforming strictly with historical accuracy, except that it must be admitted that the term 'seaplane' had not, at the time of its emergence (summer 1913) superseded 'hydro' or 'hydro-aeroplane', or even the grossly confusing 'hydroplane'. (This last expression was, in truth, applicable only to high-speed skimming craft, or very fast motor-boats, with which Tom Sopwith was familiar). More than this: by correctly introducing the type of Sopwith aeroplane now under scrutiny by the distinctive name of its engine one feels wholly vindicated in one's continuing emphasis on the

A salty study of an Anzani Tractor Seaplane in motion. Even so distantly viewed, the ten-cylinder radial engine—with a frontal exhaust-collector ring that would not have disgraced a radial installation of twenty years later (this picture was made in 1913)—is prominent. (Original Sopwith print uncaptioned, but numbered 36).

Almost directly comparable with the preceding study of the same aircraft is this revealing close-up—restful, too, though the Anzani engine is running. Note especially the centre section, braced top-wing extensions and underhung portion of rudder. (Original Sopwith print uncaptioned, but numbered 32).

significance of powerplants in its maker's history; for here we meet the first of the fixed, star-shaped air-cooled radials to come into general use—remembering, of course, that, using a three-cylinder fan-form 25 hp Anzani, Louis Blériot had long since flown the English Channel (25 July, 1909). The particular type of engine in the seaplane now studied had ten cylinders, arranged in two rows; and it weighed 363 lb (165 kg).

But first to describe the brand-new Sopwith aeroplane itself—for such it was considered, notwithstanding a resemblance to the 'Three-seater' landplane; and, indeed, as we shall see, there once existed a landplane version (or *con*version) of the basic type now considered. Thus we must proceed with special care, though emboldened by Mr R. J. Ashfield's own description of the particular waterbird now in our sights as the '100 Anzani Tractor Seaplane.'

This actual mention of the engine's nominal output—100 hp—can be directly linked with the fact that the new Sopwith aeroplane was sustained on the water, and hampered in the air, by three weighty and clumsy floats (two main, one tail, as was then usual). Even so, the description 'clumsy' is here applied only in a general, and not in a particular, sense; for with his motor-boat, balloon (and now fast-growing aeroplane) experience Tom Sopwith had a special concern with weight and drag. Thus the float landing gear and associated considerations are rightly our own first concern also—with the ensuing contemporary account meeting the case perfectly:

'Two main floats with spring suspension are fitted, in addition to a single tail-float. The 100-h.p. Anzani drives a propeller of approximately 9 ft. diameter, covered with thin copper to prevent splintering on the waves. The span of the top plane being approximately 56 ft. the floats are widely spaced, 10 ft. 3 ins. apart. There is, in consequence, no necessity for wing-tip floats. The main ones are mounted on inverted V-struts. As in all the other models [*sic:* meaning 'current Sopwith models'] balanced ailerons are fitted . . . '

The account now under quotation went on to make a point which the present writer has already advanced, by noting that 'the subject of floats' was 'an interesting one' which had 'obviously been studied by Messrs Sopwith with the greatest of attention'. These remarks were amplified—or, as will be seen, in some degree skirted—as follows:

'The finished product is of pleasing appearance externally, and, of far greater importance, the work and constructional design leave little to be desired. Of the hydrodynamic and aerodynamic qualities in the design we are not prepared to speak, for, by doing so, we should set ourselves up as authorities on a subject which is as little understood as was aerodynamics generally at the period when the Wright brothers first commenced their experiments . . . '

'Each float is covered with thin Holland blind union [*sic*], which is glued on and varnished, and through which the wood can clearly be seen. Three inspection covers are fitted, the interior edges of the interstices for which are padded in order to render them watertight. The hull [*sic:* obviously meaning the float-structure proper] is built up in two $\frac{1}{8}$-in. thicknesses of cedar, the first skin being diagonally built up with 4-in. strips, while the outer is composed of similar strips running longitudinally. In addition to the outer layer of fabric, another one is placed between the two layers of wood. The interior is coated with black varnish—a suitable combination of gas-tar and naphtha.

'The floats are fitted with laminated steel springs. Four of these springs are attached to each float, the extremities of the front one being rigidly fastened to clips screwed onto a 1 in. by $1\frac{1}{2}$ in. vertical strut within the hull. The rear spring, on the other hand, is free to move, this arrangement being the result of a problem which offers some serious consideration. The solution has been effected in this manner: the apex of the front spring is connected to that of the rear by means of a radius rod. There are, of course, two of these—one on each side of the float. The only result of the flattening-out of the front spring is to slide the rear one backwards, the enormous compression stresses which would otherwise arise on that portion of the float between them being, in consequence, avoided.

'The bottoms of the floats are convex, with a camber of $1\frac{1}{3}$ in. An inspection of the interior shows the interesting manner in which the step is rendered sufficiently strong to withstand the by no means inconsiderable strains to which it is subjected. The bottom consists, in part, of a number of 'ribbons', or minor longitudinals. Those running along the bottom of the rear half of the float are continued past the step until they die off where they meet, and where they are attached to, the ribbons from the bow. These are themselves continued to the upright portion of the step, on which they abut, the consequence being that a triangular girder is formed.'

The foregoing account of the floats etc. is quoted at length in the knowledge that some readers will not only be instructed thereby, but amused into the bargain ('gas-tar', indeed . . .). Yet this same account gives eloquent support for the concern expressed in June 1912 by George Holt Thomas: 'It has always seemed to me that too little attention has been paid to the flying part of the hydro-aeroplane machine, *i.e.,* to the planes of the waterplane. What I mean is this; no matter how good the floats may be, an efficient waterplane can only be evolved by using an efficient aeroplane. The floats should be regarded as a landing chassis and a landing chassis only . . . ' Holt Thomas was speaking in the context of the 'lifting' floats developed by Henri Fabre; so let us now see what the gentleman of the gas-tar etc. had to say concerning 'the planes of the waterplane' as built by Sopwith in 1913. Thus:

'Balanced ailerons now take the place of the warping action on the wings. This is highly to be commended, for it is really strange that the warp on biplanes has been tolerated for as long as it has been. The twisting strain on the spars is by no means pleasing to the eye of the engineer, and it seems that the righting couple produced by the pilot is infinitely more nearly instantaneous with the use of ailerons than is the case with warp. The reduction of speed, if any, is infinitesimal, so it is difficult to see where the point in the use of warping wings on biplanes comes in. Each wing-section [*sic:* presumably meaning wing-panel] can, in consequence of the use of ailerons, be built considerably stronger—not only this, but another addition is employed to increase the strength of the wings. This is in the shape of a number of rectangular struts between the front and rear spars at each point where the interplane struts are attached. They might well be called distance-pieces, for their only use is to relieve of compression strains the various ribs, which, being of curved shape, have not the power of opposing these stresses to as great an extent as does a member subjected to direct compression. The four tips of the main planes, and the outer extremities of each member of the tail planes, consist of circular-section steel tubing. Attachment of the fabric is effected by sewing, the "bag" thus formed being slipped on afterwards.'

Following such an intimate insight into 1913-style aircraft construction and terminology (wherein—whatever the non-attributable authorship—the mind of Fred Sigrist in particular may be perceived) it remains to add the following points: that the attachment of the floats (each of which had five watertight compartments) to the bottom wings, and not to the fuselage, was a point that commended itself to adoption for the torpedo-dropping Type C; that the three-bay wings had a 'clear-view' centre section and extensions to the upper mainplanes (the term 'extensions'

The curious landplane (No. 58?) referred to at the end of the present chapter—with the Anzani engine recognisable, though covered. The Naval gentlemen add tone to the scene, but fail to obscure a main wheel and an unwheeled skid-tip.

was applicable whether this feature was premeditated or an afterthought); that the ailerons—of which the quoted observer rightly made much—were fitted on all four wings (hence his emphasis on 'biplanes'); that the ten-cylinder two-row Anzani radial engine was installed very neatly indeed, complete with exhaust-collector ring in front; and that the petrol tanks were pressurised by an air-driven pump.

Having named Fred Sigrist in connection with the aircraft generally, if one person in particular can be associated with the design of the floats it was, apparently, Sidney Burgoine, an experienced boat-builder.

Three Sopwith Anzani Tractor seaplanes (numbers 58, 59 and 60) were delivered to the Admiralty, the first being formally accepted at Calshot in June 1913. No.59 went to Cromarty and No.60 to Great Yarmouth. At least one of these machines had a 'wireless' installation, the current for which was provided by a dynamo, chain-driven from the engine-starter shaft at 3,400 rpm. A cone-clutch was fitted to 'disengage the magneto when necessary'.

That one of these Anzani-powered biplanes (apparently No.58) saw service in the early weeks of war, when the 'Eastchurch Squadron' went to France, seems fairly certain. Two, indeed, may well have survived. A photograph herewith shows just such an aeroplane (No.58 ?) in appropriate company, fitted with a wheel-cum-skid landing gear instead of floats, though without the little auxiliary wheels at the front of the skids as on the original 'Three-seater'. Side windows, such as were characteristic of the last-named type, were, however, a feature.

Anzani Tractor Seaplane

(100 hp Anzani)

Span (approximately) 56 ft (17 m). Other data lacking.

Bat Boats

The principal authors who inspired some of Britain's aircraft pioneers—Tom Sopwith by no means least among these latter—were Jules Verne, H. G. Wells and Rudyard Kipling. And here one might add that C. G. Grey, as editor of *The Aeroplane* continued these writers' work ('For their work continueth', as Kipling declared in *Stalky & Co*) if only because so many of his compositions were fanciful (or fictitious) as well as being breezy (or blustering). So greatly influenced was Grey himself by Kipling that 'R.K.' was quite often quoted as 'C.G.G.'; but indubitably it was Kipling's story *With the Night Mail*—published as a separate title in the USA, with special illustrations, though familiar on both sides of the Atlantic as a component of the book *Actions and Reactions*—which provided the name for Britain's first successful flying-boat and the title for the present chapter of this book.

The true nature of Kipling's fictitious 'Bat-Boats' is conveyed in a page from *Actions and Reactions,* which calls for no comment here, except to re-emphasise Tom Sopwith's love of motor-boat racing. But, just as the bibliography of Kipling's tale can prove confusing, so is it important at this early point to make it clear that there were two distinct forms of the Sopwith Bat Boat flying-boat, and that, following marine practice, these were called by Sopwith Bat Boat I and II respectively. The Navy (in the manner wherein they styled the 'rig of the day') sometimes referred to them as No.1 and No.2.

Just as the Hawker Hart and Hornet caused a buzzing at the Olympia Aero Show of 1929, so did the joint appearance of the Sopwith Three-seater and Bat Boat (retrospectively called Bat Boat I) at the corresponding show of February 1913.

As first exhibited at Olympia the Bat Boat I (for so we shall call it) was an altogether trimmer craft than its successor, which, nevertheless was a far nearer approach to the big, successful and multi-engined British flying-boats that followed it from other works.

Though its outrigger tail is somewhat obscured by the co-starring Three-seater (at Olympia in 1913) the original Bat Boat I nevertheless displays its bow-mounted elevator—later removed.

In truth, the Bat Boat I itself was not a wholly Sopwith product, for the hull was built by Saunders of Cowes—a name that was to be sustained in the RAF by the sturdy Saunders-Roe London of 1934. As Harald Penrose (a boat-builder himself, as well as a gifted author and eminent test-pilot) remarks in Vol 1 of his splendid Putnam trilogy *British Aviation; The Pioneer Years 1903–1914:*

'The sea had long been the passion of Sopwith and Sigrist, and since they had just sold their first aeroplane to the Admiralty, it was natural that they thought in terms of marine aircraft as the opening venture of the new Sopwith aeroplane company, which was rumoured to be backed by the millionaire Barnato Joel, who had married one of Sopwith's sisters. Not only had Tom Sopwith raced speed-boats, but he was a client of the redoubtable boat-builder Sammy Saunders, of the neatly trimmed white beard and powerful personality. Grandson of the founder, he had transferred the family business in 1901 from Goring-on-Thames to Cowes, establishing the "Saunders Patent Launch Building Syndicate", and registered it in 1908 as S. E.

Saunders, Ltd., to exploit his patented system of *Consuta* laminated-strip planking cross-sewn with copper wire to give far greater strength for weight than hitherto available. In developing high-powered racing boats, the new company had experimented with many hull forms, plain and stepped, as well as a sidewall vessel some 35 ft. in length with air-lubricated bottom. Recently Curtiss in the United States had developed his simple single-pontoon biplane into a more capacious hull in which pilot and passenger were seated. The idea attracted Tommy Sopwith, and he discussed it with Sammy Saunders' hull designer Sydney E. Porter, who had started with him in 1903. Already he had evolved for Sopwith the very successful *Maple Leaf* stepped hydroplane, and he saw no difficulty in designing a similar *Consuta*-sewn single-stepped cedar hull, 21ft. long, with V entry, and side-by-side seating immediately above the step.'

Here, then, we have the essence of the Sopwith Bat Boats' history, related with multi-professional authority; and it remains to add the aeronautical appurtenances.

Mounted amidships on two pairs of struts, somewhat above the hull (which, in its bare form, weighed a mere 180 lb) was a two-bay, equal-span unstaggered wing cellule; and set high between the wings was a 90 hp Austro-Daimler six-cylinder inline water-cooled engine driving a pusher propeller. The hull being short—only 21 ft (6.4 m) overall—the tailplane and elevator, together with a deep single rudder, were carried clear of the water on converging tail-booms; but in addition to the rear tailplane and elevator already mentioned there was an auxiliary elevator, strut-mounted over the bow of the hull. Wing-warping was used for lateral control.

'The wing-tip floats' (declared one enthusiast) 'are constructed of copper plate, corrugated in order to give additional strength. A really most original point in their construction is the fact that each is equipped with a bicycle-valve in order that, should they become dented in any way, they can be blown back to their original shape by means of the ordinary pump! This is really worthy of a patent.'

For use by the Naval Wing of the RFC the Admiralty bought a specimen of the type described and used it for experimental work at Calshot, early modifications including the removal of the bow elevator. (No self-respecting sailor would put to sea with such an appendage just where the figurehead should be—or so it was said), and hardly less noticeable was the replacement of the original deep, unbalanced rudder by a larger surface, horn-balanced at each end. Later this gave place to a rudder of roughly oval form.

The Bat Boat I (with single rudder, and no bow-mounted elevator) was a new sight for British eyes.

With (temporary) human figurehead taking the place of the bow elevator, Bat Boat I in 'Mortimer Singer' trim. (Sopwith caption: '7—100 hp Green Bat Boat (Mortimer Singer) — Jan. 1913'.

To render the Bat Boat 'all-British', and thus allow it to compete for the Mortimer Singer £500 prize for the first such aircraft of amphibious form, the engine-bearers were modified to accept a 100 hp Green water-cooled unit and—of greater technical significance—two wheels were fitted, one on each side of the hull and capable of being raised clear of the water as required. On land, the hull sat tail-down. To absorb the extra power of the Green engine, the propeller diameter was increased to 11 ft (3.3 m) and twin rudders, below a new one-piece elevator, were associated with a modified tail-boom assembly. Instead of the earlier bracing cables, a pair of sturdy struts ran down to the hull from the new engine-mounting, and a further improvement was the fairing-in of the bottom-wing/hull junction round the supporting struts. The wing-warping system now gave place to ailerons, but the original pattern of wing-tip stabilising floats (cylindrical, with pointed ends) remained unchanged.

The demands imposed by the Mortimer Singer prize performance were very stringent and somewhat bewildering; but on 8 July, 1913, carrying Lieut Spenser Grey as official observer, Harry Hawker completed the specified tests in 3 hr 25 min, thus winning the £500 prize and an important place in British aircraft history. In securing these distinctions Spenser Grey did not lend a hand, as might have been expected of a sailor (even though an official observer) but a foot—to kick the wheels down for each landing at Hamble, the reason being that after take-offs from the Solent they had failed to drop into position when released.

Thus, although it bore a general resemblance to the slightly larger Supermarine Walrus of the Second World War, the Bat Boat was far more deserving of the description 'primitive' that has been too frequently applied to the 'Shagbat', or Walrus—which had, in any case, a full-length hull.

Still, the original Sopwith Bat Boat represented a truly significant accession to the development of British Naval flying. That Naval pilots flew the machine with and without the bow elevator seems certain; and, in his book already referred to, Harald Penrose has shown a photograph of it upside down on land and with the elevator prominent, though much the worse for wear following an incident which Mr Penrose records as follows: 'It was wrecked at the end of August [1913]'—the Austro-Daimler engine having by that time been re-installed, and the wheels removed—'after it had been moored for the night, because the sea was too rough to beach the machine at Calshot. Next morning heavy seas were breaking over the boat, eventually filling it, aided by the wash from passing steamers. Coastguards attempted to get the craft ashore, but in the process it struck a submerged groyne and was holed and turned over. The Admiralty ordered a replacement.'

As a lover of the sea, T. O. M. Sopwith must have had a special fondness for this stately seascape—featuring the Bat Boat I and a battleship bedecked. (Original Sopwith print uncaptioned, but numbered 38).

This mishap notwithstanding, the Bat Boat which bore the Service number 118, and which was generally regarded as the 'original', though clearly much rebuilt, was sent to Scapa Flow when war broke out for Fleet-patrol work (after being present at the Spithead Naval Review in July 1914)—and though it suffered gale-damage on 21 November, 1914, it was not officially written off until March of the following year.

That this pioneering Bat Boat I was a proud possession of the Royal Navy (if sometimes fractious and fractured) is clear, not only from its presence at the 1914 Spithead Naval Review, but from its use for experiments involving a little searchlight in the bows (searchlights by that time having become as much a part of a warship's equipment as were guns)—and also by some semblance of armament itself. As I recorded in my *Armament of British Aircraft 1909–1939*: 'The first flying-boat of this type was used for armament experiments with which the names of Lieut A. W. Bigsworth and Sub-Lieut J. L. Travers are particularly associated. The dropping of darts and practice bombs was preceded by the discharge of potatoes, Naval ratings observed the fall of shot. Data on bomb-aiming were thus accumulated.'

Comparative studies of the Bat Boat I with twin rudders and movable wheels—or retractable landing gear.

Even so, I feel that the Bat Boat's significance in armament development may have been much overplayed by reason of the delightful circumstances attending this episode, for by 1914—contrary to widely held opinion—a great deal of experimental, as well as theoretical, work had been done in Britain with a variety of weapons and gear—bombs and bombsights included.

The second—and seemingly separate—example of the Bat Boat supplied for British Naval service was No.38, which, at one stage at least, was distinguished by a triangular fin ahead of a single ellipsoidal rudder. But such was the extent of modification and rebuilding, and so great the perils of confusion that existed in those times (and have since been multiplied) that firm identities are exceedingly difficult to establish. In any case, the Sopwith Bat Boat II—as we shall call it for consideration now—was a very different aircraft, and was used not only by the British, but by the German Naval Air Service.

The fact just stated, though doubtless already known to many readers, has never, in the present writer's view, been accorded due prominence; for if ever the heartcry that has echoed down the years and through the wars—'Whose side are we on, anyway?'—clamoured for renewed expression it is surely here. Indeed, the instance of the German Bat Boat II must rank almost equally with 'Kestrels for German prototypes' in the 1930s and 'Nenes for Russia' in the later years. True, the aircraft itself probably had little influence on German design or policy; true likewise that such anomalies recurred, as the present writer can attest with warm personal feeling. Yet, whatever the facts of such matters, and the pretexts advanced in extenuation (notably continuance of business contacts until a few weeks before the 1914 war) there is something clammy in any transaction whereby a threatening Power can acquire, on the very eve of conflict, a prime example of a prospective opponent's technical potential.

In essence, the Bat Boat II was not only a larger and more powerful development, but differed quite strongly in appearance from its precursor. This was immediately evident on the first public showing—at Olympia in March 1914, less than five months before Britain declared war on Germany. The differences, moreover, were more than superficial, for the new and stronger hull had been made not by Saunders on their patented system at Cowes, but by Sopwith themselves at Kingston-on-Thames. The entire hull-structure was deeper, and suggestive of the sturdiness that was in fact conferred by a double skinning of mahogany on a framework of ash stringers. As on the earlier boat, there was a single step, though the planing bottom was flatter, and, for better water-clearance, the bottom wing (which was staggered appreciably behind the top one) had quite a sharp dihedral. The outboard stabilising floats were of a new design, with a rectangular instead of a circular section, and similar to those of the Type C torpedo-dropping floatplane.

The Bat Boat II had a Sopwith hull (as well as superstructure), and this hull is well shown—with sheet-brass channels screwed on to the sides, to deliver air to the 'vented step'—in the close-up here.

The uppermost of the three-bay staggered wings had strut-braced extensions (again, as on the Type C) and—unlike the lower wing—carried ailerons. The interplane struts were of spruce, and spruce was also used for the wholly new tail-boom structure, the side-struts of which were raked to conform with the staggering of the wings. Atop the convergence of the upper booms was a tailplane/elevator assembly of very deep chord (far more so than formerly) with raked tips matching those of the mainplanes. There was no fin, and the rudder was ellipsoidal.

One especially remarkable feature of the new, Sopwith-built, hull was the 'vented step', and hardly less remarkable, the means whereby air was led to it. 'The method of leading air to the step', commented one marine-minded observer, 'is very ingenious. Instead of doing this by leading tubes through the interior of the boat, which necessitates piercing of the bottom, the same results have been obtained by sheet brass channels screwed to the sides of the boat.'

Much of the interest in the new Sopwith flying-boat was nevertheless concentrated in the powerplant, which resembled the earlier scheme only in driving a pusher propeller and in being associated with forward-running struts between the engine-bearers and the hull. The engine itself was a 200 hp Canton-Unné (Salmson) water-cooled two-row radial—a form somewhat difficult to comprehend these days—with a broad frontal radiator instead of the earlier side-mounted layout. For this impressive engine (concerning which more will be said in connection with the Type C) a compressed-air starter was provided in the side-by-side two-seat cockpit. To deliver power for a wireless transmitter (note how Sopwith were meeting, and even anticipating, Service demands, though there was no provision for armament) a Motosacoche motor-cycle engine could be installed forward of the passenger's seat, and put in gear by hand.

In addition to the German Bat Boat II, which was actually being flown over the Baltic by German Naval pilots before war came, a similar flying-boat (understandably known as the 'Circuit Bat Boat') was constructed for the 1914 *Daily Mail* 'Round Britain' contest, this machine being chiefly distinguished by a 200 hp Sunbeam engine; by the mounting of the bottom wings a little above the hull, instead of being directly attached; and by an increase in petrol tankage to give an endurance of 5 hours. C. Howard Pixton would have been the pilot, but the war caused cancellation of the contest. It was reckoned that the Sunbeam-powered machine was about 5 mph (8 km/h) faster than the Canton-Unné version.

Bat Boat I

(90 hp Austro-Daimler or 100 hp Green)

Span 41 ft (12.5 m); length 32 ft (9.7 m); wing area 422 sq ft (39.2 sq m). Empty weight 1,200 lb (544 kg); maximum weight 1,700 lb (770 kg). Maximum speed 65 mph (104 km/h).

Bat Boat II

(200 hp Canton-Unné)

Span 55 ft (16.8 m); length 36 ft (11 m). Empty weight 2,300 lb (1,043 kg); maximum weight 3,180 lb (1,443 kg). Maximum speed 70 mph (112 km/h).

Circuit Seaplanes

Before identifying and describing the two distinct Sopwith types to which this chapter is devoted it will be helpful to outline the circumstances that led to their construction.

As early as May 1910 that most air-minded of newspapers the *Daily Mail* (with an eye as closely fixed on circulation as on circuit-flying) had offered £10,000 to the winner of a 1,010-mile 'Circuit of Britain' contest, specifying thirteen compulsory control stops and five days for completion of the flight. The contest did not, in the event, take place until July 1911, when thirty aircraft were entered. Among these no Sopwith was numbered, and as things transpired the Cody biplane was the only British aircraft to stay the course—and this machine came fourth. (Sopwith was a great admirer of Cody, as were so many of his contemporaries).

In March 1913 the same newspaper offered £5,000 to the winner of another Circuit of Britain, the main conditions being that the aircraft must be a 'waterplane' entirely of British design and construction, and that, starting and finishing from the mouth of the Thames (as befitted an 'all-British' event), the machine should fly—in 72 continuous hours—not only round England, Scotland and Wales, but to within one mile of Kingstown Harbour, Ireland. At the same time a second *Daily Mail* prize, of £10,000, was offered to the first pilot to fly across the Atlantic Ocean, again within 72 hours, from any point in the USA, Canada or Newfoundland, and with no restriction on the nationality of entries. Of Sopwith's Atlantic aspirations, more in a later chapter; so for the moment we have to record that the dates eventually fixed for the 1913 Circuit of Britain event were 16—30 August. This was to be a most exacting affair, with 1,540 miles (2,478 km) to be covered in nine stages. Three of the four entries—Cody's 'Circuit' Waterplane; the Short S.68 of Frank McClean and Gus Smith; and the Radley-England Waterplane of James Radley and E. C. Gordon England—were withdrawn. This left the Sopwith entry only, which, contrary to many expectations, turned out to be not a Bat Boat, but a new type of floatplane, clearly developed from the Anzani Tractor seaplane already described, though having a 100 hp six-cylinder all-British Green engine and four-bay wings of equal span.

A magnificent-looking aircraft, the 1913 Circuit Seaplane owed much of its appearance and performance to its 100 hp Green engine.

Although built hurriedly, this fine new machine was of very handsome and businesslike appearance, its lines being enhanced by the fine aerodynamic entry afforded by the slim Green water-cooled engine, the radiators for which were disposed as large flat surfaces, one on each side of the fuselage between the wings. The centre section was left uncovered, and in the definitive (floatplane) version—for the aircraft was also built with a twin-wheel/twin-skid landing gear—the gap thus left had 'endplate' fairings. The primary object of the gap was to allow the crew to get out smartly in a crash, though some later Sopwith aeroplanes had 'fancy' centre sections for other reasons—notably clear view. The two seats were in tandem; construction was of wood, with fabric covering; the two main floats were of lenticular form; and the tail float was cylindrical.

The 1913 'Circuit' event turned out to be a sad affair all round. Cody had been killed at Laffan's Plain on 7 August, when his 'Hydro-biplane' broke up in the air (this likewise had a 100 hp Green engine); the Radley-England was without a suitable engine of any kind; and the Short was not ready in time. Although the 100 hp Green engine was almost invariably described as 'reliable', poor Cody had little chance to find out (in his particular installation the propeller was driven by a chain), and Sopwith, as intimated, was using an advanced radiator system. In any case, Fred Sigrist had plenty to occupy him in connection with the powerplant; and Harry Hawker, who was the pilot, and had his compatriot H. A. Kauper as passenger, fainted from inhaling exhaust fumes after leaving Yarmouth. Sopwith arranged for Sydney Pickles to take over as pilot; Pickles tried to start again, but from a sea so choppy that water got in to the tail float and elevator. Then the machine went back to Cowes, where longer exhaust pipes were fitted.

The contest now having been re-convened by the *Daily Mail* for 25 August, and Hawker having now recovered, the 'two Harries'—Hawker and Kauper—left the Solent at 5.30 am in calm and mist, and by the end of that day had set a new record for over-water flying. They alighted at Beadnell, Northumberland, at 7.40 in the evening—this notwithstanding an unscheduled alighting, occasioned by a burst exhaust pipe which had heated water-connections and boiled the water away, the radiator system being refilled at Seaham with sea-water. Thus the Green product, as well as the Sopwith, was able to continue next day, when Oban was the night-stop. At 5.42 am on the following day (27 August) this splendid outfit was once more getting under way; but a waterlogged float obliged a return for repair. In spite of this the new Sopwith pressed on to Larne (Antrim, Northern Ireland); but after flying on nearly to Dublin Hawker decided to alight to adjust valve-springs on the engine, his foot slipped on the rudder-bar, and the Sopwith fell into the sea off Loughshinny, a few miles north of Dublin. Hawker was unhurt, Kauper broke an arm; but 1,043 miles had been covered in 20 hr flying time, and the *Daily Mail* awarded Hawker a special prize of £1,000 for his determination.

While recognising that the effort just described was very much a British affair, the present writer nevertheless ventures upon a little Empire rebuilding by noting what must be one of the most eloquent, though concise, tributes on record—not only to the Sopwith aeroplane concerned but to its crew and their homeland. Thus a solitary resounding line in the ten-volume Angus & Robertson *Australian Encyclopaedia:* '1913—H. Hawker and H. A. Kauper—Sopwith seaplane—*Daily Mail* circuit of Britain—The *two Australians* crashed after 1,043 miles.' Only the final italics are the present writer's own.

Technically, the significance of this flight was in its demonstration of the long-distance capabilities of British seaplanes (just as the Tabloid was to show at Monaco what they could do in the way of speed); and a suitably impressed British

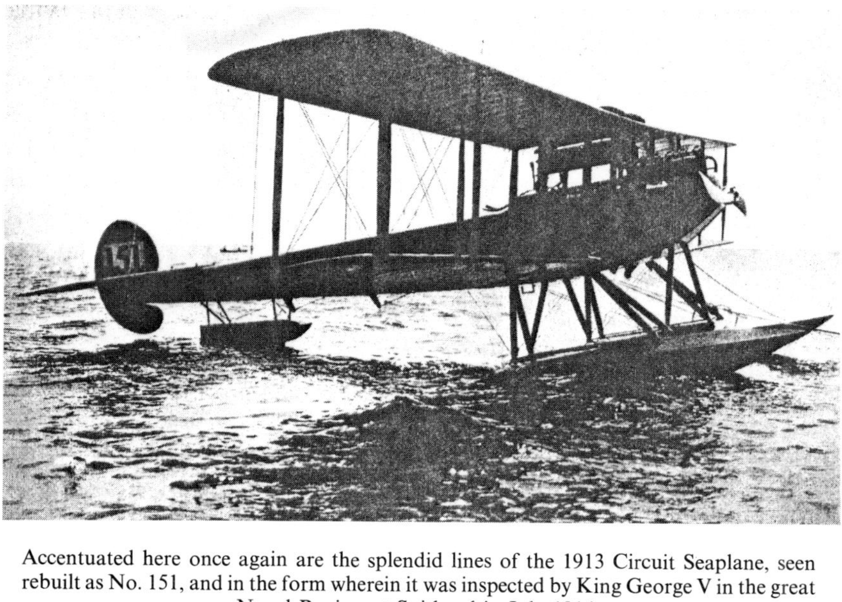

Accentuated here once again are the splendid lines of the 1913 Circuit Seaplane, seen rebuilt as No. 151, and in the form wherein it was inspected by King George V in the great Naval Review at Spithead in July 1914.

Admiralty ordered a rebuilt example of the 1913 Circuit Seaplane which, in company with a Bat Boat I, made a brave show at the Naval Review of July 1914. By that time the comma-shaped rudder bore the number 151, and this very machine was later in service with No. 4 Wing, RNAS. It was reported at the time that No. 151 was flown on the Cuxhaven Raid of Christmas Day 1914 by Flt Cdr R. Ross; but though Robin Ross participated in that raid, some doubt exists concerning the identity of his aircraft—especially so as all the seaplanes concerned were otherwise declared to have been Shorts, and also in view of Sir Arthur Longmore's testimony that Robin Ross later flew the (presumed) Sopwith Type C, a somewhat similar machine, in torpedo-dropping tests at Calshot.

That the airframe of the 1913 Circuit Seaplane (as we have already named the type concerned) has not been described in detail is due not only to the fact that few details have survived, but to the concentration of interest in its powerplant. This being so, it may be noted that the 100 hp Green six-cylinder engine, first publicly announced near the end of 1912 as having 'lately been placed upon the market by the Green Engine Syndicate, to whose specifications the engines are built in Great Britain [the country was always emphasised in connection with Green engines] by the Aster Engineering Co.' weighed 442 lb complete and delivered its 100 hp at 1,150 rpm. 'Although the dead weight per horse-power is not of the lowest', it was observed, 'compared with rotating [*sic*] engines for instance, the economy in fuel and lubricating oil reduces the total load to be carried by a machine destined for extended journeying well below that of less efficient types.'

That for 'extended journeying' a Sopwith/Green combination could indeed place Great Britain in the forefront was surely established by the effort of the Sopwith 1913 Circuit Seaplane. A landplane version of the same machine was first tested by Hawker on 4 October, 1913—at Brooklands, it is hardly needful to add. On this occasion the rudder bar once again enters the story; for, finding himself caught in a down-current soon after take off, and realising the inevitability of a crash, Hawker deliberately removed his feet from the bar in order to brace himself. Thus, as the ensuing sideslip finished abruptly (near the Weybridge—Byfleet road) he sustained only fairly minor injuries.

Although the Green-engined landplane is said to have been repaired for competition flying, there could be confusion here with the Green-engined variant of the Three-seater, referred to in the appropriate chapter. (For an exposition of the complicated competitive scene towards the end of 1913 the reader is commended to Peter Lewis' Putnam book *British Racing and Record-Breaking Aircraft*—pages 76/77 especially).

By what must surely be the ultimate in paradox, the later Sopwith 1914 Circuit Seaplane existed in name only—and even so, apparently, as a landplane! Powered by a 100 hp Gnome Monosoupape engine, this was a trim-looking tandem two-seater, with the widely spaced cockpits having head-fairings between and behind; two-bay staggered wings, of equal span and aileron-equipped; and a twin-wheel landing gear, incorporating also two skids. The ailerons had inverse taper (increasing in chord towards the tips) and were interconnected by struts. Respecting engine installation and landing gear at least, this aeroplane, which was constructed to drawings marked 'D3', resembled the Tabloid; but its real significance was that (the 1914 Circuit contest having been abandoned by reason of the war) the design was developed into the Folder Seaplane (Admiralty Type 807) which is separately described. There was, in any case, very strong Admiralty interest not only in the contest itself, which was to have started from the Admiralty yacht *Enchantress* (so closely associated with Winston Churchill) but in the individual entries. Totalling nine, these included a Bat Boat II—to be flown by Howard Pixton, as Sopwiths' second string—and, of all things, a German D.F.W., Beardmore-built.

The nominated pilot for the Sopwith 1914 Circuit Seaplane was Victor Mahl, by this time prominent in the Sopwith team, and who himself tested the aircraft (as a landplane, at Brooklands) on 15 July, 1914. Of this aeroplane these brief particulars were given:

'Immediately behind the engine are situated the petrol and oil tanks, whilst an additional supply of petrol is carried in another tank behind the passenger's seat. This is situated sufficiently far forward to provide a good view in a downward direction, whilst from the pilot's seat, placed as it is in line with the trailing edge of the lower plane, which has been cut away near the body, an excellent view is obtained in a downward and forward direction. By cutting away the trailing edge of the centre portion of the upper plane, the pilot is enabled to look upwards and forwards, so that it would appear that the arrangement of the pilot's seat and the staggered planes is such as to give the pilot, as nearly as possible in a machine of this type, an unrestricted view in all directions.

'The main planes are of the usual Sopwith type, and are very strongly built. Compression struts are fitted between the main spars in order to relieve the ribs of the strain of the internal cross-bracing. Ailerons are fitted to the tips of both upper and lower main planes, and are slightly wider than the remaining trailing portion of the wings in order to render them more efficient. The ailerons are operated through stranded cables passing round a drum on the control lever in front of the pilot's seat. The tail planes are of the characteristic Sopwith type, consisting of an approximately semi-circular tail plane, to the trailing edge of which is hinged a divided elevator. The chassis is of a substantial type, and the two main floats are sprung by means of leaf springs interposed between the rear of the float and the rear chassis struts, whilst the floats pivot round their attachment to the lower end of the front chassis struts. The floats are spaced a comparatively great distance apart, in order to render the machine more stable on the water. A tail float of the usual type takes the weight of the tail planes when the machine is at rest.'

Its obvious superficiality notwithstanding, the foregoing quotation is, in fact, quite significant—especially respecting the attention paid to field of view, for this was manifest likewise in the Type 807 (Folder) and the Two-seater Scout, both Admiralty types. What Sopwith were clearly trying to do was to reconcile tractor performance with pusher visibility—an 'unrestricted view . . . as nearly as possible in a machine of this type'. But as the war was to prove (the most notable instance being the D.H.4) widely spaced cockpits, especially with petrol tankage between them, were not a paying proposition, though Sopwiths' preoccupation with field of view continued undiminished.

Although it has been stated that by the beginning of August 1914 larger vertical tail surfaces had been fitted to the (intended) 1914 Circuit Seaplane, and although such a modification was commonly associated with the fitting of floats—out of consideration for side area—there is no firm evidence that floats were installed.

1913 Circuit Seaplane

(100 hp Green)

Span 49 ft 6 in (15 m); length 31 ft (9.4 m); wing area 500 sq ft (46.5 sq m). Maximum weight 2,400 lb (1,090 kg). Cruising speed 65 mph (105 km/h).

1914 Circuit Seaplane

(100 hp Gnome Monosoupape)

Span 36 ft 6 in (11.1 m); length with float landing gear 30 ft 10 in (9.4 m). Performance data not established.

Underwing bomb-carriers, revised nacelle—and especially a two-wheel landing gear—characterise this Sopwith Gun Bus (probably Robey-built).

Pushers and Gun Buses

Just as it has been thought fit to accord a single chapter to the Bat Boats, their notable variations notwithstanding (for in truth they represented a species rather than a type) so, now, we consider those wholly individual Sopwith products generically called 'Gun Bus' though including also the 'Greek Seaplane' and the 'Pusher Seaplane Gun-carrier No.127'. (Though the Bat Boats were themselves of pusher form they had, as we have seen, a very strong marine individuality of their own).

The true chronology and lineage of the family of floatplanes and landplanes now to be studied is indeterminate and unimportant; but they were all of 1913/14 vintage—even though production of the last Robey-built landplanes was still in hand late in 1915.

At this early point in our account it must be remarked that as long before the war as August 1913 (one whole year, that is) there was a seemingly firm report of a Sopwith type then known as 'the 80-ft. span machine', and as mentioned in the following context (the rendering being a precise transcription):

'The dimensions of the floats for the "gun 'bus" are:— Length, 16 ft.; beam, 2 ft. 9 ins. For those on the machine of 80 ft. span the length is 20 ft. and the beam is 3 ft. 9 ins. It is interesting to note that, although the total weight of the latter machine in working order is somewhere in the region of $2\frac{1}{3}$ tons, the loading per square foot, owing to the enormous span, is very nearly as small as that on the average Brooklands "box-kite". We await its trials with more than ordinary interest, for with its engine of 240-H.P. and its span of 80 ft. something distinctly unusual in the way of weight-carrying and *décollage* from the water should be seen.'

Here, then, we have two different (or differentiated) aircraft: one of unstated span, referred to as the 'gun 'bus' and having floats measuring 16 ft (4.9 m) in length; and another, called 'the machine of 80 ft. span' (24.4 m), having much bigger floats—20 ft (6.1 m) long, and a whole foot (0.3 m) broader in beam—and powered with an engine of unusually high output (240 hp) which, jointly with the great span (and implicit aspect ratio and area) was expected to confer extraordinary lifting capacity.

There is now a great temptation to affirm that the smaller machine of this pair was the gun-carrying floatplane bearing the British Service number 93 and having an Austro-Daimler engine of 120 hp, while the huge weight-lifter with double the power was another gun-carrier, numbered 127.

Temptation having thus been recognised, it is permissible to advance the reasonable assumption that such an affirmation would be approximately true, and accordingly to restate the following entry from *British Aircraft Armament 1909-1939*, made under the heading 'Sopwith Pusher Seaplane Gun-carrier No.127'. Thus:

'The identity and significance of this historic aircraft is apparently now established for the first time, the significance being that it was armed with the 1½-pdr Vickers gun before that weapon was transferred to Short S.81 No.126. First, there is the testimony of Sir Arthur Longmore that "one of our Sopwith pusher seaplanes" (at Calshot before the 1914 war) carried a 1½-pdr gun weighing 265 lb, with which Lieut R. H. Clark-Hall conducted many successful tests. Second, it was stated on the occasion of the Naval Review in July 1914 that a "Sopwith Gun Carrier" with 200-hp Salmson (Canton-Unné) engine was unable to fly because of tail alterations. On this same occasion the Short S.81 No.126 was present carrying a 1½-pdr gun and it was remarked: "The gun on the Short is the biggest weapon yet used in aircraft. It was first used on the Sopwith, and later was used to test the Short's ability to stand the recoil."

'Aircraft No 127 is on record as being a Sopwith with 200-hp Canton-Unné engine, and it may be supposed that this and the Short machine were ordered as a pair for trials with heavy guns. That No.127 was of the well-known Greek Gun Bus type is certainly open to question, having regard to the fact that this was a much smaller machine than the Short No.126 . . . and there can be little doubt that No.127 was the Hydro Biplane Type S of 80 ft span, already associated by J. M. Bruce with a quick-firing gun. Thus No.127 must take its place in history, not only on account of its big gun, but as the largest British aeroplane of its time.'

The foregoing extract is quoted not with any motive of vindication but to stress these points: (1) That Sopwith appear to have built 'the largest British aeroplane of its time'—a feat which, in itself, should go far towards justifying the issue of this present book. (2) The company's early use of increasingly powerful engines—now in the 200 hp+ bracket. (3) This same company's equally early involvement with the development of aircraft armament. (The stressing of any aeroplane to withstand the recoil of a 1½ pdr gun, especially when this was exerted in any of several directions, was a job demanding mathematical skill, as well as structural ingenuity). (4) The inception of peculiarly Sopwith armament innovations—all of which will later have their places in these pages, and best exemplified, perhaps, by the classic '2 × 0.303 in.' installation on the Camel; by multi-gun combinations on the Dolphin and Snark; internal stowage on bombers ranging from 1½ Strutters to the Cobham; and pioneering work in the highly specialised field of torpedo-dropping.

Such swiftly mounting experience—and, even more important, determination to lead and be seen to lead (for the haze around the '80 ft span machine' may well have arisen from 'security' rather than obscurity) must inevitably have aroused commensurate interest overseas; and although the German Bat Boat has already had its place, the first Sopwith export order for aeroplanes in quantity came from

Greece. This event was signalled in *Flight* of 10 July, 1914, by the printing of four fine photographs, though these featured personalities rather than aircraft and were accompanied by the merest note under the old 'rag-bag' heading *Eddies*. Thus:

'Some details are to hand regarding the work of Capitaine de Freigate Collyns P. Pizey, who before his appointment to the Marine Royale Hellenique, was so well known to our readers from his connection with the Bristol Co. A very flattering report is given of the work of the Anzani-engined Sopwith "pusher" seaplane, which in one month was flying for some 40 odd hours [*sic*]. Five Greek officers have been trained on it, and they are now ready for solo flights. This is probably the first time that naval officers have been taught to fly directly from the sea without first doing land flying.'

Another Sopwith 'first'?: eloquent testimony, surely, not only to the tractability of the first (dual-control trainer) Sopwith Anzani-engined 'Greek Seaplane' (as we shall call the type concerned) but indicative also of the distinction now arising between 'sea' and 'land' flying. The following facts must, however, be appended:

The man behind the scenes here was not, in fact, Collyns Pizey, but Rear-Admiral Mark Kerr, an air-minded British Naval officer who had been appointed in 1913 as an advisor to the Greek Government. Though in British naval-air history the name of Mark Kerr is inevitably overshadowed by that of Sir Murray Sueter there was certainly liaison between the two men concerning the 'Greek Seaplane', not least respecting coast-defence in narrow waters—a problem facing Greece as well as Britain. It was, as matters transpired, through long Anglo-Greek associations, as much as by a twist of fate, that in April 1929 the Hawker company (Sopwiths' successor) received an order for six Horsley torpedo-bombers, as adopted for coast-defence by Britain. Six, likewise, was the number of 'Greek Seaplanes' ordered as gun-carriers, though these were additional to the single machine supplied as a sample and a trainer—and clearly the one referred to in the quoted *Flight* report.

Viewed in flight, the Anzani-engined 'Greek Seaplane' (as the type is identified in the text) assumed a certain gracefulness that was less in evidence on the water.

The fourth machine of the main Greek order for six gun-carriers—to which type the designation 'S.P.Gn.', signifying Sopwith Pusher Gun—[carrier] was seemingly applicable—was already on test at Woolston, Southampton, by late June 1914; but when war came in a few weeks' time the entire batch of Greek Seaplanes was taken over by the British Admiralty. To these the Service numbers 896—901 were apparently allocated, though the reported fitting of 200 hp Canton-Unné engines in the aircraft so numbered adds confusion—especially so as C. F. Snowden Gamble recorded in his splendid book *The Story of a North Sea Air Station* that Nos.897, 898 and 899 were at Great Yarmouth in September 1914, powered by 100 hp Gnome engines. More comprehensible, having regard to later developments yet to be recorded, is the classifying of these three examples as 'bomb droppers' as well as 'gun machines'.

The installation of the Lewis gun on this Sopwith-built Gun Bus—and features of the gun itself—are specially mentioned in the text. Though the installation of the Gnome engine in this aircraft is obscured, the landing gear is sufficiently prominent for 'Jack' to climb the rigging!

No less comprehensible, in the early months of war, would be conversions of seaplanes to landplanes, and vice versa—though equally so the description Gun Bus (with initial capitals) could be strictly applicable only to RNAS landplanes of the general type now to be considered. (This having been said, it must still be emphasised that the term 'gun 'bus'—being a generic and colloquial one—had been used in print, as already exemplified, as early as 1913 to describe a Sopwith aeroplane).

Now to extricate the Gun Bus (with initial capitals to connote its RNAS landplane status) from its near-relations. This aeroplane was a four-bay biplane with a four-wheel landing gear, the wheels being in two close-set pairs mounted one

Robey-built Gun Bus No. 3833, showing the angular fin.

on each side of a cable-braced skid, which was itself attached to the outer ends of the lower centre section, and thus reminiscent of the original 'hybrid' tractor. Another 'family' resemblance was the forward raking of the struts between the four converging booms, on which the monoplane elevator/tailplane assembly was fixed. The single, comma-shaped rudder had a horn-balance portion wholly above this assembly, and the engine was a Gnome. Nacelle-shape, nevertheless, may well have varied; but a splendid photograph survives (and is now reproduced) showing not only the general shape—with a 'cuniform' nose—but also the installation of a Lewis gun. The gun is seen to be of land-service pattern (with the characteristically massive cooling jacket, or, more strictly, radiator by reason of its internal finning) and is secured to a cranked-pillar mounting. Covering the ejector slot of the gun is a deflector/collector assembly—to restrain and contain the spent cartridge cases; and this component may well have been made up at Hendon, where this particular early Gun Bus was stationed. Certainly it is of unfamiliar pattern—and equally certainly it would be unsafe to regard the Gun Bus shown as representing 'standard' practice,

Second from left in this group with a Robey-built Sopwith Gun Bus is Harry Hawker. The picture shows not only his 'flat hat' but the flat radiator for the 150 hp Sunbeam engine, of the type seen in the subsequent view.

for the times demanded what the Navy traditionally knew as 'make and mend'. That the Gnome Monosoupape engine of 100 hp was generally fitted seems probable; and this was one important respect in which the early RNAS Sopwith Gun Bus differed from a developed, or 'intermediate', model. Sopwith-built 'intermediate' Gun Bus aircraft were officially styled Admiralty Type 806, though the actual Service numbers of the aircraft concerned were 801—806. The designation Gun Bus No.2 may well have applied in this instance—but though convenient and explicit would not strictly conform with the Navy's system of numbering (say) lieutenants ('Number One' or 'Jimmy the One' being the classic instance).

This 150 hp Sunbeam engine displays characteristics discernible in the preceding picture.

In the Admiralty Type 806, with the design of which Herbert Smith and R. J. Ashfield were closely concerned, the engine was a water-cooled 150 hp Sunbeam (retrospective class-name Nubian) installed at the rear of a wholly new nacelle, which was slightly raised above the lower centre section instead of being directly attached as formerly; the elevators were stronger; the lower centre section was reduced in chord; and the wing trailing edges were given little cut-outs to afford easier acceptance of the tail booms, which were subject to distortion in the air.

A word about the engine seems in order, for the 150 hp Sunbeam was a fairly obscure powerplant, and one with particular RNAS associations. Like the Rolls-Royce Hawk, it seemingly achieved its greatest distinction in airships.

That these particular specimens of the Sopwith Gun Bus were built by Robey & Co, of Lincoln, is surely not in doubt. Numbered 3845 (foremost) and 3846, they are on their way for testing at Bracebridge.

Perhaps the most noticeable feature of the Admiralty Type 806 was its new two-wheel cross-axle landing gear, with revised attachments and bracing. Though the track of this new gear—the wheels whereof were rubber-bound to skids as formerly—was relatively narrow, this was greatly accentuated by the 50 ft (15.2 m) wing span.

What may be considered as the definitive version of the Gun Bus (for, as noted at the outset, production was still in hand late in 1915) was built not by Sopwith themselves, but by Robey & Co Ltd, of Lincoln; and that the order called for only thirty machines, and had been placed (in May 1915) with an engineering company that dated far back into Victorian times may at first seem strange. It could, however, be regarded otherwise. In the first place, the order might be compared with at least two other obscure contracts for pushers—specifically the D.H.1As from Savages Ltd, of Kings Lynn, Norfolk, and the Vickers F.B.9s from Wells Aviation Co Ltd, of Chelsea, London (the latter, for some mysterious reason, having different 'RAFwire' lengths from F.B.9s built by Vickers themselves). A second point is that the Robey-built Sopwith Gun Buses had tangible Sopwith associations: Harry Hawker himself tested the first machine of the batch—No.3833 (the total order was 3833—3862, from 3850 onwards being delivered as spares); the first two examples at least were sent to Brooklands; and the drawings used by Robey were the responsibility of Sopwith's 'super draughtsman' R. J. Ashfield. One test pilot, additional to Harry Hawker, was the man who in later life became Col The Master of Sempill, Baron Sempill—a very eminent figure in British aviation and who will later be named again in connection with the Sopwith Cuckoo.

Whatever name or designation they may have gone by, it seems tolerably certain that the Robey-built Gun Buses were constructed with bombing in mind; for not only was the pilot moved to the front cockpit in the lengthened nacelle, but there were underwing carriers (set close inboard) for four—and latterly, with reduced petrol capacity, six—65 lb bombs.

Hardly surprisingly, some of these pushers did service as trainers—at Hendon and at Eastchurch (from which latter base Tom Sopwith had set out in the old Howard Wright pusher back in 1910 to win £4,000).

Gun Bus (as built by Robey)
(150 hp Sunbeam)

Span 50 ft (15.2 m); length 32 ft 6 in (9.9 m); wing area 474 sq ft (44 sq m). Maximum speed 80 mph (128 km/h).

Tabloid

This tiny biplane was Sopwith's historic essay in building an aeroplane primarily for high performance (especially respecting speed) both for sporting and military-scouting work. In its first—two seat—form it appeared in November 1913; and when officially tested at Farnborough late in that month (the actual date quoted by Sopwith being the 29th, on which day also Hawker 'buzzed' a crowded Hendon meeting) showed a speed range of 36.9—92 mph (59—148 km/h) and an initial rate of climb of 1,200 ft (336 m) per minute—and this when carrying two men and petrol for 2½ hours' flying. Here, then, was yet another example of what Sopwith—Britain's new, up-and-coming aircraft constructor—could achieve with low power, by virtue of excellent aerodynamic and structural design; for the reader may now recall the earlier rhetoric of *Aeronautics* concerning the Three-seater: 'How many constructors would undertake . . . ' etc.—and stressing the low power required.

Like the Three-seater, the Tabloid had a Gnome engine of 80 hp only; and one of the few criticisms that could be made of the little newcomer was that the two occupants, seated side by side, with the pilot to port, had a poor field of view—especially upwards, for the cockpit was under the centre section, and the wings, though short in span, were deep in chord, and initially had no trailing-edge cutout. The depth in chord was due in part to the use of warping for lateral control; and another notable feature of the controlling surfaces was the absence of a fin ahead of the aerodynamically balanced rudder.

Although construction was conventional in the conventional sense (wooden, wire-braced, fabric-covered) it was sufficiently ingenious in detail to be lengthily

Engine cowlings on Tabloids were modified according to period and operating conditions. On this, the first of the type, the engine was almost totally enclosed, though with an exhaust outlet at the bottom of the cowling. Note the metal windshield—a feature later developed as the Camel's 'hump'.

Rear view of the first Tabloid. Note, among other features, the absence of a trailing-edge cut-out.

described in two consecutive articles in *Flight* during February 1914. These articles were not only partly written, but illustrated also, by the present writer's one-time editor C. M. Poulsen; and though no specific reference could be made at the time to the Tabloid as such, 'C.M.P.' assured me once that what he had to say and show in those early efforts was essentially applicable.

'We are indebted', the first article acknowledged, 'to the Sopwith Aviation Co., Ltd., for their courtesy in giving us every information and in placing their extensive works at our disposal. It would have been difficult to find a more representative

Harry Hawker himself demonstrates, on a single-seat Tabloid, how the trailing-edge cut-out could prove serviceable to a photographer as well as to a pilot.

firm' (the declared topic of the articles being 'An Aeroplane in the Making') 'for at their works at Kingston, where the Sopwith Aviation Co. employ some 150 workmen, in addition to a large staff of draughtsmen, all the most up-to-date machinery and latest labour-saving methods are employed, whilst the workmanship of their machines has already established for them an enviable reputation, equalled only by the excellence of their design.' Poulsen then went on to say a great deal about woodwork, which was always his particular interest (one remembers him talking of his days in American forests); but perhaps the most relevant paragraph in his first instalment—especially in view of the many stories told (with picturesque variations) about Sopwith aeroplanes being 'chalked out on the floor' was this:

'The method of constructing the fuselage is similar to that employed in building up the wings. The longerons, struts and cross-members are all cut to shape on the spindle machines, and, this being done, they are taken to the erecting shop, on the floor of which are secured some thin strips of wood bent to the shape which it is desired to give the fuselage. The longerons are placed on the floor, and kept to the desired shape by wood blocks pressing them against the strips on the floor. The struts with their respective sockets are then put in place, and the two sides of the fuselage made rigid by means of diagonal cross-bracing. Each of the bracing wires has incorporated in it a wire strainer in order to allow of each bay being tuned up separately. When the two sides are thus finished they are raised up on edge and connected by the cross-members and by diagonal bracing, and the whole structure is then adjusted, or, as it is called, "tuned up". The next job to be done is that of putting the engine bearers, tank supports, pilot's and passenger's seats, controls &c., into place, and when this is accomplished the fuselage is ready to be covered with fabric. In the Sopwith machines, this is done by cutting the fabric to shape and, passing it around the fuselage, lacing it along one longeron. The fabric is then doped similarly to the wings, and the fuselage is ready for the wings.'

Here, I think, we have a singularly interesting study in philology (apart from one's deliberate suppression of italics for the words fuselage and longeron); for the expression 'tuning up' was to endure only in respect of engines, whereas the early wire-braced airframes were quite literally tuned-up like musical instruments (flying, in any case, then being more of an art than a science). Further respecting terminology and nomenclature, one is here constrained to quote a more famous editor than C. M. Poulsen, in the person of C. G. Grey, who declared of the Sopwith aeroplane which now concerns us: 'It was nicknamed the Tabloid, and those universal benefactors Burroughs Wellcome & Co. objected to the use of their registered trade name—thinking that it was poking fun at them. I took much pains to show them that it was a compliment, and we went on using it.'

Ironically, in the context of the 1919 Transatlantic attempts, it was the victorious Sir John Alcock and the hapless Harry Hawker who pronounced the 'Tabloid' first-aid kit to be 'the only possible medical equipment for airmen'; and as early as March 1914 Burroughs, Wellcome and Co. had been advertising their 'No.706, or The Aviator's Case' as being no larger than a cigarette case.

Exploitation of the Sopwith Tabloid's military potential as a fast scout dated from April 1914, when the first of a small batch for the RFC was finished. By 26 May the sixth was at Brooklands for test. These Service Tabloids were single-seaters, with a fin and plain rudder and finer nose lines. It was, in fact, these very features that—apart from a float landing gear—had distinguished the specimen to which attention is now directed, and which was to prove the most famous Tabloid of all by winning for Great Britain what came to be regarded as almost the Grail of airmen—

If this evocative picture shows—as it appears to do—Pixton's Monaco Tabloid being tested on the Thames, then it also depicts that historic Schneider Trophy winner with a strut-mounted tail float.

the Schneider Trophy. With this achievement the name of the pilot concerned—C. Howard Pixton, or 'Pickie' to Tom Sopwith—has become so closely identified (and rightly so) that the essential contribution made by Harry Hawker to design, demonstration and development are sometimes overshadowed. This being so, before concentrating once again on Pixton's resounding feat at Monaco, the following facts must have their place: First, the actual design, or basic concept, of the Tabloid owed much to Hawker personally, though Sopwith and Sigrist also had their say; second, it was Hawker who, on the very day that the Tabloid demonstrated its speed range and climb at Farnborough (29 November, 1913) had circuited Hendon before a crowd of 50,000 at 90 mph; third, Hawker himself had taken the first machine to his native Australia (surely the classic instance of 'local

Surely one of the most enticing 'Wish you were here' postcards ever printed (for, like the preceding Tabloid floatplane picture, it was indeed produced in postcard form) this view calls for little remark beyond affirming that the Sopwith caption reads: '43—100 hp Seaplane. Winner of Schneider Cup at Monaco.—April/1914.'

boy makes good') for demonstrations—and by the time of his return on 6 June, 1914, the appearance of the machine had been transformed by the stripping-off of much of the fuselage covering and the fitting of a plain V-type landing gear. Thus, it was during Hawker's absence that the seaplane which won the 1914 Schneider Trophy was built—and that this was one of a batch of twelve that had been ordered as single-seat scouts for the RFC has been affirmed by Sir Thomas Sopwith himself. 'It was decided to modify one to compete', he said, adding: 'In its original form this aeroplane had one central float which was installed too far aft. Three days before we were due to ship the aeroplane to Monte Carlo it had not flown. Howard Pixton was the pilot and on the first attempt to fly, at Hamble, the machine cartwheeled over on to its nose and sank. At daylight next morning we salvaged the aeroplane, took it to Kingston by road, sawed the single float into two, built two new sides and installed a twin-float chassis. We then took the aeroplane to Teddington and without permission flew it off, this time successfully. From the time it was at the bottom of Hamble river until it was airborne again was less than three days.'

A simple V-type landing gear and stripped rear fuselage gave a distinctive appearance to the modified version of the first Tabloid that Harry Hawker demonstrated in Australia, and seen here after its return.

To the foregoing recollections Sir Thomas added: 'It is interesting that up to the time that I received the first contract for the Tabloids for the army, none of my aeroplanes, and, so far as I know, no one else's, was ever stressed. All of them were built by eye and we had no idea of the factors—except that they were more than one! I have always maintained that if an aeroplane looks right, it generally is right, although at the same time this must not be carried too far.'

Here, unquestionably, we have 'The Skipper' speaking (for as such I have heard his co-directors address him)—the lover, co-designer and steersman of high-speed surface and skimming craft; and in warning that the eye alone could be deceiving he could well have had in mind (for example) the Royal yacht *Victoria and Albert* which, though one of the loveliest-looking vessels of all time, heeled over when being first undocked because of mistakes in calculating weight-distribution. Thus, the first Tabloid seaplane was in Royal company.

Stimulated by his love of yachts and all the best that Monte Carlo means, Tom

Whatever doubts may be entertained concerning the photographic authenticity of this picture of Sopwith's 1914 Schneider Trophy winner—or for that matter the elegance and strict authenticity of the lettering—there can be no doubt regarding the tail-float attachment. (The smaller lettering in the Sopwith inscription reads: '49—100 hp Sea Scout—Winner of Schneider Cup.').

Sopwith was to experience, on 20 April, 1914, one of his long life's greatest joys; for on that day—flying a Tabloid having a triangular fin ahead of a plain rudder; two short strut-mounted main floats; a tail float with faired-in attachments (earlier, as a photograph suggests, this float was attached by an unfaired set of struts); a specially tuned 100 hp Gnome Monosoupape engine ('the first 100 hp Mono ever to come to this country' as Sir Thomas once averred); 'SOPWITH' in seemingly huge capitals on its tiny fuselage (conforming with a similar proclamation on the 1913 Circuit Seaplane); and the racing-number 3 on its rudder—C. H. Pixton won the 1914 Schneider Trophy Race at Monaco. At an average speed of 86.78 mph (140 km/h) he covered the 150-nautical mile course in 2 hr 13.4 sec. Then he carried on for two extra laps—making thirty in all—at 92 mph (148 km/h) to establish a new world speed record for seaplanes.

Invited by Jacques Schneider himself to celebrate on the best the Principality could offer, he wondered if he might have a bottle of Bass.

With Pixton perched on the port float, and propped against the wing, this view of the Tabloid on floats at Monaco is more familiar than those earlier reproduced, but is nevertheless valuable for comparison with the Naval Schneider seaplane later illustrated and described.

Except for the French engine it was Britain all the way; and even the French engine had received the British treatment in the matter of installation and cowling, with a neat and sturdy fore-and-aft mounting (or nose-bearing mounting as it was sometimes called by Sopwith) which afforded not only stiffness but a very clean aerodynamic entry. As the engine-makers explained: 'The 100 h.p. Monosoupape engine' (for such was the unit fitted in the Tabloid racing seaplane instead of the standard 80 hp Gnome) 'is supported in the machine by two bearer plates, both upon the long end [*i.e.* rear end] of the crankshaft. That is, the standard practice is to let the engine overhang its bearers. Where it is desirable to have a more rigid fixing, or fixings spread over a greater length of longitudinals, a third support is added between the propeller and the engine. With such a fixing the standard pattern of short nose has to be replaced with a long or medium nose, and a ball bearing is interposed between the nose and support.'

The 100 hp 'Mono' engine (for so the French power unit was known for short, otherwise as the Monosoupape, or Single-Valve, Gnome) normally whirled round at no more than 1,300 rpm., and after arrival at Monaco it was decided by Victor Mahl that the 1,350 revs allowed by the Lang propeller then fitted were too high, and would lead to overheating. Whereupon an Integral propeller of coarser pitch was substituted.

Although, as earlier noted, the basic concept of the Tabloid was a joint effort, it is worth giving a final note about Pixton's seaplane, written by C. M. Poulsen, whose intimate knowledge of Sopwith design and construction has already been instanced. 'C.M.P.' was writing on the occasion of 'Uncle Fred' Sigrist's retirement in 1940, in the particular context of this seaplane, though probably having in mind the landplane original also. 'It is interesting to record', he said, 'that this machine was designed entirely by Sigrist, and he was in charge of construction.'

After the blue of the Mediterranean—meanwhile, back at the (Brooklands sewage) farm . . .

From the Sopwith Scouts—as the military Tabloids were now sometimes called—R. H. Barnwell of the Bristol Flying School (who is not to be confused with his brother F. S., and who had taken over as temporary test pilot) was getting splendid results. These single-seaters, as already noted, had a tail fin and a plain rudder, and though they retained the skid/wheel landing gear, the skids were longer than those on the original Tabloid and were necessarily supported by longer, and more sharply raked, struts, one specimen at least having three instead of two main struts. Conversely, the nose and other features had been cleaned up, and more than one form of engine mounting had been tried.

Though as yet un-numbered, this Service Tabloid—with Brooklands track behind, proclaims its maker's name below the blank oblong on the rudder. (Sopwith caption reads: '62—"Tabloid"—80 hp Gnome.'

Barnwell's best performance was, perhaps, not a test flight in itself but a delivery flight to Farnborough on 22 April. There a new speed range of 39.6—94.9 mph (63.6—153 km/h) was established. Pixton was quickly home from the racing to carry on the good work on behalf of 'the military'—and fortified not only by a good lunch stood by the Aero Club but by having heard the Marquess of Tullibardine affirm at that lunch that Sopwith was one of the world's foremost aircraft constructors.

The summer of 1914 saw the delivery of one of the first military Tabloids to No.7 Squadron, RFC; and thereafter the technical, as distinct from the operational, history of the type became obscured by the fog of war—especially so as the RNAS, as well as the RFC, employed it. Four were shipped in crates to Boulogne as early as 19 August, 1914, and two of these were flown by Lieuts Gorden Bell and Norman Spratt. This last-named officer has achieved an unusual kind of fame for having forced an enemy aircraft to land by manoeuvring in 'an aggressive manner', though he carried with him on that occasion a few steel darts; and he is also remembered for his tentative trailing of a grenade on a cable. A grapnel and a pistol have also been named among Spratt's weapons; but although he has been credited with firing thirty shots from a 'revolver', he must have been as dextrous as he was ingenious to reload about six chambers several times over.

In any case, air fighting was only just beginning in those times, and Tabloid production ended in the spring of 1915. Though the redoubtable 'Sammy' (Cdr C. R. Samson) seems to have approved a top-wing mounting for a Lewis gun, the two most famous Tabloids that came to his squadron at Antwerp—Nos.167 and 168 (which, with No.169, made up a brave little trio)—were destined to establish Sopwith and the RNAS in the low-level bombing, or strike, business. The importance of this function was implicit in the following letter received by Sopwith from the Director of the Admiralty Air Department on 28 December, 1914—alluding to some stirring events of the previous October:

'Gentlemen, With reference to the recent attack on the German air [ship] sheds at Cologne and Düsseldorf, carried out by Sqn. Cdr. Spenser D. A. Grey and Fl. Lieut. R. L. G. Marix, you may be interested to learn that the machines used were your Sopwith Tabloid aeroplanes.'

Sad would it be indeed if we were to leave the Tabloid landplane—a classically clean biplane if ever there was one—still hampered by the wheel/skid landing gear that was a feature of early-production specimens. Happily, the facts are otherwise, for when it returned to England from Monaco Pixton's racing seaplane was further modified for sporting use by fitting one of the neatest V-strut gears (two wheels, 'split' half-axles, no skids) that could have been desired. In this form it was intended to participate in the 1914 Aerial Derby—with 100 hp Gnome Monosoupape engine, and with R. H. Barnwell as pilot. A second Tabloid, from the RFC production batch, flown by Pixton and powered by an 80 hp Gnome, was also entered, but fog obliged both Tabloids to retire. Respecting speed one notes with special interest this passage in C. H. Barnes' book *Bristol Aircraft Since 1910* (the 'modified Scout' referred to being, of course, a Bristol product, and the Tabloid's spritely rival). Thus: 'On 14 May, 1914, Busteed put the modified Scout through an A.I.D. performance test at Farnborough and recorded a speed range of 97.5 m.p.h. to 40 m.p.h.; he then flew to Brooklands, where he gave a spectacular demonstration and in a handicap race was beaten only by seconds by Harold Barnwell in the 100 h.p. Gnome-engined Sopwith Tabloid.'

Concerning the very neat V-strut landing gear earlier mentioned, it is helpful to have the following description, written in 1915, because this general form of gear,

At first glance, Tabloids Nos. 326 and 394 were seemingly identical; but scrutiny of the wheel/skid landing gear proves otherwise—the latter having extra struts (or a V-type gear plus skids).

utilising steel tubes for the main members, was to become a veritable Sopwith 'trademark', being associated in particular with the 1½ Strutter. Having stressed the quality of simplicity, the 1915 description added: 'The axle is divided in the centre, where it is pivoted between two transverse members joining the apexes of the chassis struts. In order to prevent a downward movement of the centre of these members a single wire is taken from this point to the bottom of the body. Transverse rigidity is established by diagonal cross bracing between the front pair of chassis struts. This undercarriage, it should be pointed out, is not that fitted as standard on the Sopwith scouts. It was, in fact, quickly produced for the last Aerial Derby.'

Quickly though this form of gear may have come (and here, if one is not grossly in error, the hand of Sigrist is once more discerned) long was it to remain. 'Split-axle', though technically correct in a descriptive sense, is nevertheless misleading (having regard to later usage of the term) and 'stub axles' were the main transverse components.

Fine man, fine aeroplane: Victor Mahl with a Tabloid having the 'racing', or V-type landing gear, fabric-faired wheels and one of several forms of engine cowling tried on the Tabloid. A spinner adds an extra sporting touch.

It was not, in any case, sheer speed which was to distinguish the Sopwith fighters of later years so much as their powers of manoeuvre (Spratt's aggressiveness has already been instanced)—combined, of course, with effective armament. To emphasise the manoeuvrability factor (and remembering what construction T. O. M. Sopwith had placed on the term 'factor', as noted earlier) one could adduce no more convincing or entertaining instance than the following report of Brooklands activities, rendered as early as April 1914:

'Mr. Pixton was out on the "tabloid" Sopwith biplane; but *the* flights of the afternoon were by Mr. Barnwell on the Sopwith "tabloid" on which he achieved the distinction of being the first Brooklands airman to "loop the loop", after having quickly climbed to 4,000 ft. In the strong sunshine and at the height the evolution was carried out it was difficult to follow every detail of movement, but to most it seemed that the machine, after turning vertically upwards, fell to one side, and then, turning over, completed the loop prior to planing down. Mr. Tom Sopwith ran out to greet Mr. Barnwell and to ask him exactly what he had done, to which question the aviator called out: "That's just what I've come down to ask you!" His idea of his movements, however, coincided with the above opinion, so, apparently quite satisfied as to what he had to do the next time, he immediately restarted, and at his second attempt made three very good loops at a height of about 3,000 feet. On his return to *terra firma* he had a warm reception by the crowd near the "Blue Bird". It is worthy of note that nobody has previously attempted the feat on a machine approaching the speed of the Sopwith "tabloid". Mr. Barnwell seemed to think nothing of the feat he had accomplished, merely remarking after his second attempt: "I began to wonder where the world had got to that last time!"

That this little effort was indeed 'worthy of note' would not have been disputed by Pup and Camel pilots in the war then drawing near; nor was the rendering 'tabloid Sopwith' unknown in Service circles, for about six months before the war Lieut Col F. H. Sykes—named as a captain in the chapter on 'Other Men's Aeroplanes' was reported as having affirmed: 'Experience places the value of aerial reconnaissance beyond a doubt, not only in calm weather, but practically in all weathers; reconnaissance will however be opposed, and that in the air. Anti-aircraft guns will assist, but probably insufficiently. This fact being granted, there is the obvious necessity for different types of aircraft . . . Does the advanced cavalry require one type? The Headquarters of an Army another? Flanking divisions a third? Will there be a battle squadron? A fast scout flotilla? A squadron to hunt down and destroy airships and to attack aircraft bases? A low-flying armoured destroyer of ammunition parks and supply trains? A heavy transport convoy craft? A breakdown and repair craft?

'For all these duties, slightly different types and qualities are required. Even now, one can hardly imagine the tabloid Sopwith taking the role of a Sikorski argosy.' [*sic*].

An interesting study, this, not only in the coming needs of air warfare (which Sopwith was to do so much to meet) but in aircraft nomenclature also.

With the quality of manoeuvrability one was careful to associate a little earlier 'effective armament'; and though the full significance of the qualification 'effective' (as distinct from merely 'heavy') will later be emphasised—especially in regard to the 1½ Strutter and Camel—the moment now arrives to proclaim the Tabloid as an initiator of the 'deflector' propeller—one, that is, which had an arrangement of steel plates to protect it from damage from bullets fired straight ahead through the plane of the revolving blades. That this same innovation is sometimes ascribed to the type of aeroplane called the Sopwith Gordon Bennett appears to be amply justified, and

Though sometimes called a 'Sopwith Gordon Bennett' (see separate chapter following) this particular Tabloid variant was acquired by the Admiralty and was distinguished not only by a 'racing' landing gear and liberally ventilated cowling, but by a Lewis gun fixed on the starboard side and firing ahead by virtue of a deflector propeller.

is explained in the 'Gordon Bennett' chapter which follows the present one; but for the present at least this particular armament fitment is associated with the Tabloid. Relevant in any case is a minute written by Winston Churchill in April 1915 calling for a single-seat Naval aircraft 'with a Lewis gun firing through the deflector propeller.' (It is not, however, suggested that the device was in any sense a product of 'Winnie's toy-shop'—as a certain exciting, dangerous and unofficial emporium came to be known one war later).

Whatever its origins (probably French) one deflector scheme had associations with The Integral Propeller Co Ltd of Euston Road, London. A stout bracing-rod for the channel-form bullet-deflector was a prominent feature.

Beyond recording that between October 1914 and June 1915 Tabloid production totalled 36 machines (all Sopwith-built); that the later examples had ailerons instead of wing warping; and that allocated Service numbers included 123, 124, 167—169, 326, 394 and 1201—1213, little remains to be said of this historic little biplane; for though four were sent to the Dardanelles aboard HMS *Ark Royal* (later renamed *Pegasus*) early in 1915, though two were at the Isle of Grain and one or more served at Great Yarmouth, these achieved little operational distinction.

These facts notwithstanding, we must not omit to note that *Ark Royal* was commanded by that pioneer of aircraft armament Cdr R. H. (later Air Marshal Sir Robert) Clark-Hall, who himself reported that the 'four land machines' (the Tabloids) had so small a wing span that they could be hoisted in and out 'with their wings spread' (meaning fully rigged, for the wings did not fold). These aeroplanes, said Clark-Hall, could be flown off the ship's deck, but could not land back on to it, and they could alight on the sea only with great risk. One wholly fascinating observation by 'Clarkie', was that they could be set ashore, for service from land bases, 'without difficulty in the ship's boats'.

Shades of Sir Francis Drake and company; yet the armament of these particular Tabloids was no mere musketoon, but four or more 20 lb bombs, or quantities of steel spikes weighing 50 lb per thousand.

So, from Kingston-on-Thames and Monaco, it was out to the Dardanelles . . . Yet, later in the war came a reminder—a somewhat poignant one today for lovers of Sopwith 'original' material—of Pixton's historic peacetime victory, on the strength of which some observers steadfastly declared that it 'had proved the biplane to be as fast as the monoplane' (the dubiety of which would have been less in 1914 than now). The reminder just mentioned came after the Sopwith Sports of 1917, concerning which it was announced: 'The gate receipts totalled £52. 7s. 6d., the tickets sold realised £46. 16s. 6d., while the draw for the beautifully made model of the Schneider Cup winner brought in £16. 3s. 6d.'.

Where—many readers will join the writer in wondering—is that model today?

Tabloid

(80 hp Gnome)

Span 25 ft 6 in (7.7 m); length 20 ft 4 in (6.2 m); wing area 241.3 sq ft (22.4 sq m). Empty weight 730 lb (331 kg); maximum weight 1,120 lb (508 kg). Maximum speed 92 mph (148 km/h); endurance 3.5 hr.

Gordon Bennett

There are, one feels, two justifications for this present short chapter—both of them aeroplanes. The first of these machines, as shown in the accompanying photographs, was very distinctive, not only in matters of detail wherein it differed from any known form of the Tabloid, but in palpably basic features also. That this particular aeroplane had a special claim to the name 'Gordon Bennett' or 'Gordon Bennett Racer'; that it was taken over by the Admiralty as No.1215; and that it had associations—at Hendon—with Lieut Spenser Grey are indisputable facts; but concerning the second aeroplane mentioned some haziness persists, though in having tucked it away under the 'Tabloid' heading one feels relief and confidence in equal measure. Let us then recognise what the photograph showing it fitted with a deflector propeller proclaims it to be—a Tabloid, or a very close derivative. But equally let us admit the possibility of its having been intended (in one form or another) as a standby, or second string, for the 1914 Gordon Bennett Aviation Cup.

On the purely sporting aspects of the matter once again we seek counsel from Peter Lewis, writing of 1914. Thus: 'Racing had been extremely popular for some three years, but relatively few attempts had been made in Great Britain to design small machines exclusively for the purpose.' (Here we shall do well to remember Sir Thomas Sopwith's own assurance that Pixton's victorious Schneider Trophy mount of 1914 was originally a Tabloid from an RFC production batch). But to allow Mr Lewis to finish: 'Among the efforts just before the war began was the Sopwith single-seater derived from the Tabloid with the express intention of competing for the 1914 Gordon Bennett Aviation Cup. It used an 80-hp Gnome equipped with a deep-chord cowling blended into a finely conceived circular fuselage but was never raced and, instead, saw war service at Hendon in the RNAS as 1215.'

Sqn-Cdr Spenser Grey with the Gordon Bennett—the aircraft displaying its converging central struts, and even the mottled 'engine turning' round the shapely cowling of the 80 hp Gnome.

The most noticeable differences between the special Gordon Bennett machine and a typical Tabloid were the slender fuselage, with convex side-fairings running back to the deep-chord tailplane, and blending forwards with a particularly smooth cowling having an annular air-intake of very small diameter, behind the propeller spinner; the vertical tail-surfaces—alarmingly small in area, even though disposed both above and below the fuselage (the triangular fins seeming almost negligible ahead of the deep-chord, shallow and rounded rudder); and the slimness of the V-form landing gear struts.

Yet there were distinctions also in the wing cellule which, though generally of Tabloid form, had no stagger, and was braced by notably slender steel-tube struts—those immediately forward of the cockpit converging sharply upwards in side elevation, though the rear struts were single, and not inverted-Vs as on the 1915

Of undeniably sporting, if somewhat perilous, mien (by reason of its almost negligible tail): the Sopwith Gordon Bennett No. 1215.

Martinsyde two-seater, which had a somewhat similar arrangement. Behind the cockpit was the merest suggestion of a head-fairing, and another—probably more significant—subtlety was the blending of the lower wing-roots with the rounded fuselage, somewhat as on the Gloster IV racing seaplane of 1927.

It appeared, indeed, that the only discernible feature having some possible commonality between Gordon Bennett No.1215 and Tabloid (or Tabloid-derivative) No.1214 was the low-drag landing gear; and that 1215 should be appreciably the faster of the pair—having a reputed speed of 105 mph (169 km/h) with a Gnome engine of only 80 hp can well be understood.

Less comprehensible—having regard to the fact that the 1913 Gordon Bennett contest had been won, by a Deperdussin monoplane, at 124 mph (200 km/h)—is that any hope whatever should have been entertained for competitive success, though a more powerful engine may well have been in view.

Churchill

Although the names 'Sociable' and 'Tweenie' have both been justly applied to this enlarged development of the original Tabloid two-seater, it is important to note these facts: (1) the term 'sociable' was in common use during 1913/14—from which period the aircraft dates—for any aeroplane having side-by-side seats: (2) that 'Tweenie' was a colloquial description signifying that, in the Sopwith hierarchy, the type came somewhere between the Tabloid and the Three-seater (also remembering that a 'tweenie' in the purely domestic sense was a 'between-floors' maid—in this instance, it may later be considered, a maid-of-all-work). The name Churchill is nevertheless adopted here not only because it is equally 'correct', but because of its particular associations with 'a former Naval person'—and in the knowledge also that this is the name by which it would be best remembered by the British people who made it.

In any case, that in December 1913 Winston Churchill, as First Lord of the Admiralty, dictated its general specification (wherein dual control was a dominant consideration) is both plausible and laudable—especially so having regard to this 'one-off' aeroplane's future military service. And while apportioning personal credit we must reintroduce Lieut Spenser Grey, who not only took delivery of the machine in February 1914 and acted as aeronautical mentor to the First Lord, but took this same machine to war.

Unlike the Tabloid two-seater, the Churchill had two-bay wings, staggered, and carrying inversely-tapered ailerons. The fuselage was long and slender, thus giving the aeroplane an overall resemblance to an Avro 504, notably by reason of the long lever-arm afforded for the 'comma'-form balanced rudder. There was no fin, and the landing gear was of the twin-wheel/twin-skid type that was common on Tabloids. The 80 hp Gnome engine originally fitted was later replaced by a Gnome of 100 hp—presumably Monosoupape, though at what stage is uncertain. By March 1914, it does seem sure, however, that the number 149 had been painted on the rudder; and by this time also Spenser Grey—flying solo—had taken the machine to 10,600 ft. (3,140 m).

Clearly, here was an aeroplane fit for military service, and this it was to see in September 1914, crewed, in particular, by Sqn-Cdr Spenser Grey and Lieut Newton Clare. Dates and details of raids at this period are difficult to determine with absolute accuracy; but that Spenser Grey 'lost' a bomb from No.149 seems well-founded—this projectile supposedly having vibrated off the 'pipe-rack' holder provided for it, together with an additional number of other bombs.

C. G. Grey (who, though not related to Spenser Grey, knew him well) related the circumstances which followed 149's return to Antwerp (having got itself 'completely lost') in the following terms—which the reader is invited to interpret as he will:

Called in this book the Churchill, though otherwise known as the 'Sociable' or 'Tweenie', this Sopwith aeroplane, with its two-bay wings, long slender fuselage, and scuttle-fronted side-by-side seats, was not only distinctive in appearance, but distinguished 'in the field' likewise (one particular field being shown in the next picture).

'When Spenser Grey and Newton-Clare [*sic*—the latter name was variously rendered, as, for that matter, was the former, though Spenser was positively spelt as given here] landed they saw a vacant space in the pipe-rack which showed that one of the bombs had vibrated itself off. They only hoped that it had fallen in Germany and not in Belgium.

'After dinner they were sitting in the lounge of their hotel—war was a comfortable game in those days—when an excited Belgian staff-officer dashed in and told them that a complaint had come from the Dutch Government that one of the Allied aeroplanes had dropped a bomb in the city of Maastricht, and had blown up a school and some houses and had killed a lot of women and children, and that the Dutch Government were seriously contemplating declaring war on Belgium. Spenser turned to Newton-Clare and remarked 'That must have been a damned good bomb.'

The Churchill No. 149 (the '4' whereof is clearly visible on the rudder) with Sqn-Cdr Spenser Grey, Lieut Newton Clare, and personages who may not be unconnected with the affair of the 'lost bomb'—as recounted in the text.

On this Churchillian note our brief appreciation of Sopwith tractor biplane No.149 might well be concluded—except, perhaps, to remark that the Navy of the period would have been well pleased to possess a bomb of the potency attributed.

Yet even now one footnote remains to be added, and this on the personal authority of Sir Thomas Sopwith. Once again, Spenser Grey is the dominant figure, though in this instance accompanied by Jerry Aldwell, a Naval engineer officer. The pair, Sir Thomas recalled, were flying the Churchill at Eastchurch when they got into a spin and went into the ground. They had been flying at about 90 mph (149 km/h) in this aircraft which—Sir Thomas' figure, a shade high though it may seem—stalled at about 50 mph (80 km/h). The stick then being pulled hard back, the spin occurred, the outcome being not only the sudden finish mentioned, but a better understanding of the nature of a spin and its avoidance.

Thus in several ways, the Churchill may be reckoned a more useful product of 'Winnie's toy-shop' than some that came in later years. Moreover, it is not uninteresting to note that the affair of the 'lost bomb', as earlier recounted, occurred in or about September 1914, and that on 3 October there arrived at Antwerp from Dunkirk the very man whose name heads this chapter.

Type C

The designation of this large, experimental torpedo-dropping floatplane may—as has sometimes been suggested—have been a sequential one (earlier designs having supposedly been nominated A and B) or the C may have connoted Calshot, for this was the Naval air station with which the aircraft had special associations. In any case, it will be well to preface this study of the machine—to which the Service number 138 was allocated—with a sketch of torpedo-dropping experiments in the months before the First World War.

Although we have the personal testimony of the Italian General Alessandro Guidoni that as early as 1912 he had been ordered to help Pateras Pescara (then a lawyer) to build a torpedo-dropper which Pescara had proposed to the Italian Navy, and that after preliminary experiments with his 'faithful Farman' he (Guidoni) went on to build a special monoplane—having two 200 hp Gnome engines in tandem and hydrofoil floats—from which, in February 1914, he dropped a torpedo weighing 750 lb (340 kg) we also know that in the very same month Winston Churchill, Great Britain's First Lord of the Admiralty, said: 'The objectives of land aeroplanes can never be so definite or important as the objectives of seaplanes, which, when they carry torpedoes, may prove capable of playing a decisive part in operations against capital ships.'

In June 1914 Churchill said that the development of a 'torpedo-carrying seaplane' would greatly affect the value of 'this type of aircraft' (presumably seaplanes generally) and added that he hoped that it would very shortly be possible for a seaplane 'to fire a 14-inch torpedo.' This weapon would be sufficient, he expected, to sink a small cruiser, a destroyer or a transport. By that time, said Churchill, torpedo-dropping had already been practised from what he called a 'mock-up machine'—adding that torpedo officers expected a high degree of accuracy, and that an order had been placed for 'a machine to carry an 18-inch torpedo.'

Well before these predictions and disclosures by Churchill—as early, it appears in fact, as the summer of 1912—a paper discussing the torpedo-dropping problem had been written by Lieut D. Hyde-Thomson and placed before the Captain of HMS *Vernon* (the Navy's establishment responsible for specialist training and development in all matters concerning torpedoes). At that time, or somewhat later, a particular interest in Hyde-Thomson's ideas was shown by Capt M. F. Sueter (later Rear-Admiral) Director of the Air Department of the Admiralty, which led to the application for a secret patent—jointly in the names of Sueter and Hyde-Thomson—early in 1914. By that time, it is tolerably clear, Sopwith, as well as Short, were regarded as intended, or potential, builders of torpedo-dropping seaplanes; and, having regard to several contradictions respecting early British torpedo-dropping work generally, one quotes directly the following declaration by Sir Arthur Longmore:

'I was commanding the experimental [Naval air] station at Calshot from January to September 1914, and it was there in July 1914 that for the first time a 14-in torpedo was taken into the air and dropped successfully. Indeed, I did it. The actual seaplane was a 160 Gnome Short, the preliminary experiments on the water having been carried out for some months previously in a Sopwith taxiplane, a clipped-wing seaplane. Shortly afterwards the war broke out and what may have been really only a "stunt" experiment was regarded as an operational possibility. In August 1914 a 200 h.p. Canton-Unné Sopwith seaplane regularly went into the air with a torpedo and a considerable amount of target practice was carried out . . . '

Sir Arthur gave the following additional facts in a letter to the present writer some years ago: 'It seems probable that confusion has arisen in some accounts of the first drop by reason of the fact that there was at Calshot at the time a Sopwith Canton-Unné *taxi*-plane (centre section only) which was used for developing the release mechanism and for registering the behaviour of the floats, undercarriage and general construction under taxying conditions on the water. Lieut. Robin Ross frequently handled this taxiplane. Lieut. Hyde-Thomson was specially appointed to Calshot to take part in the development of the torpedo carrying . . . Shortly *afterwards* a Sopwith Canton-Unné floatplane was delivered at Calshot fitted for carrying the 900 lb. 14 in torpedo (the same as I had previously dropped). With this machine I and my other pilots frequently flew with the torpedo and did a few successful runs at a target.'

These matters having thus been presented by one of the principal figures concerned, and taking no further cogniscance of the 'Sopwith Canton-Unné *taxi*-plane' mentioned by Sir Arthur (especially as a similarly rigged Borel floatplane has also been associated with Hyde-Thomson's early work) we may proceed to examine the Sopwith Type C—or, as a reasonable supposition, what Sir Arthur Longmore called the 'Sopwith Canton-Unné floatplane', which, although frequently associated with the year 1913, seems more properly to date from 1914.

Understandably—having regard to the difficulty of lifting even a 14 in torpedo, and the early prospects for the airborne launching of one of the heavier 18 in patterns—this was a large and powerful machine. Its size, in fact, is implicit in the sheer breadth and rigging of its biplane wings, which might well have been similar to those of the 'machine of 80 ft. span' mentioned in the chapter on 'Pushers and Gun Buses'. In any case, they were not only of four-bay construction, but had the additional distinction of strut-braced top-wing extensions. Strut-connected ailerons were fitted on all four wings.

The Sopwith Type C experimental torpedo-dropping seaplane, as evident here, was a very large aircraft, and had an engine installation of uncommon interest. The 'half-hoops' seen above the floats were probably for springing, and not associated with the torpedo gear.

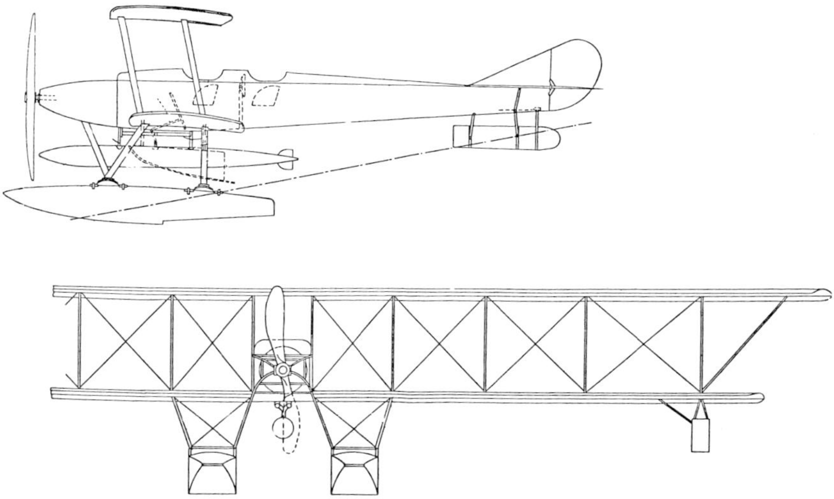

While not corresponding precisely with the characteristics of the Type C as shown in the photograph, these drawings clearly depict a machine of the same general type, though their primary purpose was to show how the torpedo was slung. Note especially how the typically Sopwith method of attaching the floats to the wings gave a (literally) clear advantage for torpedo-dropping.

The only known surviving photograph is reproduced, and it would be pointless to add the conjectural to the obvious. Certainly, the features commanding most attention (apart from the impressive size) are the powerplant installation and the gear for carrying and releasing the torpedo. The photograph shows—or at least suggests—side-mounted radiators for the 200 hp Canton-Unné (Salmson) water-cooled radial engine. Probably by reason of the name 'Canton', this engine has sometimes been described as having Swiss origins, though this is evidently not so, the two inventor/patentees concerned—MM Canton and Unné—having been French, and their truly remarkable engines being products of the Société Anonyme des Moteurs 'Salmson', of Billancourt, Seine, France, or of the Dudbridge Iron Works Ltd, of Stroud, Glos, England.

A point concerning the powerplant that may be of more than trivial interest is that in the line-drawings reproduced (and prepared with the primary object of showing the torpedo-stowage on an obviously Sopwith-type aircraft more or less identifiable—except, perhaps, for the slightly staggered wings—with the Type C) the representation of the engine is apparently tentative. There could, of course , be several explanations of this; but one that instantly comes to mind is the following: Whereas the 200 hp Canton-Unné engine was normally mounted conventionally (that is, vertically)—driving the propeller directly or through a shaft (the latter apparently being the case in the Type C as shown in the photograph)—certain engines of the same family were constructed with the cylinder-axes horizontal and the crankshaft-axis vertical, the propeller then being driven at right angles by means of bevel gearing. It was contemporarily stated: 'This horizontal disposition is often adopted in dirigibles, and occasionally in *large sea-planes* . . . ' (present author's italics).

Whatever the facts or intentions, however, the 200 hp fourteen-cylinder water-cooled radial engine of the Type C, with its shaft drive and side-mounted radiators, is certainly worthy of this special note.

In one form of torpedo gear apparently designed for the Type C (and evidently that which the line-drawings show) the projectile was to have a single-point suspension from a shackle on a longitudinal beam, and was to be steadied against swaying by fittings that eventually became known as 'crutches'. Release—effected from the pilot's seat—was to be mechanical.

Having introduced these notes on the Sopwith Type C with reference to early Italian experiments in torpedo-dropping it seems fitting to append the following:

In 1913 it was reported—apparently on good authority—that three 'Sopwith type hydro aeroplanes' had been ordered by Italy; and though these were termed, in one context, as 'torpedo air-craft' it must be recognised that this particular expression was in fairly common international currency at that period to connote heavier-than-air machines of the 'fighting', 'scouting' or 'bomb-dropping' persuasion, as distinct from lighter-than-air craft. Whatever the implication, no Sopwith aircraft of any kind are known to have been supplied to Italy before the war of 1914.

Yet even so, though Erskine Childers' story *The Riddle of the Sands* had first been published ten years earlier, those same ten years were fraught with many mysteries of fact as well as fiction. One doubtless unrelated, but nevertheless curious, fact was that among Erskin Childers' brother-officers in 1914 was Lieut Robin Ross, earlier singled out for particular mention by Sir Arthur Longmore.

Folder Seaplane Type 807

In concluding the earlier chapter dealing with the 'Circuit Seaplanes' a special emphasis was placed on field of view—this same consideration having led Short Brothers (who were Sopwith's chief rivals in the Naval-aircraft business) to prepare a more-or-less contemporary design for a tractor floatplane wherein the pilot and observer were seated almost entirely forward of the unstaggered wings. The engine in this Short aeroplane was to be mounted in the fuselage, in line with the wing trailing-edges, and driving the nose-mounted propeller by means of a shaft running beneath the two seats. The wings would fold, and, except for the special provisions just mentioned, the machine would have appeared to be similar in all essentials to the familiar Short 'Folders' that saw much service. The occupants were to have—for a tractor layout—a matchless field of view, especially for reconnaissance, defensive armament having been little considered (except in its theoretical and experimental aspects) until the coming of war.

Sopwith's partial solution to the field-of-view problem in their own 'Folder' seaplane, or Admiralty Type 807, was to place the observer under the centre-section leading-edge and the pilot under the trailing-edge. For the observer there was a small cutout; for the pilot a large one. Thus, although there was a general relationship between the 1914 Circuit Seaplane, the drawings marked 'D3' (to which allusion has been made in the context of that seaplane), and the Admiralty Type 807 now under study, the most significant development was folding wings.

Having already named Short Brothers as Sopwith's chief rivals in the Naval-aircraft business, it is fitting now to proclaim Short's undisputed leadership; and

The particular aircraft which led to the official styling 'Admiralty Type 807' for the class more generally known as 'Sopwith Folder' is here portrayed in a position and condition which afford their own explanations. Note the top-wing overhang, float arrangement, tail design and cockpit spacing.

this can best be done by quoting what Oswald Short himself had to say—many years before his death in 1969. Thus: 'My late brother, Horace, designed and built the first seaplanes with folding wings at Eastchurch in 1912. A patent was taken out for the invention. At a later date Mr. T. O. M. Sopwith wrote to me, saying he wanted to build an aeroplane with folding wings, and what royalty payments would

Another aspect of Sopwith Folder No. 807, showing not only the folded wings but the partly metal-sheathed propeller (for protection against water-damage).

we ask for. I replied £15, and received a cheque for that amount. Strange as it may seem, that was the only sum Short Brothers ever received for that lasting invention from an aircraft constructor.'

That the wings of the Sopwith Type 807 had no stagger—though its near-relation the 1914 Circuit Seaplane did incorporate this feature—is probably due to the wing-folding requirement; and this consideration, jointly with that for a good field of view, must have dominated the design. Though the Type 807 appeared in more than one version (or with relatively minor alterations) this fact is of secondary importance.

Twelve Sopwith Folders were built—in two batches, numbered respectively 807—810 and 919—926, the first aircraft being delivered in July 1914. All were constructed by Sopwith, and all initially had the 100 hp Gnome Monosoupape engine, installed in a Tabloid-like cowling. Clearly, these seaplanes were underpowered, especially when carrying bombs or wireless equipment in hot climates. More, indeed, is known concerning the operational performance of these aircraft than can be stated with confidence respecting their technical characteristics.

Folder No. 920 (nearest camera) shows distinct differences in wing cellule as compared with the preceding side view of No. 807.

The following facts, however, are relevant: Although—typically—the span was about 36 ft (11 m) both this measurement and the overall length may have varied quite widely. Recorded differences in span could well be accounted for by the extent of the very marked overhang on the outer portions of the upper wings, which were wire-braced not only to kingposts above, but to the lower wings also. Ailerons were fitted to the upper wings only, though it is certain that the lower wings were of at least two different spans. Length might have varied because of differences in tail design, two known forms whereof are confirmed. Engine installations likewise may well have varied, some cowlings being more bull-nosed than others—or even removed altogether—as in the instance of No.920 when operating at Niororo Island in the *Königsberg* affair (to improve cooling and save weight). Conversely, the floats were sometimes deliberately filled with water—though only when an aircraft was not in use—to keep them from warping in the heat.

Certainly the airframes—equally with the engines—were prone to get 'under the weather'. Take-off was generally poor (when it was possible at all, in a hot climate or a choppy sea), and the sprung floats of No.922 at least were criticised not only for their fragility, but for imparting 'bounce' at take-off. Rate of climb and ceiling were rendered in expletives rather than figures. And as though the take-off was not a sufficient problem, the airframe came 'unstuck' when not required to do so, the trouble here being the type of glue applied in England.

Although no guns were carried, bombs most certainly were—seemingly on the float cross-bracing struts; and in this regard one may now enlarge on the well-known instance of one of the 'anti-*Königsberg*' Sopwith Folders refusing, day after day, to leave the water with a full complement of crew and petrol—plus two 50 lb and four 16 lb bombs. Less understandable than the Sopwith's stubbornness in this instance is the fact that it should have been carrying 50 lb bombs at all, even though by early 1918 the Bomb, H.E.R.L., 50 lb., Mk.IV was apparently in current use, and a Mk.II pattern was evidently still being issued at about the time of the Armistice. Conceivably the bomb carried by the Sopwith Folder was an adaptation of a 50 lb lyddite-filled artillery shell, as used in the Boer War. In any case, the Mk.IV 50 lb bomb had a particular association with internal (vertical) stowage in the D.H.9—a much later aircraft than the Sopwith Folder, though fittings, described as a 'bomb band and lug', were available late in the war for horizontal stowage. A particular point is made of this fact because, whereas the early Sopwith seaplane may well have been the first aircraft to carry the 50 lb bomb operationally, then among the last may well have been the Camel 2F.1s which raided the Tondern Zeppelin sheds in July 1918.

Whatever the facts concerning armament, were one asked to enumerate the truly significant events in the development of British Naval aviation, one would certainly include among them the delivery for our first 'custom-built' seaplane-carrier HMS *Ark Royal* (the description of the ship is, I think, a just one, though the craft had first been intended as a merchantman) of Sopwith Folders Nos.807, 808 and 922—to Blyth, Northumberland. This was at about the time of *Ark Royal*'s commissioning, in December 1914, and the pilots employed by Sopwith to check these aircraft out—

In this view of Folder No. 920 the uncowled engine (increasing cooling airflow and decreasing weight) is even more in evidence than the vertical tail surfaces, which differ from those seen in the side view of No. 807.

after the machines had been assembled in dockside fish-packing sheds—were Harry Hawker and Ronald C. Kemp.

Caution having earlier been exercised in noting that 'Clearly, these seaplanes were underpowered', that 'the lower wings were of at least two different spans' and that 'engine installations may well have varied' (though meaning, in that connection, installations of the 100 hp Gnome Monosoupape) one final qualification may now be added: in at least one superficially similar Sopwith seaplane operated in the Dardanelles an installation could possibly have been made of a (presumably more powerful) water-cooled engine, this having a frontal radiator and thus giving the appearance of an unequl-span Type 860—two examples of which type did, in fact, go to the Dardanelles—though having no underwing floats for lateral stability. Fin area certainly seems to have been greater than on either form of the 'Folder' (Admiralty Type 807) illustrated in this present chapter, and the 'superficially similar' machine just mentioned could indeed have been a form of the Type 860, for this, as later noted, had—initially at least—wings that were arranged to fold.

Although tabulation of data for the Folder Seaplane Type 807 would be futile, it is worth remarking that, while the type was serving in the Dardanelles (having been taken there in *Ark Royal*, which left Sheerness on 1 February, 1915) it was reported to have a fuel capacity for 4½ hrs' flying 'when equipped for scouting only'. It was further reported that, although not fitted with a 'long distance wireless installation' it could carry 'a light W/T transmitting set'. This set was 'effective up to 10 miles' and (it was added) 'works on a 700 foot wave which can be received by T.B.D.s' (torpedo-boat destroyers).

Type 137

That some close association existed between this curious and obscure 'one-off' Sopwith type and the 1914 Circuit Seaplane is apparent from its general appearance—the non-folding wings (lightly staggered, and having inversely tapered ailerons) being the most obvious similarities. To a marked degree, however, the true derivation of the machine (numbered 137, and bearing the Admiralty type-designation that heads this chapter) is concealed by the unequal-span wings, with strut-braced upper extremities; by the bedazzling—even hypnotic!—effect of the huge roundels painted on the under-surfaces of the upper wings, as well as the lower ones; and—most of all, perhaps—by the entirely different engine installation.

The engine was, in fact, a water-cooled 120 hp Austro-Daimler, the deep frontal radiator for which appears to have been outsize—projecting, as it did, high above the engine itself. The bizarre appearance thus conveyed (which could, of course, have been accounted for by any of several considerations, among them water-clearance) was heightened by the echelon arrangement of exhaust ports in the heavily louvred side-cowlings. A long fore-and-aft member lower down and in line with the two cockpits was apparently a foot step. Behind and above the engine was what appears to have been a tank, substantially oblong in form; and on the rearmost inboard bracing strut was a wind-driven pump, or the like.

Here it is pertinent to note that the '120-hp Beardmore Austro-Daimler Aero Engine' (Beardmore having also obtained a licence to build the German D.F.W. biplane) was being promoted before the war by the Austrian Daimler [*sic*] Motor Co Ltd, of Great Portland Street, London—partly on the strength of Cody's

A puzzling machine from many aspects was the Sopwith seaplane shown here, which bore the Admiralty type-number 137 and was powered by a 120 hp Austro-Daimler engine.

success in the 1912 Military Trials, when using an engine of this general type. Further, a number of Royal Aircraft Factory designs were prepared round the 120 hp Austro-Daimler, and this same engine was, in fact, installed in the standard R.E.5 (to which—on reflection—the Sopwith Type 137 bore a certain resemblance). Conversely, however, the R.E.5 installation was a distinctly 'fancy' one, with open-fronted cowling and an internal radiator set far back in the fuselage.

Whatever its functions—intended or realised—No.137 survived until some time in 1915, when it was overhauled by Pemberton Billing Ltd at Woolston, Southampton.

Type 860

The primary points to note concerning this big production-type torpedo-dropping floatplane of late 1914 are the following:

(1) It bore a near relationship to the experimental Type C, and, especially when provided (as it was in some instances) with strut-braced overhang on the top wings, resembled that aircraft very closely.

(2) It was more or less contemporary with the Short Admiralty Type 184 and the Wight Admiralty Type 840, and was intended to meet similar requirements.

(3) Like the Short and Wight types just named, it was designed specifically for the new 225 hp Sunbeam engine—a powerplant so important in the development of British Naval aircraft that the Short Type 184, upon which it was decided to standardise (especially after the torpedo-dropping requirement became secondary, following initial successes by Shorts during the Dardanelles campaign of 1915) was familiarly known as the 'Short 225'.

(4) An installation of a Sunbeam engine, though one of lower output (150 hp) was made in the Admiralty Type 806 Gun Bus, as noted in a preceding chapter.

(5) The wings—initially at least—were arranged to fold.

The earliest Service numbers known to have been allocated to Type 860 seaplanes were 851—860 (ten aircraft), and of these No.854 was being tested, by Victor Mahl,

Sopwith Admiralty Type 860 seaplane No. 851, showing wings of unequal span, with top-wing kingposts, and strutbraced top-wing extensions.

over the Solent at the beginning of 1915. Twelve more (Nos.927—938) were also ordered, and—except for numbers 933, 934, 936 and 937—duly delivered, thus giving the RNAS a known total of eighteen Type 860 seaplanes—all Sopwith-built. Beyond the facts that the type was used in the Dardanelles and was flown from the Isle of Grain, however, little is known of its Service history. Thus it is worth noting that—in particular for Short Type 184s—demands for 225 hp Sunbeam engines (later named Mohawk) must have been heavy; and here too it is especially relevant to note the following recollection by Rear-Admiral Murray Sueter, who, as Captain Sueter, had been Director of the Air Department of the Admiralty before the 1914 war. This officer said:

'After the war broke out, we required all Mr. Sopwith's efforts and those of his factory to produce high performance machines, then just beginning to show some promise. But Hyde-Thomson and myself [the name Hyde-Thomson will be remembered from the chapter on the Type C] were quite determined to succeed with a torpedo machine. So I sent for that fine pioneer seaplane constructor, the late Mr. Horace Short. When I explained my requirements to him and the great weight that had to be lifted with a 225-h.p. Sunbeam engine . . .'

But the successful outcome of that meeting—the historic 'Short 225'—is well enough known; and having now re-emphasised the Navy's special interest in Sunbeam engines we may proceed with our study of the Sopwith Type 860 torpedo-dropping seaplane, which appears to have continued in service (in however lowly a role) until 1916.

Here, once again, we are involved with the Sunbeam story, for one of the most arresting visible features of this big Sopwith was the immense solid-looking block, towering not only above the engine but the top wing also. This was not, in fact, the radiator—in the familiar Short-style location—but the exhaust manifold. The Sopwith's radiator was positioned in the nose, just behind the propeller (sometimes two-bladed, sometimes four-bladed).

Although existing photographs show clearly that Nos.851 and 859 had wings of unequal span—the strut-braced upper-wing extensions having additional top-surface bracing from kingposts—and although these particular machines were characterised also by the elegant 'Sopwith' tail surfaces (much as on the production

Tabloids and Folder Seaplane) No.928 or 938, here depicted on the water, had wings of equal span and a much larger fin, no longer triangular, but curved.

In the Sopwith tradition by this time established, the two main floats were sprung, and like the tail float (seen well-nigh submerged) were carried on struts of great height. Being attached to the fuselage, and not to the wings, the main alighting gear, in the form depicted, would appear to have been less favourable to torpedo dropping than that of the Type C; though the point is by no means conclusive, having regard to the astonishingly low-slung torpedo stowage on the Short Type 184. Aiming the torpedo must, in any case, have been a truly hit-or-miss business, for the pilot occupied the rear cockpit. Defensive armament could well have been

Sopwith Admiralty Type 860 (No. 928 or 938) with wings of equal span and vertical tail-surfaces increased in area.

intended or improvised, jointly with the top-wing aperture over the front cockpit (especially so as Owen Thetford's Putnam book *British Naval Aircraft since 1912* records that the Type 860 was used on patrols in home waters during 1915 and 1916); but—as with many other points concerning this Sopwith type—there is no certainty in this regard. More positively it can be recorded that Nos.851 and 852 were not written off (in the clerical sense) until March 1917, and that Nos.931 and 932 were at the same time reduced to spares in the Supermarine works (successor to Pemberton Billing) at Woolston, Southampton.

The name Pemberton Billing having now been mentioned twice (formerly in the context of the Type 137) it is interesting—though not necessarily significant—that those well-known Sopwith characters Howard Pixton and Victor Mahl were both present at early tests of the P.B.9—the 'seven day 'bus'—concerning which aeroplane some mysteries persist. Mahl, in fact, made the first flight—shortly whereafter the little single-seater was seen at Brooklands.

Whatever its vices or deficiencies, the Sopwith Two-seater Scout was of trim appearance, even when fitted (as in this instance) with a bomb-carrier, more or less between the rear legs of the heightened landing gear. The same aeroplane is shown in the next picture.

Two-seater Scout

'Spinning Jenny' was the name by which this generally unpopular biplane became known to its RNAS crews—a name which first came prominently to public attention (as, indeed, did the aeroplane itself) during the 1950s by reason of recollections then aired by J. C. Brooke, Sqn Ldr RAF (Ret).

While based at Killingholme, close by the River Humber, in Lincolnshire, this former RNAS officer had experienced the propensity of the particular Sopwith aeroplane now reviewed to spin at the least provocation. From his first—unpremeditated—spin on the type (No.1055 was the specimen concerned) he regained control, and next day did two deliberate spins, though recovering from both only after a height-loss of about 1,000 ft. A hardly less valuable contribution to aeronautical history (for the deliberate spins must have been among the earliest) was also made during the 1950s by Sqn Ldr Brooke's declaration in these terms:

'As regards performance, the acceptance test in those days [1915] consisted of a climb to 3,000 ft. and stay there for twenty minutes. The climb used to take about 20 minutes, and the top speed low down was about 55 to 60 knots. This climb was not quite as good as the Curtiss J.N.4s, which were also at Killingholme, but the speed was slightly better.'

Of the Two-seater Scout's propensity for spinning there will be more to say later; but concerning all-round performance the quotation just given enters this book at an opportune time. In the first place it emphasises the deplorable performance of a supposedly operational two-seater aircraft on the power of a 100 hp Gnome Monosoupape rotary engine (for this was the powerplant indeed); second, in the phrase 'about 55 to 60 knots', the very word 'about' is fully as eloquent as the range of the figures themselves; and third, if the trouble is taken to convert those same figures thus—55 kt = 63 mph = 100 km/h, and 60 kt = 69 mph = 111 km/h—the futility will be apparent of claiming anything like 'exact' performance data for low-powered heavily laden aircraft—perhaps improperly rigged, and suffering from an overheating engine (not to mention the ravages of time or weather).

As for citing the Curtiss Jenny (quite coincidentally named, it seems) as a pacemaker or yardstick, it might insult most readers to remark that this type was, supposedly at least, a trainer pure and simple (though in truth it was neither pure nor simple); nevertheless, it might surprise those very readers to know that the Curtiss designer concerned—B. Douglas Thomas—had worked for both Avro and Sopwith, and that the basic Model J design had been started in England.

That the Sopwith 1914 Circuit Seaplane had been built to drawings marked 'D3' has already been recorded; and that the Folder Seaplane and the Two-seater Scout (which latter type appeared in March 1915, a few months later than the Folder) bore some kinship with the general form of aircraft so designated—widely spaced cockpits with headrest fairings and a low-powered rotary engine being earmarks of the breed—is evident. Apart from the obvious distinctions in landing gear, and non-folding wings of equal span (36 ft, 11 m) as seen in the photographs, the Two-seater Scout had a shorter fuselage (apparently omitting one bay) which could well be

Historic, rather than technically informative, is this picture of 'a former Naval person' (Winston Churchill) inspecting Sopwith Two-seater Scout No. 1062 at Hendon in 1915. The camouflaging of the wings is the more curious because the fuselage and tail have not been similarly treated.

accounted for by its smaller side area, as compared with the Folder Seaplane, with its floats. Strut-connected ailerons were fitted on all four wings, and the tail resembled that of the Folder. The main landing gear was of simple V-strut form (made taller when bombs were carried, as later described), and though the tailskid was sturdy and tall, the overall appearance of the Two-seater Scout must be accounted trim, and worthy, perhaps of a more powerful engine (the spinning proclivity notwithstanding). Weight is lent to this reasoning as the type is known to have gone to war not only with assorted small arms (though not machine-guns)—and bombs, carried either additionally or as an alternative load.

It is important here to note that during 1915 bombs were regarded as anti-airship, as well as anti-terrestrial, weapons; and thus the Sopwith Two-seater Scouts that were based not only at Killingholme, but at Hendon, Chingford and Great Yarmouth also, may indeed have been fulfilling the contemporary function of 'scout'—in the sense that their purpose was air fighting rather than reconnaissance, or 'scouting' in the Boer War tradition.

In general form—and especially in being strut-attached far below the fuselage—the bomb installation resembled that made on some Sopwith Schneiders (*see later chapter*). The bombs themselves—which were thus to some extent between the rear legs of the heightened landing gear—could well have been of the pattern called 'small petrol bomb', the 16 lb carcass incendiary, the 16 lb H.E.R.L. or the 20 lb Hales H.E. The familiar '20 lb Cooper' had not at that time arrived, although an early form of Cooper fuse was designed for anti-aircraft work, and its designer had early associations with F. Marten Hale. Should a bombsight of any kind have been fitted this might well have been of the 'Lever' type, then used by the RNAS.

The last of twenty-four Sopwith Two-seater Scouts (Nos.1051—1074) delivered to the Service just named arrived at the Royal Naval Air Station, Chingford—on the north-eastern outskirts of London—in June 1915. This fact emboldens one to record that the name of W. R. D. Shaw, of Chingford Road, Walthamstow, was linked with a scheme for fitting an aeroplane, generally resembling the Two-seater Scout, with a tailplane having 'a negative dihedral angle to prevent a spin or nose-dive due to side-slip when banking'.

'Spinning Jenny' or no, we may allow the last word to a writer in *Flight* (March 1915). Thus, of 'Hendon last Saturday': 'The first visitor to arrive in the afternoon was Mr. Harry Hawker on a tandem two-seater Sopwith tractor biplane. This new machine differs considerably from the usual Sopwith biplanes, and I can best give an idea of it by saying that it is an intermediary between the Scout and the larger two-seater' (whatever this may have signified—probably 'between the Tabloid and the two-seat form of the Three-seater'). 'The planes are not staggered, as in nearly all other Sopwiths, but as the passenger sits well forward in the body he is on a level with or slightly ahead of the leading edge of the lower plane. The chassis is of the V-type, and is built of steel tubes throughout . . . Mr. Hawker tells me that she climbs exceedingly well' [good for Harry, though Sqn Ldr Brooke's opinion clearly differed here] 'besides being easy to handle and comfortable to fly. For those who had not had the opportunity to see Hawker's piloting for the past few months, it was quite a treat to watch him coming in from Brooklands travelling at a great pace, and to note that his piloting has lost none of its brilliancy since the days of looping and race meetings.'

Alas, the 'looping and race meetings' were now fast receding, and Neuve-Chappelle and Ypres were more in people's minds than the Hendon in its earlier, enchanted days. Already Two-seater Scouts were being fitted with what was then known as 'bomb-dropping gear'.

Schneider and Baby

The three-float single-seat Sopwith seaplane which, in November 1914, was put into production for the RNAS was—as the commonly used name 'Schneider' proclaimed—very much the same aircraft as the floatplane Tabloid with which Howard Pixton had made history at Monaco in the preceding April (see under 'Tabloid')—still with the relatively powerful 100 hp Gnome Monosoupape engine (though not specially tuned) on the 'nose-bearing' mounting.

By noting that the name Schneider was 'commonly used' one has in mind the quite extensive currency—in official and private documents alike—of the appellation 'Schneider Trophy [or Cup] Seaplane'. For was it not vastly glamorous (and good for recruitment) to be able to assure one's girl-friend that one was flying a 'real racer' of such high renown rather than just 'a little seaplane'—especially so as the Tabloid was quickly to become more of a hack than a hacker-down of Zeppelins, or anything so fierce. And here it may be emphasised that the later 'improved' Schneider that was officially called the Baby and will be described under that name, continued to be loosely referred to as a Schneider; and as J. M. Bruce recorded plaintively in his *British Aeroplanes 1914-18* 'There is much confusion in most of the records of the exploits of both types.'

Though initially the Schneider retained a triangular fin, as on Pixton's racer, this surface was later much enlarged in area (as was common with floatplanes) then being curved in contour; and this development was commonly associated with the fitting of ailerons instead of a wing-warping system, though the latter was used on the early Schneiders. An especially noticeable difference—even on the early Schneiders—was the additional diagonal strut in the float-attachment assembly (a feature that was to recur in the Schneider's lineal descendant the Hawker Nimrod, though in that instance respecting the centre-section struts and not those which attached the floats).

To accommodate a Lewis gun, firing upwards and forwards over the propeller, an aperture was made in the Schneider's centre section; but the nature and disposition of armament varied widely, and deflector plates were sometimes fitted to the propeller. Further, it must be noted, although rifles, carbines, shotguns and pistols were quite commonly carried aboard aircraft in 1914/15 (the Sopwith Two-seater Scout, for instance, sometimes had a shotgun with a chain-shot load, this last comprising lumps of lead linked by a steel wire to rip open fabric, leaving the wind to do the rest) with the Schneider a shotgun-related weapon of a particular type was associated. This weapon was made by Holland & Holland, who also developed the Paradox gun that was familiar to early fighting aircrew; but though a 12-bore, the particular weapon now in mind (the 'Aero gun' as it was called) had a recoil-damping device, and a muzzle-shield to prevent an inrush of air during loading. The gun was fixed to the Schneider (or was intended to be) aligned so as to fire chain-shot or buckshot clear of the propeller.

For carrying anti-aircraft or anti-terrestrial bombs, arrangements were likewise varied, though a common fitment seems to have been the '20 lb. C.F.S. carrier', taking four 20 lb Hales H.E. bombs, or four 16 lb H.E.R.L. bombs, or four 16 lb carcass incendiary bombs, or four 6 lb 'small petrol bombs' (as they were called). The carrier was generally suspended on struts well below the fuselage. One known photograph is especially interesting because the carrier is apparently capable of taking eight 16 lb or 20 lb bombs—though significantly (for the Schneider's lifting capacity was strictly limited, especially in adverse climatic or marine conditions) no gun is fitted and four bombs only are in place. These bombs are on the central group of crutches. Resulting from the width of the carrier the supporting struts are apparently toed out, instead of sloping inwards as for instance on the Schneider seen in another known photograph—showing a Schneider with experimental Linton Hope floats. Certainly it is worth noting that one possible load for a Schneider was four 16 lb bombs plus four incendiaries of unrecorded pattern—though conceivably 15 lb Carcass; nevertheless, the four 16-pounders alone would correspond roughly with the weight of the single 65 lb H.E.R.L. bomb that was otherwise (and doubtless somewhat later) carried on the Schneider. This last-mentioned bomb was of a type and calibre commonly used for anti-submarine work.

Steel darts would have been an alternative offensive load.

From many an old Naval-flying hand the very name 'Sopwith Schneider' could well bring instant recollections of one of the best-loved aircraft carriers (though strictly speaking she was a 'seaplane carrier') that the Service ever had—the *Ben-my-Chree*—formerly an Isle of Man packet boat and generally associated with torpedo-dropping Shorts in the Dardanelles campaign. Yet Sopwith Schneiders too made history (technical, if not operational) from this same vessel; and even as early as 11 May, 1915—months before the droppings in the Dardanelles but during an actual operation off the German coast—a Schneider (which could have been No.1444 or No.1557) was used for an attempt to fly-off from a dismountable forward platform. For this purpose wheels were fitted to the floats, but the pioneer effort—like the Schneider itself—was grievously marred by the wrecking of the engine-starting gear. This took the form of a crank-handle in the cockpit; and this fractious handle not only broke the pilot's wrist, but removed some of the instruments from the sparsely furnished dashboard.

Early Schneiders had a triangular fin, as witness these three revealing but contrasting views. Top, No. 3726 with warping wings; lower, No. 3717 takes a solid-tyred delivery ride, with its own pneumatic beaching or launching chassis (note also the underfuselage bomb-carrier; opposite, both types of fin exhibited aboard submarine E.22.

The first successful take-off of this kind by a Schneider was made not from the *Ben-my-Chree* but from *Campania*, the pilot being a Flt Lieut W. L. Welsh, who as Air Marshal Sir William Welsh held some senior posts in the Second World War—notably that of AOC-in-C Flying Training Command, 1941-42. With *Campania* making 18 kt into a 13-kt wind, on 6 August, 1915, Welsh's Schneider (No.1559, fitted with a jettisonable wheeled dolly) was airborne after a run of 113 ft (34 m).

These early deck-operations followed some catastrophic experiences when operating the Schneiders directly from the water: as, for instance, on 4 July, 1915, when three factory-new examples which had been hoisted-out from HMS *Engadine* had their plywood floats smashed while on the water, and only one aircraft could be salvaged. Meanwhile four Zeppelins prowled around.

The Navy's notions for off-shore Zeppelin interception were ingenious and manifold. Thus during 1915 there was a plan to patrol 50 miles out with shallow-draught paddle steamers bearing four Schneiders apiece, though by the end of March 1916 only the *Killingholme* and *Brocklesby* answered to this programme. Light cruisers of the North Sea Patrol were other largely unsuccessful carriers for Schneiders (as, indeed, were Yarmouth trawlers); and though on 2 June, 1915, the cruiser *Arethusa* made a determined launch against a Zeppelin, the pilot failed to attack before returning because of a misunderstanding.

The Schneider's delicacy in a seaway notwithstanding, one feels wholly justified in recording its pioneering take-off from a ship's platform before the sorry tales of battered floats and shattered hopes; for Sopwith and the Navy were together launching not merely a few frail floatplanes but (by the use of shipboard platforms) a novel method of air war—and this not in some quiet haven, but in face of the enemy. Even more than this: Schneiders were carried by and launched from a submarine—this in April and May 1916, when *E.22* went to sea with a pair of the little seaplanes on her deck, whence they flew home to Felixstowe. Needless, perhaps, to add, *E.22* did not submerge with her strange deck-cargo in place; though clearly presaged here were the catapult-launched Parnall Peto trials from the watertight hangar of *M.2*—trials that began in the late 1920s, though they ended disastrously in 1932.

Except that it was never catapulted, and never achieved a reputation for robustness, the Schneider did, in fact, perform just about every type of operation that was undertaken by the postwar Fairey Flycatcher, one of the most versatile single-seaters ever built; and if one adds to its distinctions those that are separately credited to its landplane twin the Tabloid, then the Sopwith contribution to the development of one of the most 'difficult' of all Service aircraft—the naval fighter—is at once apparent. Furthermore, the aircraft of that class today lack the buoyancy conferred by even partially waterlogged plywood floats.

Operational, as distinct from mainly technical, successes were achieved by Schneiders in the eastern Mediterranean and the Red Sea—these sometimes involving gunnery spotting or reconnaissance; and even as late as 21 November, 1916, a Schneider flown by F/Sub-Lieut A. F. Brandon shot down an enemy aircraft which had attacked the Mudros airship hangar.

All the Schneiders were built by Sopwith themselves, the first order being for twelve (Nos.1436 to 1447), followed by a batch of twenty-four (Nos.1556 to 1579) and another of a hundred (Nos.3707 to 3806).

The essential difference between the Schneider and its derivative the Baby (Sopwith built a hundred Babies, numbered 8118 to 8217) was the very one which rendered these generally similar aircraft readily distinguishable to the eye: namely, the fitting of a more powerful engine—the 110 hp or 130 hp Clerget—which was housed in a wholly different cowling, of open-fronted inverted-U or so-called 'horseshoe' form. Though armament varied, the most interesting innovation was the fitting, on some aircraft, of a synchronised Lewis gun, firing over the cowling just described. The gun was mounted either on the centre line, projecting backwards through the windscreen, or was offset to starboard—and, distinct from these installations there were instances of a Lewis gun mounted on the starboard side of the fuselage and of the gun being attached to the port centre-section struts, turned on its side, and firing upward at about 45 degrees to the line of flight.

Whether or not any meeting ever took place between T. O. M. Sopwith and Col Isaac Newton Lewis (the developer, if not strictly the inventor, of the Lewis gas-operated drum-fed machine-gun) one cannot be certain, though Lewis and Geoffrey de Havilland most certainly met. The matter is, in any case, of little consequence; for although the initially Belgian-sponsored Lewis gun was manufactured and adapted for aerial use by BSA in Great Britain (and demonstrated by that company at Bisley, Surrey, just a few miles from the Sopwith works at Kingston in the same county during November 1913) official interest by the British flying Services was firmly established when war came.

The Lewis gun, though an admirable weapon for free mounting (as in the classic instance of the 1½ Strutter) proved intractable in attempts to synchronise it; and though George Constantinesco (his Christian name was rightly rendered thus) was one of several men who designed synchronising gears of various types to meet the case, he was likewise eager to render the difficulties clear. These difficulties were inherent in the gun's method of operation—with the striker carried on a post which travelled backwards and forwards in a slot in the breech-bolt, with firing taking place initially from the 'open-breech' (or 'open-bolt') position under the influence of the 'return spring', which was such a prominent and important component of the gun.

These inherent difficulties notwithstanding, mechanical gears of both Scarff and Hazelton types (Martinsyde designed an electrical gear, as they also did for the Vickers gun) were installed on Sopwith Baby floatplanes, including the Blackburn, Fairey and Parnall-built versions or derivatives. On some Blackburn-built

Schneider/Baby comparison: Top, one of the last Schneiders, No. 3804, showing all the main identifying features of the type—especially the engine cowling; lower, a typical Blackburn-built Baby, with characteristically cowled Clerget engine and upward-firing Lewis gun.

examples—and possibly others with synchronised Lewis guns—the spade grip of the gun was adapted to serve as a protective pad (pilot-protective, that is); and in some instances firing may have been initiated from the spade grip rather than from the pistol grip, as was the case in free installations.

Like the Schneider, the Baby was used for varied duties, and for anti-submarine work it could carry two (instead of one, as on the Schneider) 65 lb bombs, in tandem under the fuselage. However, the Lewis gun and ammunition weighed 55 lb, and only one 65 lb bomb could be carried in addition. Though there seems to be no confirmation of the possibility, it may well be that a single 100 lb H.E.R.L. bomb was lifted, if only experimentally, though identification in photographs could prove difficult as both the 65 lb and 100 lb bombs were identical in length (as well as being of thin-case type, and commonly employed against submarines). Four 20 lb bombs would be another possible load.

One positively identified armament installation involved the fitting of Le Prieur (or possibly Brock 'Immediate') rockets to the interplane struts, and one batch of Blackburn-built Babies initially carried Ranken Darts to the exclusion of other armament. Designed late in 1915, these darts were explosive anti-airship weapons.

With the heavier armament loads (together, perhaps, with such prudent provisions as a sea anchor, a caged carrier-pigeon and emergency rations) a fully-fuelled Baby was demanding a lot even from a 130 Clerget engine, and special lift-increasing modifications made by Blackburn and Fairey will later be described. Meanwhile some Baby achievements must be placed on record.

Both in Home waters and the East these little single-seat seaplanes were operated much as were their precursors the Schneiders, and were likewise concerned in some early shipborne operations. Two Babies, for example, were hoisted-out from HMS *Vindex* in the Horns Reef area on 25 March, 1916, though one was lost—which was the more regrettable because the intended target (a supposed Zeppelin base at Hoyer) was found in fact to be at Tondern. This being so, no fewer than eleven Babies were hoisted-out from *Vindex* and *Engadine* off Sylt on the following 4 May. Four of these sustained broken propellers; one was overturned by a destroyer's wake; and three had engine failure. Of the three that got away one crashed after striking a destroyer's wireless aerial, one returned with engine trouble and one reached Tondern. Alas, its two 65 lb bombs missed their target.

From *Campania* (as with Welsh's Schneider) take-offs were made using wheeled dollies; and when this same vessel put to sea for the Battle of Jutland (fought 31 May—1 June, 1916) her aircraft included not only three Babies, but four Schneiders also, while *Engadine* bore—equally abortively—two Babies of her own. To back-up the land-based fighters at Dunkirk four Babies were transferred there from *Vindex* on 24 June, 1916, and nearly a year later (May 1917) these were supplemented by another nine—though not for long, for two months later Pups replaced them.

Bombing raids by Babies were quite frequent in the Mediterranean area, both from shore stations and the *Ben-my-Chree*. Three Babies from this famous ship, for instance, attacked the Chikaldir railway bridge on 27 December, 1916, and in February of the following year six Babies were allotted to the Otranto seaplane base. In the Aegean three flew as fighters from Thasos, and—such was technical and operational progress—that in November 1917 the seaplane carrier *Empress*, operating in the East, carried not only four Sopwith Babies but two of the newly derived Hamble Babies also—a fact that conveniently focusses attention on production and development.

With expanded production and extended capability both in mind, a sample Baby was supplied to the Blackburn Aeroplane & Motor Co and another to the Fairey Aviation Co; and some Blackburn-built examples were, in fact, engaged in certain of the operations already recorded—including fighter patrols from Dunkirk. As late as 20 January, 1918, two Blackburn-built Babies from Imbros tried unsuccessfully to bomb the German cruiser *Goeben*.

Concerning Blackburn's production effort A. J. Jackson wrote in his *Blackburn*

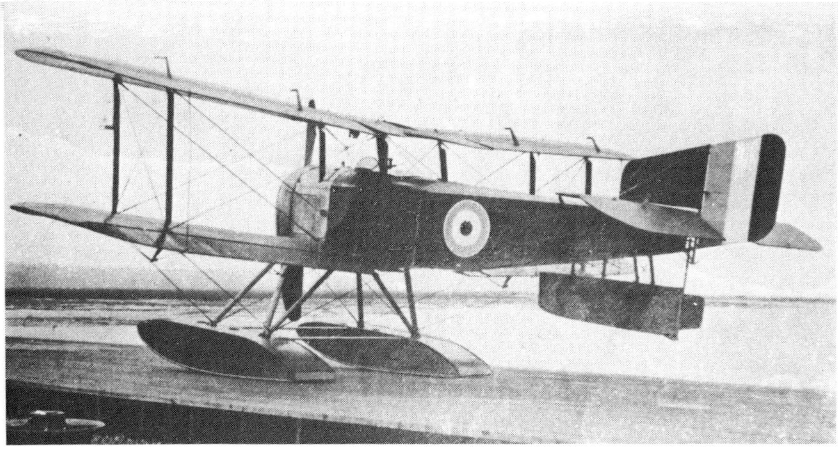

Top, Fairey Hamble Baby, with its distinctive fin and rudder and camber-changing flaps clearly in evidence; bottom, Parnall-built Hamble Baby, distinguished by its 'Sopwith' tail, and with the camber-changing flaps especially prominent.

Aircraft since 1909: 'The Blackburn Baby seaplanes (as they were called) were built in the Olympia Works, Leeds, commencing with a prototype machine, N300, and 70 subsequent aircraft all with 110 hp Clerget engines. Ten of these, N1030—N1039, were fitted with experimental mainplanes of modified section. Later both sub-contractors were made responsible for modifying the design to take the 130 hp Clerget, after which Blackburns built 115 machines with this engine. These were in two batches, the first, N1410—N1449, being armed with Ranken anti-Zeppelin darts.'

RNAS/Sopwith/Bentley associations were close, and N1410 *et seq* were originally intended to have the Bentley A.R.1 engine, at first called Admiralty Rotary, but engines of this type were not available in time. Blackburn production did, in fact, total 186 Babies—N300, N1010—N1039, N1060—N1069, N1100—N1129, N1410—N1449 and N2060—N2134.

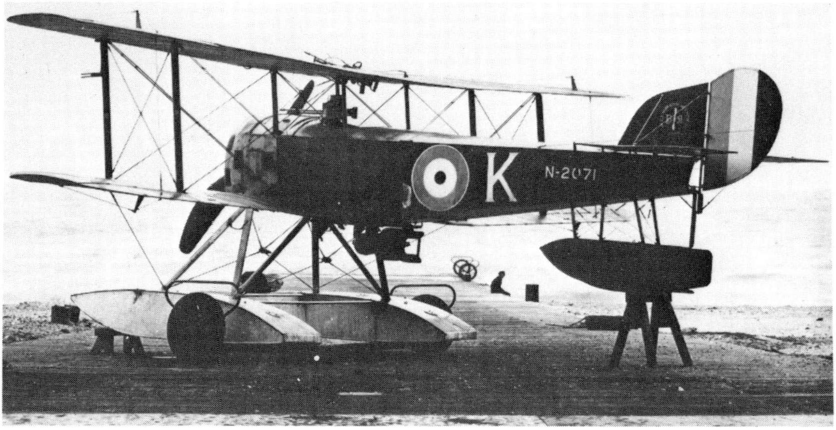

Blackburn-built Babies: Top, N2071, with two Lewis guns (one synchronised; one-upward-firing) and a 65 lb bomb; bottom, N2112, with unoccupied bomb-carrier.

In a purely technical sense Fairey's effort was altogether more ambitious, and although quite properly known as the Fairey Hamble Baby—being virtually a new type, and thus more justly renamed than was the 'Blackburn Baby'—must have a brief note in this Sopwith book (and not merely because Tom Sopwith knew Hamble well, and was an ardent yachtsman, as was Dick Fairey).

The salient novelty in the Hamble Baby (50 built) was the use of the Fairey Patent Camber Gear first incorporated on the converted Sopwith Baby No.8134—and production machines were further distinguished by newly designed floats and a characteristic square-cut Fairey tail, this last feature contrasting strongly with the new, rounded, wingtips, Thus was this Sopwith derivative a true forebear of the Fairey Flycatcher, already named in this chapter.

Parnall-built Hamble Babies and their skid-equipped landplane derivatives the Hamble Baby Converts retained the Sopwith-style tail.

Yet another derivative of the Sopwith Baby was the one-off Port Victoria P.V.1, with wings of higher aspect ratio, heavily cambered and heavily staggered.

Sopwith production of the Baby was, as already noted, the entire first batch of a hundred (Nos.8118—8217) all of which were delivered between September 1915 and July 1916. Of these the first five were non-standard, in retaining the 100 hp Gnome Monosoupape engine, as in the Schneider; the others were fitted as standard with the 110 hp Clerget. Babies Nos.8128 and 8129 were delivered to the French; ten Blackburn-built Babies were supplied to the Norwegian Naval Air Service; to Canada—for embarkation in ships of the Royal Canadian Navy—went Nos.8125, 8197, 8204 and 8209.

Finally, attention is invited to the 'mystery monoplane' in the *Apocrypha* to this volume.

Schneider

(100 hp Gnome Monosoupape)

Span 25 ft 8 in (7.8 m); length 22 ft 10 in (6.9 m); height 10 ft (3 m); wing area 236 sq ft (21.9 sq m). Maximum speed at sea level 87 mph (140 km/h.); service ceiling 8,000 ft (2,440 m).

Baby

(110 hp Clerget)

Span 25 ft 8 in (7.8 m); length 23 ft (7 m); height 10 ft (3 m); wing area 236 sq ft (21.9 sq m). Maximum weight 1,580 lb (717 kg). Maximum speed at sea level 92 mph (148 km/h).

Baby

(130 hp Clerget)

Span 25 ft 8 in (7.8 m); length 23 ft (7 m); height 10 ft (3 m); wing area 236 sq ft (21.9 sq m). Empty weight 1,226 lb (556 kg); maximum weight 1,715 lb (778 kg). Maximum speed at sea level 100 mph (161 km/h); climb to 10,000 ft (3,050 m) 35 min; alighting speed 45 mph (72 km/h); endurance 2¼ hr.

1½ Strutter (land versions)

In more ways than one the 1½ Strutter was the most significant of all the Sopwith 1914–18 aeroplanes—not in the military and technical senses alone, but in its international acceptance as a novel and uncommonly efficient airframe (dependent though it was to remain throughout its life upon French-designed engines). In Great Britain, its country of origin, it had the additional significance of being the first Sopwith type to be sub-contracted on a truly vast scale.

Here one draws attention to a parallel which, though quite remarkable in its closeness, seems to have passed unnoticed in the annals of the air: that is, the relationships which undoubtedly existed between this classic type of two-seat tractor fighter and the Vickers Gun Bus—the classic type of two-seat pusher fighter. This is true not only in the military sense, but also in the international sense; for the Gun Bus, like the new Sopwith, was licence-built in France and was dependent on French rotary engines. In the purely Service sense the parallels are even stronger, for Admiralty interest was very marked in both instances—not in the airframe alone (for both machines embodied novelties—the Gun Bus, for example, in its partly metal structure) but in the development of suitable armament. Both types, moreover, were tested and developed at Brooklands, and both were at one time

Though bearing no maker's caption, this splendid study of an early 1½ Strutter, 9382, was nevertheless clearly taken on Sopwith's behalf, and though no guns are mounted (some early 1½ Strutters were apparently delivered without the Vickers gun, and mountings for the Lewis gun varied) all salient features of the airframe are admirably shown. In the background is an early Martinsyde G.100—apparently 7263 or 7283—with its top-wing gun mounting sticking up like the proverbial sore thumb.

intended to have a Smith radial engine; but even in 1915 it was declared of the Gun Bus: 'A most painstaking, and often disheartening, series of experiments were made with a view to making this pusher type machine the equal of the tractor type of aeroplane, and a measure of the success obtained can be gauged by the fact that it has now flown in a wind against which scarcely any headway could be made.'

Certainly, compared with the 1½ Strutter the Gun Bus was to make very little headway of any sort at all.

Although in its definitive military form (and it will soon be clear that 'definitive' is used here solely for convenience) the design of the 1½ Strutter is generally ascribed to late 1915, the time might be more truly set much earlier in that year, or even in late 1914; and that 'two-seater scout' was a description already current in the early summer of 1915 for the 'Sigrist Bus' (to use a familiar appellation) will be clear from a quotation that shortly follows. First it must be explained, however, that in the course of this quotation (describing the setting-up of a new British altitude record by Harry Hawker on 6 June, 1915) the qualified barograph reading of 'about 20,000 ft' proved inaccurate, thus accounting for the apparent discrepancy concerning Norman Spratt's unofficial performance on a B.E.—for Hawker's true height was 18,393 ft (5,606 m). Thus the quotation ran:

'An altitude record is of distinct value for several reasons. First and foremost it connotes reserve of power. Secondly, provided it is made with a standard machine, not specially designed for the purpose, with neither pilot nor engine doped,' [a portent for Olympic Games to come?] 'it affords a direct standard of efficiency. Neither of these conditions was fulfilled in the case of the world's record held by the Germans. On July 14, 1914, Oelerich reached a height of 26,200 ft. in Germany, but his machine had been specially designed for the occasion; it was a 120 h.p. Beardmore-A. D.-engined D.F.W. biplane, and both pilot and engine were doped.' (*Sic:* Does one detect here a tang of wartime propaganda, especially as the figure quoted for the German effort appears to be on the low side?).

'Hawker's new British record' (continued the quotation) 'though inferior in actual height, is of a very different nature. It was made' (present author's italics hereafter) '*on a standard Sopwith two-seater scout,* engined with an 80-h.p. Gnome of the ordinary type. True, the machine is slightly different in minor points of design from the Schneider Cup craft, notably in the chassis, with a single central skid and *a new diagonal strut arrangement, of which it might be indiscreet to give particulars.*'

It was then recorded that 'The first 10,000 ft. were climbed in twenty minutes—500 ft. per minute up to 10,000 ft. is *some* going' (present author's italics no longer)—'went on steadily in the clear air and the intense cold, until the barograph needle had gone right off the chart, marked up to 6,000 metres, when the revolutions fell off to 1,100, and the climbing angle became exaggerated to the point of stalling. At this point Hawker deemed it expedient to come down; switching on and off, he made a long spiral, at the rate of 1,000 ft. a minute right over the aerodrome.' (This was Hendon—the Sopwith having flown there from Brooklands that morning). 'His vol plané lasted twenty minutes from a height of 20,000 ft. Switching on again, he made two splendid circuits over the course at full speed, having, curiously enough, lost none of his touch' (Hawker's hands were 'icy cold' even in the 'sweltering heat of the aerodrome') 'and finally alighted. His sealed barograph showed about 20,000 ft, subject to official correction. Thus Lieutenant E. F. Briggs' unofficial record (14,920 ft at Eastchurch, March 11, 1914, on an 80-h.p. Gnome Blériot), and the unofficial record of Lieutenant Norman Spratt, amounting to some 18,900 ft., on a B.E., were both handsomely beaten, to the manifest delight of Tom Sopwith, who watched the while with perfect confidence.'

The speed of this aircraft was given as 'upwards of 90 m.p.h.'. Its central-skid landing gear has already been mentioned—though not the fact that this fitment was at some stage in the aircraft's career used to contain adjustable lead ballast (possibly when the machine was flown as a single-seater); but it was the 'diagonal strut arrangement, of which it might be indiscreet to give particulars' that was the essential novelty, the two halves of the top wing being braced by the W-form strut system, the outer arms (or struts) whereof reached so far outboard that their attachment-points might well have received a second pair of normal interplane struts (there was one pair in any case, further outboard) and thus suggesting the name '1½ Strutter'—for did not the French have '1½-wingers', or sesquiplanes, and were not both the Sopwith company, and Hawkers after them, much given to trying new biplane fighters with alternative single-bay and two-bay wings?

With his usual professional touch Harald Penrose thus summarised the salient facts: 'At Kingston, Tom Sopwith's great factotum, the dour determined Fred Sigrist, as a result of discussion with Hawker on the possible form of a replacement two-seater with enhanced performance and safer characteristics, modelled a new fuselage on the 807 [see 'Folder Seaplane'] using a bigger fin having a rounded nose of bent tube, and stiffened the main wing spars in order to employ a single bay with outward-raking struts, shortening the lower wing proportionately [*N.B.* The 'Sigrist Bus', unlike the 'definitive' 1½ Strutter, had wings of unequal span]. To reduce bending moments of the upper wing he used steel centre-section struts steeply sloping from the top longeron to a point well out in the spar bay, and then braced the centre-line juncture of port and starboard spars within inverted V-struts arranged like a trestle, resulting in a widespread transverse W. The machine had been growing slowly in a corner of the old Kingston Skating Rink, for Sigrist was preoccupied with production matters, and it would be another month or more before the framework was ready for covering. Meanwhile it was jocularly referred to by the workmen as 'Sigrist's Bus'.

Further, it may now be pointed out, although it is sometimes correctly observed that the French Hanriot HD-1 fighter had a similar arrangement of inboard struts, it is no less cogent to remark that these struts braced far less of the span; and it might be more valid to submit that the 1½ Strutter was a half-way step to full Warren-girder bracing—associated especially with Fiat and Handley Page—and that the centre portion of the W was to be seen on the Hawker Horsley and Hornbill. Far more remarkable, however, is the analogy of the Fiat C.R.42, the finest biplane fighter of the Second World War, an aircraft which had not only Warren-girder bracing for the outer wings but the 1½ Strutter-style W bracing inboard. (Still one is left pondering how best might be described, in fractions, the centre-section struts of the 'progressively' modified Gloster fighters of the Gamecock era—culminating in the Goldfinch, with an N system instead of the W, though full Warren-truss bracing was nevertheless used for the ribs!).

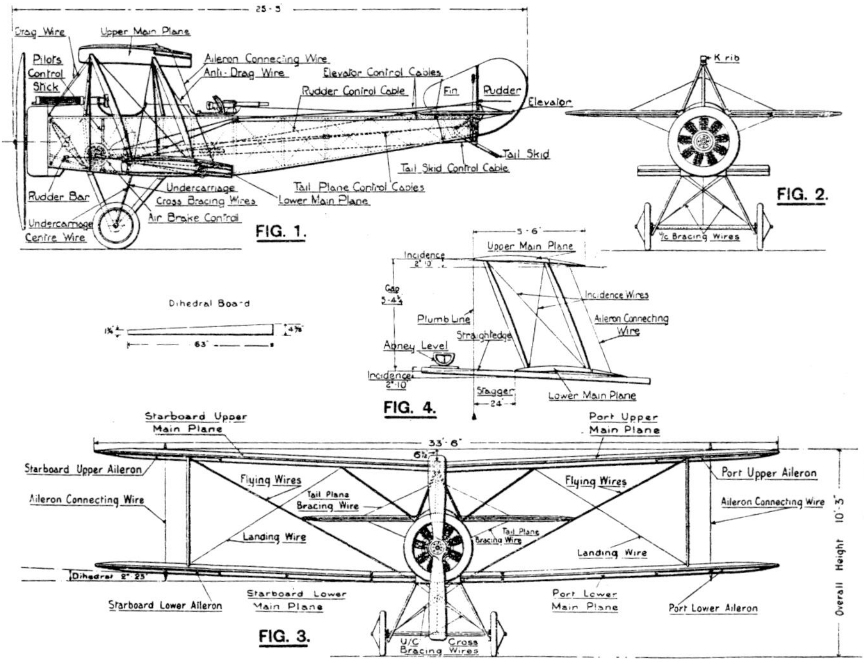

Official drawing of the 1½ Strutter.

Whatever the full implications of the matter may be (and, all things considered, perhaps the RAF's homely Armstrong Whitworth Atlas affords the closest parallel, embodying, as it did, the inverted-V struts of the RFC's 'Little Ack' and 'Big Ack' with additional long splayed-out members) in the particular instance of Sopwith's 1½ Strutter the novel wing-bracing arrangement for a biplane having no centre section was a peculiarly happy one. Certainly it differed radically from the arrangement on the Gordon Bennett single-seater (wherein the inboard struts were splayed in side elevation only) and it was one, furthermore, which helped to gain for the 'Strutter'—as it soon became known—not only military fame but its international acceptance also, the French themselves building 4,200 examples, some of which went to the Americans.

Construction on this scale in France was, in itself, some tangible repayment of the debt accumulated by British aircraft-builders over the years to the Farmans (though Henri was a British citizen until 1937), to Blériot and Voisin—and certainly not forgetting the flow of French engines, to which, as already noted, 1½ Strutters were (so to speak) to remain permanently attached.

While thus engaged in a general assessment of the 1½ Strutter one must next reaffirm one's contention that—its bombing and naval applications aside—this aeroplane is 'justly remembered as the archetype of the classic two-seat fighter (pilot with fixed gun, gunner with free gun).' As for bombing and naval use, one might with equal justification have devoted separate chapters to fighters, bombers and 'ship's aeroplanes' (as they were called), as comprising the 1½ Strutter spectrum; but one has plumped for compactness, and thus proceeds with development for land-based use, with fighting foremost—for we have already noted the early-1915 description of the aircraft as a 'scout'.

Felicitously one can record at this point a visit by Harry Hawker to Hendon on 8 April, 1916, (the best part of a year after the one already mentioned) in what *Flight* described as 'the new two-seater "bullet"'; and as at that period the terms 'scout' and 'bullet' were generically and loosely applied to the class of aeroplanes latterly known as 'fighters' it may reasonably be supposed that this aeroplane can be identified with the production-type 1½ Strutter, built to the order of the RNAS, and capable of being armed to the requirements of that pioneering Service.

This appearance by Hawker on the new Sopwith two-seater followed shortly upon another event which may now be seen to have been of no less significance in the history of the aircraft and the company alike. This event was a dinner given in the preceding January (1916) by the Sopwith employees for the directors, and graced by the presence of the Mayor of Kingston, who remarked that the occasion was something of a birthday party, for it was three years since the company had made a start at their Kingston works. Mr Sopwith declared that credit for the successful machines produced belonged to 'the fellows who had the getting out of the machines', mentioning (evidently in the order now given) Mr F. G. Sigrist, Mr H. Hawker and 'the drawing office and charge hands'. Mr R. O. Carey, in his directorial turn (and before leading the Sopwith Works Band in a rendition to which he himself added a violin solo) referred to 'the excitement created by certain newspapers about the Fokker machine'. They must realise, said Mr Carey, 'that this kind of machine was a purely defensive one, and didn't worry our people a bit' because 'this country possessed something which would successfully outstrip a Fokker any day.' He then referred to the great kindness and consideration shown by 'the Naval resident inspectors'. Thus one may reasonably suppose that the unworried people mentioned were the Sopwith staff, and that the 'something' was the 1½ Strutter, which was certainly an offensive, as well as a defensive machine; could indeed outstrip the Fokker (though by a narrow margin); and could certainly outfight it. And, in any case, the Pup was already well advanced.

A special word will be in order here concerning 'the Naval resident inspectors' to whom Mr Carey rendered such grateful tribute; for numbered among these in 1915 was R. A. Bruce who, in that same year, joined the Westland Aircraft Works (founded during that April). To many people in the aircraft industry the name of Mr Bruce—together with that of Harald Penrose—was well-nigh identified in later years with Westland, and Bruce's eventual appointment as the company's managing director was a measure of his calibre. In this present instance it is fitting to note that the Westland-built Sopwith 1½ Strutters were among the better-built examples.

An unfamiliar version of the 1½ Strutter—on skis in Russia after the Armistice.

With fighting as our foremost concern in considering the 1½ Strutter, we turn attention first to armament and then to the airframe—which embodied unorthodox features apart from the peculiar strut-arrangement. Thus, respecting armament, of particular importance were the guns, the mode of their installation and their manning.

The pilot's fixed Vickers gun was mounted on the centre line of the fuselage, with the firing lever (when the Sopwith-Kauper synchronising gear was fitted, at least) projecting from the back of the gun. A Sopwith patented padded windscreen was fitted (to protect the pilot when he was taking aim) and plain open sights, or ring-and-bead sights, were mounted on the gun itself. This was the normal arrangement, though at least one early specimen had the Vickers gun mounted on the port upper longeron, with the breech casing lying under the built-up coaming of the cockpit. The ammunition (as was then standard for the Vickers gun, as for the Maxim gun from which it had been developed) was fed from the right, and a metal plate on the left, forward of the feed block, served to prevent the canvas belt from twisting in the slipstream before re-entering the protection of the fuselage. (The Prideaux disintegrating-link metal belt had not yet arrived). On some early examples the rear of the gun was partly faired.

The 1½ Strutter being primarily an RNAS aircraft (the first contract having been placed by the Admiralty in time for deliveries to start in February 1916 and the first operational machines being in action with No.5 Wing, RNAS, before April of that year was out) the entire gunnery scheme must be acknowledged largely to the experts of that Service (notably Warrant Officer Fredrick W. Scarff) but also to Sopwith themselves—particularly in the person of Harry Kauper. Both the fixed-gun and the free-gun installations at first varied quite extensively, and apart from what has already been said about the Vickers gun scheme, it must be added that the first machines for the RFC (A and B Flights of No.70 Squadron, which went to the war a few weeks after the RNAS machines already mentioned) had gun-synchronising gear of the Vickers type, though C Flight of the same unit had aircraft which had been transferred from RNAS contracts, and which were fitted with the Scarff-Dibovsky gear developed for that Service. On a small scale, gears of Ross type were fitted to 1½ Strutters; and seemingly on a larger scale the Sopwith-Kauper gear, which is now rightly our concern.

Harry (H. A.) Kauper—at one time foreman of the Sopwith fitters and already something of a hero in the earlier chapter on the Circuit Seaplanes—was the man chiefly responsible for this mechanical gear, which proved so successful that it lived on well into the Camel era, 3,950 sets having been supplied to the flying Services and 2,750 installed. The gear was developed in 1916, and there were several variations. Sopwith themselves once described it in the context of a cam operating a mechanism 'which directly caused the actuation of the firing lever of the gun', at the same time remarking on 'the difficulty of obtaining a high rate of speed with these devices' because 'the inertia of the moving parts tends to prolong the period during which the automatic gun may repeat its movements and fire again, with the resultant risk of damage to the propeller.' The essential components were, in fact, the cam already mentioned, and which allowed the actuation of a spring, which itself operated the firing mechanism. The cam was mounted on the engine 'in such a manner that it can oscillate the tappet rod in order to synchronise with the passage of the blades of the propeller past the line of fire of the gun, alternatively holding off or preventing the fire and then permitting the fire by means of the spring.'

As for the 1½ Strutter's free-mounted Lewis gun, the Scarff No.2 ring-mounting, with which the aircraft became almost identified and which Gen Trenchard quickly requested as standard on all future 1½ Strutters for the RFC, was designed by Warrant Officer F. W. Scarff of the Air Department of the Admiralty, and in its development Sopwith played some part, the nature and extent of which is indeterminate. This most famous of all aircraft gun mountings was mainly constituted by an elevating-arm, or bow, which carried a Lewis gun, and a rotatable ring. It was contemporarily described as follows:

'The elevating-arm and the rotatable ring are locked in their adjusted positions by devices which are actuated simultaneously, or in succession, to effect unlocking by a single control wire operated by a handle on the elevating-arm. When operated in succession, the ring is released before the elevating-arm. Elastic cords balance the turning moment in a vertical plane due to the weight of the elevating-arm and the parts carried thereby. The rotatable ring is mounted on ball or roller bearings.' Not mentioned in the foregoing concise description were the two upward-projecting pairs of toothed segments, or quadrants, each pair being engaged by a locking pin carried by the elevating-arm and moving into or out of engagement with the teeth of the segments. Of this basic form of Scarff ring-mounting there were several variations, not only British but foreign also, the best-known related foreign counterpart being the French T.O.3.

This last-named mounting differed very greatly from the French-designed Nieuport type used on some early 1½ Strutters, though the Nieuport mounting was itself basically of ring type. It had, nevertheless, a very distinctive appearance, by reason of its great height and peculiar form, these features resulting from a pair of upward-projecting arms, between which the Lewis gun was carried on a cross-member.

Far less obtrusive was the British Strange mounting, a cranked-pillar pattern with which was associated a crescent-shaped toothed quadrant; but neither the Nieuport nor the Strange mounting was extensively fitted on the 1½ Strutter—such was the demand for the Scarff, when it became available in quantity.

One Sopwith contribution was a special seat for the gunner, capable of all-round movement about an eccentric pivot on a fixed stand, and having means for unlocking itself when relieved of the gunner's weight, enabling it to be freely rotated in its socket. Largely negating such refinements however (which were intended to enhance fighting efficiency) was the wide separation of the two cockpits by the fuel tank—a legacy from earlier Sopwith tandem two-seaters already described.

For the Vickers gun of the 1½ Strutter the belt held 300 rounds of .303 in ammunition, and for the Lewis gun a maximum of five 'double' (97-round) drums could be taken. Early 1½ Strutters were sometimes supplied with 'single' (47-round) drums. Supplementary armament on one aircraft of B Flight, No.70 Squadron, was an 'automatic' (self-loading) pistol with oversize magazine, attached to the starboard landing gear struts to fire outside the propeller arc. Sopwith 'Two-seaters', as the 'fighter' versions of the 1½ Strutter were sometimes called, sometimes had bomb-rails under the lower wings, possible loads being four or eight bombs of 20 lb or two of 65 lb.

Two comparative views showing 1½ Strutters with different types of mounting for the Lewis gun: Above A377 with Strange mounting (inconspicious as always); below, N5624 with Scarff No. 2 ring-mounting.

Clear-view top-wing panels are seen in the two upper views here, with the gun-carrying bow of the Scarff No. 2 ring-mounting elevated on A1924 (top) and depressed on another specimen of the 1½ Strutter (centre), while in the flying view both the Lewis gun and the Vickers gun are conspicuous.

The airframe of the two-seat 1½ Strutter was a remarkable combination of convention and novelty, the latter having been already instanced by the strut arrangement, though the wooden construction (with steel tubing used for tips and trailing edges) conformed with standard Sopwith practice. Three features that command special mention were the air brakes—handwheel-actuated trailing-edge panels in the bottom centre section, pivoted spanwise to rotate upwards through 90 deg; the large adjustable tailplane (which, though having as a primary purpose compensation for gunners of different weights, was later popular with pupils, when numbers of 1½ Strutters were used for training, because it enabled the trainees to fly with their chilly hands in their pockets); and third, the use of all-steel tubular construction for the elevators, fin and rudder.

The unhappy separation of the pilot and gunner has been mentioned, but the pilot's upward view was also poor, though adequate forwards and downwards, while the siting of the gunner's cockpit, jointly with a top-wing cutout, afforded a wide field of fire for the Lewis gun. Somewhat disappointingly (for so much care had gone into the design) the 1½ Strutter was officially declared to be 'rather heavy and slow on controls'; yet, even so, this aircraft marked a new departure in military-aircraft design, and was clearly capable of development and adaptation.

After Hawker's early spectacular demonstrations (he was then flying solo) performance was never remarkable, either with the original 110 hp Clerget engine, or the later-standardised 130 hp Clerget (nor, for that matter, with any other Clerget or Le Rhône rotary, as fitted in some French-built machines), and this deficiency in performance, jointly with mediocre manoeuvrability, clearly militated against combat effectiveness—and, of course, the armament was quickly matched. Nevertheless, the type was in first-line service until late in 1917, by which time the Sopwith single-seaters had quite outclassed it in fighting performance. This being so it is the more regrettable that a project of September 1916, involving the fitting of an American-designed 150 hp Smith Static ten-cylinder radial engine, was never realised—and at that time there was no comparable A.B.C. radial available. (Clearly, a long inline engine would have spoiled the basic concept of the design).

The foregoing reference to a ten-cylinder engine is a reminder that, however confusing were the Service or makers' designations of the rotary types actually fitted, these—for British operational service at least—invariably had nine cylinders, whereas the Pup sometimes had a seven-cylinder unit. The British-built 1½ Strutters had either the 110 hp Clerget (makers' suffix 9Z) or the 130 hp Clerget (9B), but French-built examples had either Clergets of the 9B series (9Ba 135 hp, 9Bb 135 hp or 9Bc 145 hp) or the Le Rhône 9J (110 hp) or 9Jby (135 hp). French-built trainers sometimes had the 80 hp Le Rhône 9C, a type so popular in British operational Pups. The standard annular cowling on British 1½ Strutters had a segmental slot at the bottom and was of special Sopwith design (see under Pup, in Harald Penrose's account).

Development and adaptation now being our concern it is fitting to consider the special single-seat fighter version of the 1½ Strutter developed for Home Defence. In this instance the Vickers gun was dismounted, the front cockpit was faired over, the pilot was moved to the rear, and a Lewis gun was mounted over the top wing to fire above the propeller and obviate the risk of using sensitive and temperamental 'special' ammunition in a synchronised gun. Presumably to exploit the '45-degree shot' upward-firing formula, on which much investigatory work was done, one aircraft at least had two upward-firing Lewis guns in a special installation forward of the cockpit, and in another application of two Lewis guns these were carried on a twin mounting of Foster (sliding block on curved track) type.

Single-seat bomber versions of the $1\frac{1}{2}$ Strutter, with bomb doors closed and showing clear-view top-wing panels. The Vickers gun is present in both instances, and was indeed standard.

Yet of far greater significance than these specialised single-seat fighters were the no less specialised single-seat bombers (as distinct from bomb-carrying two-seaters already mentioned). In essence the bomber transformation was simple: bombs were internally stowed in a compartment which took the place of the gunner's cockpit. On typical British $1\frac{1}{2}$ Strutter single-seat bombers, four 50 lb or 65 lb bombs were stowed horizontally. Beneath the bomb compartment were four trapdoors that were opened by the weight of the falling bombs (probably with scant effect on bombing accuracy) and were closed again by shock-absorber cord. In each side of the compartment were two inspection and access panels. The Vickers gun, was retained, and occasionally a Lewis gun firing over the top wing was additionally fitted—though strictly as 'secondary' armament, for the ammunition magazine could not be changed in flight. Nevertheless, the possibility of abandoning the Vickers entirely may well have been in mind, for the later Sopwith B.1 bomber (another single-seater) had a single Lewis only.

Confusingly, the specialised 1½ Strutter bomber variant was sometimes called by the Air Ministry 'Sopwith Bomber', though any association with the B.1 later described was obviated by 'Clerget' in brackets. Early in the 1½ Strutter's career, it must be added, the two-seater form had been styled by the Admiralty Sopwith Type 9400, the single-seat bomber being Type 9700; but some single-seaters were converted to two-seaters.

In the matter of bomb-sighting the 1½ Strutter presented acute difficulties—by reason of the relatively poor view, among other considerations. Installations of the Scarff Equal Distance and of the C.F.S.4 bombsights were quite unsatisfactory, and after being posted to No.3 Wing in 1916 Lord Tiverton gave special attention to the 1½ Strutter sighting problem. One possible 'official' arrangement was the subject of a note on 'Bomb Sighting Slot on 1½ Strutter'. This note referred to two transverse wires. When the aircraft was flying level the forward one of these was at 35 deg in advance of the vertical, and for dropping 65 lb bombs the 'nearer wire' was at 17 deg. For dropping analite bombs (of the type discussed under 'B.1. and Derivatives') this last-mentioned wire was at 18 deg. When bombing from 12,000 ft or higher a reversing stop-watch could be used 'if desired'.

To all writers and readers of the history of air war the books by Norman Macmillan (notably *Into the Blue*) will remain as standard source-works concerning the 1½ Strutter in action with the RFC; but there was another author—Lord Douglas of Kirtleside (formerly Maj W. Sholto Douglas)—who gave this concise account:

'We found that our precious aircraft were much too slow when pitted against the new Albatros and Halberstadt scouts. The Sopwith 1½ Strutter was not a particularly easy aircraft to maintain, and there were a number of modifications that we had to make before the new machines could be used. After some hair-raising experiences we found that it was necessary for us, if we were to engage the enemy on anything approaching equal terms, to fly underneath his formations so as to lure him into attacking us; and then we would trust to the good shooting of our observers to pick off the Huns as they came diving down . . .'

Thus did the all-important rear gun, on its Scarff ring-mounting, come into its own; but the pilots themselves, with their single fixed synchronised Vickers gun, did good work also—poor aircraft performance notwithstanding—and among the modifications mentioned by Lord Douglas could well have been the fitting of transparent panels in the top wings to afford some upward view.

On the Naval side (and distinct from over-water work described in the following chapter) bombing was pre-eminent. Some of the first 1½ Strutters in service were, as noted, with No.5 Wing in April 1916, this Wing having been formed in the previous month for long-range bombing; and now, in its new Sopwiths, it possessed machines that were not only bombers themselves, but (especially by reason of their rear gun) potent escort fighters also. Even more, it was a 1½ Strutter of No.5 Wing which, on 2 August, 1916, was used to control a bombing formation by firing Very lights—a pathfinder, as it were, for the master bombers of a later war.

The future was presaged also in operations by No.3 Wing, when, for instance, on 12 October, 1916, thirteen of its 1½ Strutters, in company with British and French Breguets and French Farmans, set out to attack the Mauser factory at Oberndorf—not, alas, with great success.

To the present chapter, rather than the next, rightly belongs the following summary of British 1½ Strutter production—conveniently and reasonably reckoned as 'over 1,500'. French production has already been quoted—a trifle hesitantly be it now added—as 4,200 (Darracq alone built 753), and countries other than France which used the type included the USA, Belgium, Japan, the Netherlands, Romania and Latvia. Although Sopwith had a Paris office, which had a close liaison with the French industry, the French were on occasions 'difficult' partners. Official classification in France was Sop. 1A.2 (two-seat artillery observation and reconnaissance); Sop. 1B.2 (two-seat bomber); and Sop. 1B.1 (single-seat bomber), but many built in these categories were ultimately used for training.

Civil usage was fairly extensive (including a French conversion with a cabin for two passengers) but the only example on the British Civil Register was G-EAVB (C. H. Oliver, 1920).

Transfers, conversions and non-deliveries were numerous, but British production was, in essence:

Sopwith 3686 (became B394); 9376—9425; 9651—9750; 9891 (allocated to projected trainer variant with 80 hp Gnome); 9892—9897; N5170—N5179; N5500—N5549 (last 12 cancelled).
Fairey A954—A1053
Hooper A6901—A7000
Mann, Egerton N5200—N5249; N5630—N5654
Morgan A5950—A6149 (last 100 apparently not delivered)
Ruston Proctor 7762—7811; A2381—A2430; A8141—A8340; B2551—B2600
Vickers A1054—A1153; A8744—A8793
Wells A5238—A5337
Westland A1511—A1560; N5120—N5169; N5600—N5624

N.B. The special 'Ship's Strutters' F2210—F2229 and F7547—F7584 were apparently rebuilt examples.

1½ Strutter (two-seater)
(110 hp Clerget)

Span 33 ft 6 in (10.2 m); length 25 ft 3 in (7.6 m); wing area 346 sq ft (32.1 sq m). Empty weight 1,259 lb (572 kg); maximum weight 2,150 lb (975 kg). Maximum speed at 10,000 ft (3,050 m) 96.5 mph (155 km/h); climb to 10,000 ft (3,050 m) 20 min 25 sec.

1½ Strutter (two-seater)
(130 hp Clerget)

Span 33 ft 6 in (10.2 m); length 25 ft 3 in (7.6 m); wing area 346 sq ft (32.1 sq m). Empty weight 1,305 lb (592 kg); maximum weight 2,150 lb (975 kg). Maximum speed at 10,000 ft (3,050 m) 97.5 mph (157 km/h); climb to 10,000 ft (3,050 m) 17 min 50 sec; service ceiling 15,500 ft (4,720 m).

1½ Strutter (single-seat bomber)
(130 hp Clerget)

Span 33 ft 6 in (10.2 m); length 25 ft 3 in (7.6 m); wing area 346 sq ft (32.1 sq m). Empty weight 1,316 lb (597 kg); maximum weight 2,342 lb (1,062 kg). Maximum speed at 10,000 ft (3,050 m) 98.5 mph (159 km/h); climb to 10,000 ft (3,050 m) 24 min 35 sec; service ceiling 13,000 ft (3,960 m).

$1\frac{1}{2}$ Strutter demonstrating the 'skids-in-troughs' launching technique, as tried aboard HMS *Vindex*.

$1\frac{1}{2}$ Strutter (Ship's)

Although the part played by Sopwith aircraft generally in the development of British Naval flying has been extensively recorded by other writers (and properly so) the contribution made by the $1\frac{1}{2}$ Strutter in particular has not, one feels, been fully recognised. In the 'Ship's Strutter' (as the basic form concerned was sometimes known, though commonly adapted from a landplane fighter or bomber) we see not merely an aeroplane capable of operating from a platform on a ship (or from a carrier's deck) but one that could perform in the pre-eminent roles that the Navy so urgently demanded should be filled: namely those of spotting fall-of-shot for the big guns of the Fleet(s) and of reconnaissance in general—aided by 'wireless'.

Between the specially prepared shipborne versions of the $1\frac{1}{2}$ Strutter and the original form of the aircraft were analogies that were strikingly in parallel with those existing between the Hawker Hart of over ten years later and its Naval derivative the Osprey; and this was evident even in external appearance, for both could be seen with jury struts where the outer wing-panels were (on some Ship's Strutters) detachable or (on all Ospreys) foldable. Further, although in the Osprey there existed (supposedly at least) a 'fighter' element—the official classification being 'fleet fighter-reconnaissance'—the pilot's armament was only half that of the corresponding specialised land-based fighter variant of the Hart (this, the Demon, having two Vickers guns and the Osprey one only) whereas on the shipborne 1½ Strutter there was no Vickers gun at all, having regard to the pressing requirement to keep the aeroplane light while carrying its all-important wireless and other Naval appurtenances. By the same token, the Parnall Panther, which was specifically (and very ingeniously) designed to undertake essentially the same duties as the Ship's Strutter, had a fixed Vickers gun in prototype form only, this being absent on production versions—with even more weight being saved by the adoption of a special pillar mounting for the defensive Lewis gun, instead of the generally adopted Scarff ring.

To conclude the Ship's Strutter/Osprey analogies, even the apparently basic difference—the fitting of floats on the Osprey in its catapulted form—was, in fact, far from being as basic as appearances suggested; for catapults were not widely fitted to British Naval vessels until the 1930s, the two possible launching methods formerly used from battleships, battle cruisers and craft other than specialised aircraft carriers being the 'flying-off platform' built over a gun turret and rotating with it, or the technique of lowering a floatplane into the water by derrick and hoisting it back again.

Herein lies the full significance of the Ship's Strutter—it gave the Navy ears as well as eyes and claws; for on 4 April, 1918, (immediately after the formation of the RAF) an aircraft of this type, flown by Capt F. M. Fox of the new Service, and carrying an observer, W/T equipment and an Aldis signalling lamp was successfully launched from a platform on Q turret of the battle cruiser HMAS *Australia*. 'Successfully' here is emphasised, for an earlier trial from another battle cruiser (HMS *Repulse*, in March) had failed. Thus battleships and battle cruisers carried a two-seat reconnaissance aircraft (typically a 1½ Strutter, though a Parnall Panther was used, for instance, from HMS *Revenge*) on a forward-turret platform, with a single-seat fighter (generally a Camel) on a rear-turret platform. By the end of the war, in fact, the Grand Fleet had over 100 platform-borne aircraft, 22 cruisers having themselves been given platforms—though not, of course, rotatable atop the turrets of big guns.

The extent and importance of these provisions were little appreciated at the time by reason of wartime secrecy, but shortly after the Armistice a Naval officer gave this concise account: 'My ship carried one Camel and one 1½ Strutter. These were carried on our broadside turrets. The 1½ Strutter was used for spotting purposes, and the Camels in the squadron performed formation flying. The method of release is effected by a "quick-release". When about to fly off, the ship steams at about 30 deg. to the wind and the turret is trained 30 deg., thereby pointing into the wind. The clamps are taken off the ailerons, elevators and rudder, and the quick-release attached. The pilot then starts his engine, and gradually works up to the maximum revolutions. When he waves his hand, the men holding on to the leading edge of the lower plane let go and stand clear, and at another signal, usually dropping a flag, given by the executive officer, the A. M. [air mechanic] pulls sharply on the quick-release and thus frees the machine, when the pilot runs along, and takes-off from, the platform. All the time when the engine is running the quick-release alone holds the machine. A tail guide of about 2 ft. length keeps the tail from dropping at the start. The greatest length of run for taking-off is not more than 30 ft.'

This helpful little note is given here, rather than in the context of the Camel, with deliberate intent, for reconnaissance and spotting, rather than fighting, were (as noted in this chapter's opening paragraph) pre-eminent among the Navy's requirements. A brief word, too, on terminology may not be superfluous.

First, the term 'broadside turret' must not necessarily be construed as archaic or tautologous, for not all turrets during 1914—18 were on the ship's centre line. Second, the Camels' 'formation flying' may reasonably be construed as 'flew in formation (or in company) with the 1½ Strutters'—just as, in later years, the single-seat Hawker Nimrods were dependent on the slower two-seat Ospreys for navigation (and found station-keeping a hindrance to their purpose). Last, even when flying-off platforms had become obsolete on Naval ships, the essential launching technique described was perpetuated in the aircraft carriers *Furious, Courageous* and *Glorious* for launching their 'slip-flight' fighters—sometimes using a quick-release—straight out of their hangar and over the bows below the level of the main flying-deck.

Conversely, it is worth pondering the true origins of training the launching platform by a rotary mechanism; and here it may be noted that torpedo-tubes were mounted on turntables just as guns were mounted in rotating turrets. This being so one may further remark that well before the end of the 19th century a humorous fictional reference had been made to launching a flying machine from a ship of the Royal Navy by means of a veritable torpedo-tube. A relevant passage ran: '. . . the navigatin' commander give the correc' course to the torpedo lootenant, who trained the toobe by compass, an' fired . . . . (But perhaps this was a precursor of the aircraft catapult).

To the earlier-quoted account by a Naval officer of how Sopwith aircraft were operated aboard 'his ship' it may be added that a technique was also developed whereby the aircraft was launched, as it were, in reverse, the guns being elevated to give a steep run down the platform, but still into wind. A method involving downward launching from a ship had, in fact, been patented long before the war in the names of Capt F. M. Sueter, Lieut F. L. M. Boothby and Engr-Lieut H. G. Paterson, this method being applied jointly with a launching trolley, which was to drop into the water and be hauled on board again. Acceleration of the aircraft was assisted by a Wright-style falling weight. In the purely practical sense, Lieut C. R. Samson, on 10 January, 1912, had used a sloping staging, built over the forward gun turret (thus rendered unworkable) of HMS *Africa,* to accelerate his pusher Short—and secretly, it was said, 'Sammy' had done the job in the previous December.

In the preceding chapter it was affirmed that 'On the Naval side' (and distinct from work described in this present chapter) 'bombing was pre-eminent'; and this particular theme may now be taken up by remarking that 1½ Strutters were used for anti-submarine patrols not only in home waters but in the Mediterranean area also, the home-based patrols beginning in April 1917 and those in the 'Med' about two months later. Thus, on 17 September of that year (by which time the 1½ Strutter had been in service for about 1½ years) an Otranto-based aircraft of this type was claimed to have sunk a U-boat with a 65 lb bomb.

As a summation of other developments and as a basis for further comment one acknowledges this excerpt from Owen Thetford's *British Naval Aircraft since 1912:* 'At the Armistice some 170 Sopwith 1½ Strutters remained in service with the R.A.F. and nearly 40 of these were at sea with the Grand Fleet. Indeed, when, in March 1918, HMS *Furious* was made the flagship of the Flying Squadron of the Grand Fleet her complement included fourteen 1½ Strutters. For deck flying the 1½ Strutter was used both with the normal wheeled undercarriage and with a special skid undercarriage first developed in trials at the Isle of Grain; the latter version usually had a hydrovane mounted at the front end of the skids to prevent the aircraft nosing over if forced into the sea. This device, as well as the inflatable air-bags located either side of the engine, remained a feature of naval aircraft until about 1923, when flotation equipment was mounted inside the rear fuselage instead.'

To these remarks one would add that, before ditching a 1½ Strutter, it was advisable (one experienced pilot said 'necessary') for the pilot—seated as he was under the centre section—to get out on the bottom wing. Further points that one now makes revert to the earlier Sopwith/Hawker analogies and to the development of flotation gear generally. Whereas the Hawker Nimrod and Osprey had (in common with other Fleet Air Arm types of their period) air-bags of the officially-styled 'permanent atmospheric-pressure type' as part of their normal equipment, the land-based Hart bomber was not thus endowed. So, when over-water dive-bombing became a requirement, flotation gear had to be provided as an 'extra'—just as it had been on the 1½ Strutter—and although this gear took two quite distinct forms (inflatable bags in under-wing containers, or a Youngman dinghy, as fitted on some Ospreys, and housed at the root of the top starboard mainplane, between the spars) both forms were inflatable—again as on the 1½ Strutter.

Another fitment of the general type which was to reappear on the Nimrod and Osprey was the arrester hook with which experiments were made on the dummy deck at the Isle of Grain (particularly associated with Harry Hawker's compatriot Harry Busteed) in June 1918. The 1½ Strutter used was a skid-equipped single-seat bomber, though the significance of this fact should not be over-estimated, for many types of aircraft were used as experimental 'hacks', and the absence of a gunner and internal bombs may have been deemed especially beneficial respecting c.g. position and flying weight, for the hook was attached far forward under the fuselage. These considerations notwithstanding, the claim by Vice-Admiral Richard Bell Davies that the idea of 'detachable-wing 1½ Strutters' of 'the long-range single-seater bomber' type was his own (for attacking Zeppelin bases) is not to be ignored.

Clips on the landing gear spreader-bar were ultimately favoured to effect arrest, and were to be generally used on British carrier-borne aircraft pending the return of the under-fuselage hook. The pilot who pioneered this 'clip' technique has, in fact, already been named in a loftier rank: he was Lieut Col R. Bell Davies, VC, DSO, and the aircraft concerned—on 1 October, 1918—was 1½ Strutter F2211, which, when flown from and landed aboard HMS *Argus,* had a special propeller guard.

The full extent of Naval experiments with the 1½ Strutter may never be known; but one that must clearly be recorded is that wherein fore-and-aft troughs, as developed at the Isle of Grain, laid along a special deck on HMS *Vindex* received the landing-gear skids of another experimentally-employed single-seat bomber. This particular aeroplane was the Westland-built N5601, and there is ample evidence that more than one aircraft from the same batch was used for Naval experiments of various kinds—with inflatable air-bags for instance. Westland's own first major contribution to British Naval flying was, as matters transpired, an adaptation of another of their products—the Walrus, for which the D.H.9A provided the basis, but which utilised the hydrovane landing gear and inflatable air-bags as developed on the 1½ Strutter.

To the 1½ Strutter the US Navy, as well as the British, owes a debt for pioneer experiments and operations. Apart from a single specimen shipped to the USA and numbered there A5660, twenty-one additional examples were obtained from the US Army after the war. Thus the 'yellow rose' that bloomed on the turret platform of the USS *Texas* in 1919 was really planted at Kingston, Surrey, England.

At the very outset of this account deliberate allusion was made to 'specially prepared shipboard versions' (plural) of the 1½ Strutter; and any attempt at a precise definition of the common appellation 'Ship's Strutter' (or Ship Strutter) would be hazardous. However, that the Isle of Grain should have primary credit for the detachable wings, skid-type landing gears, hydrovanes, flotation gear—even, perhaps, for that all-important 'wireless' (W/T)—used by aircraft of this general denomination is likely. That conversion sets were later ordered from the Gosport Aviation Co. is hardly less credible, though this order appears to be quite unconnected with the fact that Sopwith's R. J. Ashfield had joined Gosport around the turn of 1917. With the Isle of Grain also was associated the development of a special lightweight gun-mounting for the Ship's Strutter, though whether this was of the pillar type fitted to the Parnall Panther or of a specially lightened Scarff pattern (the Scarff, in any case, taking many forms) is uncertain. With Grain development, too, was associated a Clerget engine having a nominal output of 140 hp—in which regard it may be noted that the 9BF pattern was described by Gwynnes Ltd, of Hammersmith ('Sole licencees for the British Empire') as having a stroke of 172 mm (that of the 130 hp 9B being 160 mm) and a nominal power of 150 hp. (For French ratings see previous chapter). Concerning the long-stroke Clerget it may also be pertinent to note in the present context that in February 1918 an official Information Circular was issued comprising notes on the conversion of the 9B engine to what was designated (verbatim) 'the 9B.F. (Long Stroke) Clerget Engine'. As no Vickers gun was fitted to the Ship's Strutter any increase in cowling diameter would be of small consequence.

SL.T.B.P.

Of the various Sopwith aeroplanes which (men associated with them have assured the present writer over the years) were 'chalked out on the floor of the Experimental Shop by/for Harry Hawker etc, etc' (see also under 'Bee') this little, though evidently larger, precursor of the Pup appears to have the strongest claim to that dubious distinction. That Hawker laid down the general lines of the design, by dictation if not by draughtsmanship, can be accepted; and that he used this 'light tractor biplane'—for that nomenclature is implicit in the designation given above—as a 'runabout' is wholly credible. The name 'Hawker's Runabout' appears, indeed, to have gained general currency, though Hawker used it also for aerobatics.

That the SL.T.B.P. was a larger aeroplane than the Pup—which it otherwise generally resembled except in respect of the centre section and associated struts, and in having warping wings instead of ailerons—is apparent by courtesy of these attendant gentlemen.

By any reckoning, at any period, the SL.T.B.P. would be considered a dainty creation—even though the pilot's upward view was deficient, as is plain from this photograph.

Hardly less certain is the notion that Hawker's personal ideas for a single-seat fighter were developed—possibly with the intention of having a Lewis gun on the high-set top wing—while he was flying this aeroplane during 1915. A Lewis gun scheme, in fact, may well have had a special appeal to him, particularly when flying near Brooklands; for it was in this same area (at Bisley) that the Lewis gun had first been demonstrated in England—by BSA—and where Harry's own father is said to have shot, with a rifle, for Australia. (This last asseveration notwithstanding, the name Hawker does not appear among winners of the Queen's Prize between 1860 and 1900 as, according to one account, it should have done).

Upward view for possible combat may certainly have been less in Harry Hawker's early thoughts than armament, though lightness and simplicity were targets clearly shared with Sigrist and the drawing office staff. Lightness was implicit in the installation of a low-powered engine—a 50 hp Gnome (the actual one, it was said, from the Burgess-Wright bought by Tom Sopwith in 1911), this unit being cantilevered from a single rear mounting and having a circular cowling, quite different from the 'fish-mouth' of the Tabloid. (The term 'cantilevered' here implies that the engine had a circular bearer-plate, or plates, bolted directly to an engine-bearer in the fuselage, and leaving the front end of the crankshaft unsupported—the rear end being located by a transverse member in the fuselage. This last -mentioned member also resisted any bending moments tending to pull the engine out of alignment).

Simplicity was proclaimed by adherence to wing-warping for lateral control. Of hardly less significance (the deficiency in upward view having earlier been intimated) was the sharp stagger of the wings, with evident advantages to forward and downward fields of vision. Rearward rake on the tips of the wings and tailplane were other features portending the Pup, though the narrow-span centre section was carried on vertical (not splayed-out) struts.

Concomitant with lightness and simplicity were aerodynamic cleanness and an air of daintiness—characteristics evident in the photographs, which, truth to tell, contribute most of the information which might usefully be added in this text. One exception to this honest declaration is evidence that, with an 80 hp Le Rhône or 100 hp Gnome Monosoupape engine, a second seat and ailerons instead of wing-warping, the airframe still existed in 1926.

Remarkable though the fact may seem, this head-on aspect (with the camera serving as an aiming-point for the Vickers gun, or vice versa) shows practically every basic feature of the Pup. The Sopwith caption reads: 'S.80—Sopwith "Pup". 80 hp Le Rhone—1916', and the installation of the engine named is especially well shown—likewise the centre-section transparent panelling.

Pup

A rightful and honoured tradition among the hundreds of authors who have published dissertations on the Pup is to characterise it as 'the perfect flying machine', or something closely akin, and in so doing to honour also the name of Oliver Stewart, whose first-hand knowledge of this aeroplane was so lovingly and memorably expressed in his writings to that effect. Oliver, alas, died when this present book was being planned (December 1976); yet one can still pay tribute to this old friend—and at the same time to the Pup—by declaring that he was a man by whom others were measured, and it is certainly no exaggeration to affirm that many of his ways and manners (even, it seemed, his physical characteristics) were—consciously or otherwise—reproduced in men around him. So it was with the Pup.

These matters being so, one turns not to Oliver Stewart for an introduction to the Pup, but to a quite exemplary appraisal by Lord Weir of Eastwood, Controller of Aeronautical Supplies and member of the Air Board, 1917/18; Director-General of Aircraft Production, 1918; and—likewise in 1918—Secretary of State for Air and President of the Air Council. With experienced men to advise him, Lord Weir declared soon after the Armistice:

'The characteristics of the Sopwith Pup, our first good tractor single-seater, were very light surface loading, a small but good rotary 80 h.p. French engine, and every scrap of unnecessary weight eliminated by careful design. The view, particularly overhead, was not very good, but the aeroplane was so handy fore and aft that this did not interfere very seriously with its fighting qualities. The type lasted a very considerable time before it was superseded, which, in view of the comparatively small horse-power, was remarkable. During the period in which this type was in use fighting acrobatics [*sic*] advanced to a marked degree, and in the next type [the Triplane] an effort was made to increase view and manoeuvrability.'

This last reference to the continuing need for increasing manoeuvrability gives emphasis to the fact that it was not manoeuvrability *per se* that distinguished the Pup (for that particular attribute was not to be fully realised until the coming of the Camel) though Lord Weir makes special allusion to fore-and-aft handiness, just as Oliver Stewart once mentioned 'the rather powerful and quick elevator'. The Pup's pre-eminent quality in combat was, in fact, its ability to 'hold its height' (in the parlance of those times) as now finally affirmed by Maj Stewart: 'It was this power to hold height during a dog fight that made the Pup a useful aeroplane, and it was this quality that the pilots sought to amplify. By the selection of an appropriate type of airscrew and by lightening the machine as much as possible the height-holding powers were enhanced.'

So much for the significance of the Pup as a fighting machine; and as for its history, it must first be remarked that although manifestly a very near relation of the SL.T.B.P., the traditional ascription of its ancestry to the 1½ Strutter (of which it was supposedly declared by Col Brancker to be 'a pup') is not to be dismissed. This is fairly clear from the fact that 'Pup' was the name insisted upon (if not conferred by) Service pilots, who were familiar with the 1½ Strutter though far less so with Harry Hawker's 'pup'—which he liked to take around with him not merely as a pet, but as a development vehicle for a new fighting aeroplane.

Self-evident is the absence from the Pup, or Sopwith Scout as it was first officially known (with or without initial capital for 'Scout') of the salient feature—the peculiar form of wing-bracing—which not only identified the 1½ Strutter, but whereafter that type was named. 'So' (as Oliver Stewart summed the matter up) 'I

Of such clarity is this study of a Sopwith-built Pup that the gunfiring lever (which, as Oliver Stewart recalled, 'projected back from under the rear part of the gun') is clearly seen. This might not have been so were the Sopwith padded screen installed, though fittings for this are present on the gun (a Vickers).

suppose *that* [the second of two official orders relating to nomenclature] and the perverse state of mind of the fighting forces when it came to language, both good and bad, accounts for the fact that the aeroplane has ever after been known exclusively as the Sopwith Pup.'

Yet still one has the merest suspicion that the full story of one of the most famous aircraft names in history has not been fully told; for even though Peter Lewis, in his '1809-1914' book, faithfully records that there was built, in 1909, a tiny single-seater called the Neale Pup, it is not explained by Mr Lewis that this same aeroplane was characterised not only by its name, small size and single seat but that (according to an official account) it was built for J. V. Neale at Brooklands, and that this same man later formed an aircraft company at Richmond—obtaining, in fact, a War Office contract for four machines. Both Brooklands and Richmond had strong Sopwith associations, though here, perhaps, we have a chain of mere coincidences.

Whatever the niceties of nomenclature, we now perceive how the 'thoroughbred' Pup was really a mongrel—by $1\frac{1}{2}$ Strutter out of SL.T.B.P. But that among officially adopted 'scouts' it had a character all its own, in the sharply raked tips of its wings and tailplane, was conveyed by the *official* recognition drawings.

To do full justice not only to officialdom as well as historical precision, the 'Admiralty-system' designation (for the Pup, like the $1\frac{1}{2}$ Strutter, was blooded in operations by the RNAS, and not the RFC) was Sopwith Type 9901.

Though construction was fairly conventional, typically with spruce wing-spars and ribs, ash longerons (earlier spruce) and spruce spacers—or 'transverse struts' and 'side struts', as these last-named members were formally designated—steel tubing was extensively used, not only for the wingtips and trailing edges and for the landing gear, but in the tail—the fin and rudder especially; and visually there were other strongly marked features apart from the sharply raked tips already mentioned. Most prominent among these features was the small size of the four ailerons, contrasting with the large horizontal tail surfaces (the elevators included)—the size of these last accounting for the aircraft being, as Lord Weir said, 'so handy fore and aft'. That same minister's critical remark that 'view, particularly overhead, was not very good' (this notwithstanding the more rearwardly positioned cockpit, compared with the SL.T.B.P., and the provision of a trailing-edge cutout) was in some degree met, both experimentally and in service, by the inletting of transparent panels—which unfortunately were prone to splitting—or the provision of non-standard cutouts. Such tampering with wing area was not, of course, compatible with height-holding, and there were even some misgivings at one time (early 1917) concerning the possible effect on performance of a little hole—about a foot square only—as a palliative against tail-heaviness.

Possibly in 'the perverse state of mind of the fighting forces' referred to by Oliver Stewart, the first Pup seems to have been criticised by RNAS pilots respecting not only upward view, but 'straight downwards' also, in this last respect being considered inferior to the Nieuport, though otherwise the fields of view were reckoned equal.

The Pup having been regarded above all as a 'pilot's aeroplane', and Oliver Stewart's first-hand knowledge having been so freely drawn upon, it is salutory to consider the affirmations of a pilot hardly less renowned who, though having no first-hand knowledge of the Pup in combat, has nevertheless made a particular study of its design and engineering. Thus Harald Penrose:

'The hand of R. J. Ashfield is discernible in the design of the Pup, and its derivation from the Tabloid is clear when structural drawings of the two are superimposed. The length from stern-post to engine bulkhead is the same; the fuselages have identical depth; and the lower longerons are set at the same upward angle from rear spar to tail, though in plan the Pup is noticeably narrower. Vertical spacer-struts spindled to H-section are displayed in slightly different positions from the Tabloid, but the attachment fittings appear interchangeable, though Pup metalwork had more lightening holes. On a fuselage general arrangement drawing in my possession there is an annotation by Fred Sigrist that the spruce longerons are to be changed to ash in subsequent versions. Wing construction of Pup and Tabloid show further similarities, for the chord is identical, the gap only fractionally different, and spar positions the same, but the Pup's wing stagger is seven inches greater.

'Many features show how attentive Harry Hawker was to the draughtsmen's boards. The wing attachment was his design, and had the simplicity of a flying model of that day, for the butt of each spar was reinforced with a hollow metal ferrule within the end rib, and slid on to projecting ends of the centre-section spars and the lower spars traversing the fuselage. A long thin pin, with localized increased diameter at load points, was externally pushed through a tubular guide in the leading edge and through the spar abutments before emerging through the trailing portion of the wing. An airtight chordwise join resulted, simple and quick to secure compared with separate pin joints enabling the wing cellules to be placed in position or removed while tautly boxed with bracing wires. The original patent was secret, but registered in Hawker's name as No. 113,723, and ultimately dated May 1917. So also was Patent 127,847, protecting the practical and extremely simple method of attaching the annular cowling of the 1½ Strutter and Pup, whereby an encircling groove, formed round the aft end of the aluminium cowl, engaged a similar groove in the nose structure of the aeroplane. A cable passed round the cowl groove and, tightened by turnbuckle, compressed the perimeter into the fuselage groove and locked it in position. It was a system quickly adopted by other makers.

'Because axles slung between the inverted Vs of conventional undercarriages tended to become permanently bowed after several heavy landings, another simple solution was achieved in the articulated axles devised and patented (No. 109,146) by Tom Sopwith and standardized for all his machines. Using two separated horizontal spreaders from apex to apex, he pivoted the half axles between them from a central hinge, springing the hub ends with shock absorber cords wound on studs at the bottom of each undercarriage frame—but the crux of the invention was to suspend the central hinge by cable from the fuselage in order to resist collapse as the wheels moved upwards. Compared with sleeving the axle to increase bending strength, it saved several valuable pounds.

'The Pup on early flights had a fixed tailplane like the Tabloid, but slight changes in trim with speed and pilot weight made a trimmer desirable for *finesse*. To use the nut and worm gear patented for the 1½ Strutter would be unnecessarily expensive, so Hawker devised a simple crank hinged from the stern-post, connecting it at mid-length to a vertical push-tube attached to the rear spar, and operated it with wires running from a diagonal sliding knob on the right side of the cockpit.'

Here it is fitting to note that—Mr Penrose's accurate record notwithstanding—official notification was given in January 1917 that RFC Pups would have a fixed tailplane as standard. This alteration was, indeed, only one of very numerous modifications made to meet pressing needs or personal preferences. Thus, at various times, tailplane incidence was increased; the landing gear (which had already been somewhat heightened in early-production machines) was strengthened for specialised training applications—for which purpose production was run on into 1918; the ⅓-inch plywood decking round the cockpit was variously

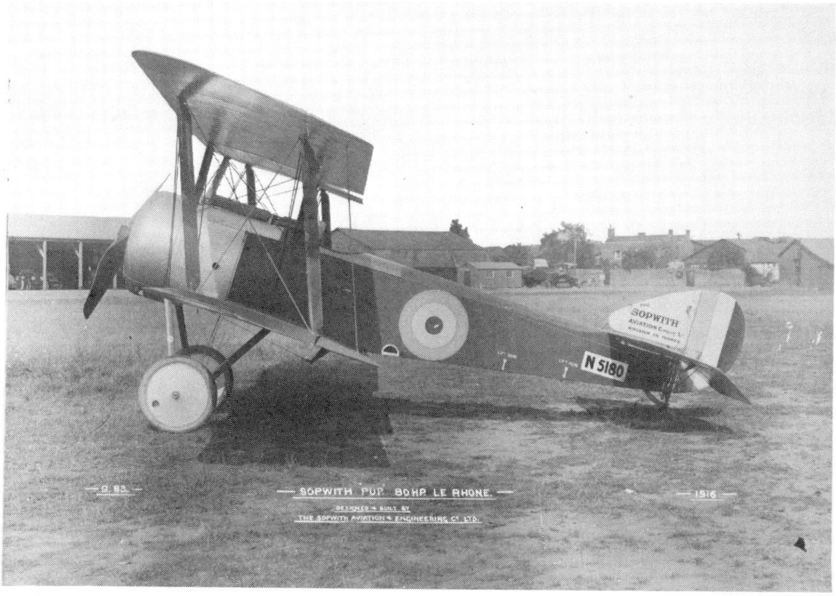

Markings and movements are illustrated by these three side views: Top, a Sopwith-built Pup wearing its Service number (N5180)—absent in the previous view of a similar machine—the picture being captioned by the makers: S.83—Sopwith 'Pup'. 80 hp Le Rhone—1916; bottom, a Standard-built specimen, darkly numbered B1704 hugely on the fin, and with port ailerons down; opposite, a Whitehead example, with its A6158 more readable, and with port ailerons up.

cut away; the Sopwith padded screen on the Vickers gun was sometimes abandoned as a further interference with fighting view (one known alternative being a Triplex screen in two halves, one on each side of the gun—though unarmed trainers, for instance, sometimes had an Avro screen); while there were variations too in stagger from the standard 18 inches—eight Sopwith-built Pups, for instance, being turned out with only a 15-in stagger. But among some 2,000 simple aeroplanes (especially when flown by pilots prone to 'perversity'!) alterations must have been vastly more extensive, though the major ones were those later instanced in the context of special Naval applications.

The first Pup was No.3691, which was cleared by Sopwith's experimental department on 9 February, 1916. This example (which had a shorter landing gear and smaller vertical tail surfaces than later versions) may have initially been fitted with a seven-cylinder 80 hp Clerget engine, though a nine-cylinder 80 hp Le Rhône was installed for CFS tests in March. Clerget-powered—initially at least—were the next five Sopwith-built examples (Nos.9496 and 9497 and Nos.9898—9900) and the first eleven built by Beardmore, to which company the initial production contract was transferred, and which was later awarded special Naval development contracts. For the apparently unbuilt N503 a 110 hp Clerget was intended.

With the 80 hp Le Rhône (as made by W. H. Allen, Son & Co Ltd, and in an annular cowling with a segmental slot at the bottom) the Pup airframe became not only chiefly associated but well-nigh identified; and there was wide agreement among Pup pilots that the '80 Le Rhône' (makers' suffix 9C) was the perfect partner—sweet-running and dependable, even when over-revved, as in chases and escapes. (Later a few Pups were fitted with the 110 hp Le Rhône 9J engine, but with this unit—which could, in fact, deliver about 130 hp—airframe-strength suffered quite alarmingly, and already the Camels were coming. The fairly common 80 hp Gnome installation was apparently confined to Pups used for training).

As soon as No.3691 went to the RNAS for service tests in May 1916 (Chingford, Grain and Dunkirk were visited before the aircraft was allocated to 'A' Squadron of No.5 Wing at Furnes, in France) it became apparent that an entirely new class of

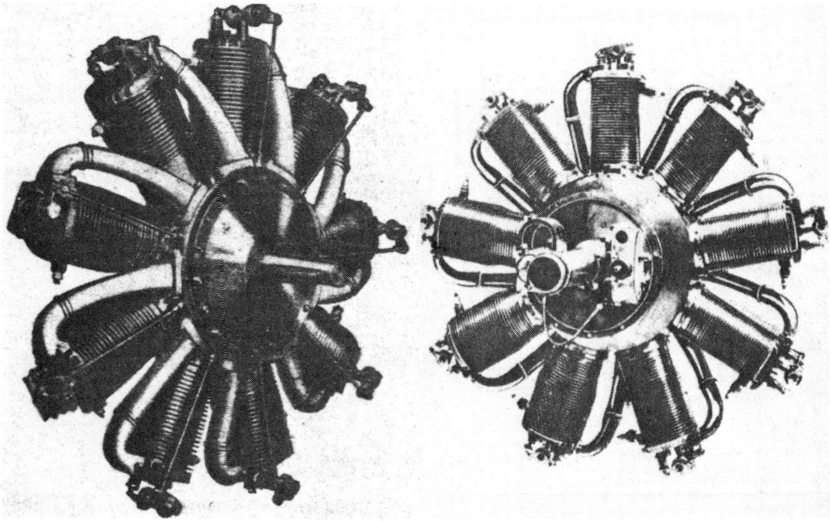

As remarked in the text, there was wide agreement among pilots that the '80 Le Rhône' was the perfect partner for the Pup, and aspects of this delightful French rotary are here presented: left, $\frac{3}{4}$ front; right, $\frac{3}{4}$ rear.

fighting aeroplane was in Britain's hands—and an ascendancy was established which lasted from the autumn of 1916 until around mid-1917. Although credit for this must go initially to Sopwith (who not only designed the Pup, but together with William Beardmore & Co built large numbers for the RNAS—the War Office orders going to the Standard Motor Co and Whitehead Aircraft) it must be recognised that ascendancy over the enemy stands to the glory of Naval and RFC pilots alike. To a member of the first Pup-equipped RFC squadron (No.54) this vivid explanation of that ascendancy is due:

'We attained 18,000 ft with regularity, and could get even higher. Our best chances came from climbing above the maximum height obtainable by the German fighters and then hoping to make a surprise attack. The Germans were always superior in level speed and in the dive, but the Pup was much more manoeuvrable and we could turn inside any German fighter of the day.' And—as clinching evidence that the Pup did indeed represent an 'entirely new class of fighting aeroplane', as already declared: 'The winter of 1916/17 was bitter in Northern France, and at 18,000 feet everything froze—the engine-throttle, the gun, and the pilot . . . The aircraft itself always behaved in a most gentlemanly way, but it needed careful handling—a dive of 160 m.p.h. was fast enough, and at 180 m.p.h. the wings were definitely flapping and a gentle recovery was essential, since to lose a wing when one had no parachute offered no future.'

Curiously, this same officer went on to describe the Pup's lightness—one of its great advantages in combat, as earlier established in this chapter—as its 'one *dis*advantage', though this was in the context of ground-handling, which, in a strong wind, entailed the calling-out of all available personnel when a patrol was landing-back to seize the wingtips before a gust could blow the aircraft over. Here we have an inter-Service parallel, as well as a technical one, with the famous picture of an RNAS Pup being literally hauled out of the air by a deck-party using rope toggles affixed to the aircraft.

The first RFC Pup squadron was, as already noted, No.54, which arrived in France on Christmas Eve 1916; but considerably earlier—by late October—one flight of No.8 Squadron, RNAS, had been equipped, and the Pup's first recorded victory had, in fact, been chalked up on 24 September, when F/Sub-Lieut S. J. Goble, of No.1 Wing, RNAS (later Chief of the Air Staff, R. Australian A.F.) shot down an L.V.G. in flames. By this time the 'image' of the First World War fighter pilot was forming—the man in the warm-lined leather coat, with silk gloves under leather gauntlets, sheepskin boots, and perhaps on exposed facial areas—whaleoil.

A consistently successful Pup pilot was, of necessity, a good marksman, for the Pup was armed as standard with a single Vickers gun only, installed as on the 1½ Strutter in conjunction with a Sopwith padded screen and—initially—Sopwith-Kauper synchronising gear, though some later aircraft had the Scarff-Dibovsky (mechanical) or Constantinesco (hydraulic) gun-gear. As was the case with the 1½ Strutter, there was a non-standard installation of the Vickers gun on the port upper longeron, and there were several unofficial—and generally unsuccessful—installations of a Lewis gun above the centre-section.

The Pup's 'perfection' was, of course, relative, and this was manifest not only in criticisms of its field of view and too-light armament (one Vickers gun being standard) but in the difficulty, in late 1916 at least, of holding the sights on-target in a fast dive by reason of a 'surging' motion in the 'up and down' plane. Another point concerning armament was that while the Pup was becoming established in service (early 1917) so, also, was an innovation in feeding the ammunition—by the use of the Prideaux disintegrating-link belt. Previous to this, sodden, frozen, swollen or damaged belts (even though these were stoutly made of webbing) had given trouble to the Army and the flying Services alike, and non-disintegrating belts had been briefly tried by the Army in France, though they were never standardised.

With the plain circular cowling that characterised the 80 hp Le Rhône and 80 hp Gnome installations (the latter is shown) and without a Vickers gun, the Pup presented an especially trim appearance.

For aircraft use the Vickers gun presented a particular problem because the used portion of the fabric belt had somehow to be stowed away snugly and safely, where it would not (for instance) seek to reintroduce itself into the gun or create special kinds of mischief to which aeroplanes were sensitive. By making the cartridges themselves form the hinge-pins, the metal links (which were expelled from the side of the gun) and the spent cartridge cases (which came out through the bottom of the gun) could be more conveniently disposed of—through chutes, though with the new form of belt the Pup's ammunition box could hold only 350 rounds instead of the specified 500. Removal of the original receptacle for the used webbing belt, however, permitted restoration of the full ammunition supply.

The summer of 1917 found the Pup outclassed in France but still an attractive proposition for Home Defence—especially with the 100 hp Gnome Monosoupape engine to improve the rate of climb and ceiling. Here the light drum-fed Lewis gun came in for special consideration as a top-plane fitment—partly to obviate firing 'special' ammunition through the propeller arc; and the great McCudden made himself a 'rough sight of wire and rings and beads' for such an installation.

Even with Vickers gun, Sopwith padded screen, and ring sight (and with ailerons awry) the Pup was still one of the daintiest of all aeronautical creations, as A7302 here proves.

The 'H.D.' Pups which had the 100 hp Gnome Monosoupape engine—an installation which had, in fact, been projected early in 1916 for the second and third Sopwith-built examples (Nos.9496 and 9497) though not then implemented—were distinguished by a longer, open-bottomed cowling, which was further characterised by four auxiliary lips to admit cooling air, these lips being set in a group round the upper-starboard segment of the rim. (There were, nevertheless, instances of Le Rhône engines in cowlings of similar form). The greater power of the 'Mono Gnome' gave an appreciably better rate of climb; and it may well have been this attribute which, early in 1918, prompted a member of *Flight*'s staff (one seems to recognise his brand of humour from having shared an office with him long ago) to comment in excruciating style on an article by Sous-Lieutenant Viallet in *La Guerre Aérienne*, this article having the title *Considérations sur les Avions de Chasse*. The comment made was that the French article dealt chiefly with 'a *pop*ular British aeroplane' styled as the Sopwith "Pop". 'Is this', enquired *Flight*'s humourist, 'because it is a machine which has given the Boches ginger, or is it because the useful

With the 100 hp Gnome Monosoupape engine, the Pup had a distinctive cowling, open-bottomed (or 'horseshoe') and with four auxiliary 'lips' round the upper-starboard segment of the nose-ring, as shown.

Modifications to Service (as distinct from experimental) Pups were relatively few, one of the best-known being the increase in tailplane incidence from 1½ to 2½ degrees when the Gnome Monosoupape engine was fitted; but the tail surfaces of the one in distress here—and not, apparently, by reason of the sad event depicted, involving the inexpertly numbered 7313 (presumably the Standard-built A7313)—were non-standard.

little scout flies upwards like a cork out of a bottle?' With a 170 hp A.B.C. Wasp radial—performance wherewith reached the calculation stage by Sopwith—it should have ascended like a cork out of a magnum; but although in postwar years a Pup in Australia was, in fact, fitted with a radial engine, this was an Armstrong Siddeley Genet of only about 80 hp—and the Pup was no longer in the fighting business.

Certainly the wartime Pup had latterly become an intercepter as well as a dog-fighter, and one must note again that by early 1918 it was still 'flying upwards', although it *had* (in the past tense) given the enemy ginger in France. Yet what it had really done was far more than has so far been related here; for though its operations with land-based units of the RNAS have already been touched upon, its contribution to 'ship flying' (as it was known) was basic, wide in scope and exceptionally interesting in the purely technical, as well as the operational, sense. That no separate chapter on a special 'ship's Pup' is appended, as was the case with the 1½ Strutter, is explained by the fact that, although such an aeroplane did indeed exist, it was a Beardmore, and not a Sopwith, development; but before passing attention is paid to this very highly specialised machine, praise must be accorded to the contribution made by (more or less) 'ordinary' RNAS Pups to the techniques of flying from ships—for these were very great indeed.

Though landing gears of various sorts were much involved in the work to be discussed (but never floats, as on the Schneider and Baby, it must be emphasised) the first point to be mentioned concerns not the under-part but the overhead centre section; for though the first Pups built by Sopwith for the RNAS (N5180—N5199) were fitted, as standard, with the same form of centre section as the company's Pups for the RFC, there was soon a special Naval requirement for an aperture to be formed between the spars—not, as was the case with some modifications, to improve the pilot's view, but to allow a Lewis gun—mounted on a tripod of steel tubes forward of the cockpit—to fire upwards, especially at Zeppelins. This feature was characteristic of the 'Sopwith Type 9901' as built for the RNAS by Beardmore, though, as the Service named was from the earliest times much concerned with the attack of airships, the Beardmore-built Pups—with or without the Lewis gun—sometimes had provision for eight strut-attached Le Prieur rockets.

By no mere carelessness or contrivance has this particular part of the narrative—dedicated, as just declared, to the techniques of flying from ships—turned towards armament. This turn is readily explained by twin considerations: first, that the Navy did not fly from ships for fun or danger-money, but to serve the Fleet; second, by its particular preoccupation with armament of many forms—best exemplified by Scarff, with his gun mountings, gunsights, synchronising gears and bombsights. Thus it is fitting now to give a note concerning the rockets just named—especially so as they will be mentioned again in the Camel context. Historically, perhaps, there is an even more compelling reason, for the Le Prieur rocket was a stick-stabilised weapon and, as such, a latter-day development of such patterns as the Congreve, deployed by the Royal Navy in the early 1800s. A development by Lieut Y. P. G. Le Prieur of the French Navy (who also designed some complicated gunsights—rivalling in complexity those devised by Scarff) the French rocket gained no high reputation, although it was officially recognised that some early specimens had been damp. Firing was electrical, from steel launching-tubes, and mean velocity about 330 ft per sec—the intended targets being, of course, Zeppelins.

But from airships we must now firmly turn to surface ships; and with those of the Royal Navy, No.4 (Naval) Squadron had been well acquainted after equipment with Pups in March 1917—undertaking not only offensive patrols (fighting

Albatros scouts, for instance) but close-protection also. From May 1917 Naval Pups flown from Walmer, between Deal and Dover (at which last-named British stronghold there were also Pups) escorted and protected merchant ships and seaplanes alike—speed-difference between the patrolling Pups, which had air-bags for emergency flotation, and even the merchantmen, being relatively small; and in July of the same year Pups superseded some Sopwith Baby seaplanes when they (the landplanes) took up station with the Seaplane Defence Flight, operating from St Pol, across the Strait of Dover.

Though the high-flying Gotha bombers were by no means easy meat even for a Pup, Flt Lieut H. S. Kerby, manning a Pup attached to Walmer, sent a straggler into the sea.

The flying of Pups from ships—with anti-Zeppelin work especially in view—dated from early 1917, when it was recommended that machines of this type should replace Sopwith Babies aboard aircraft carriers (*Campania* and *Manxman* were first proposed), cruisers and other fast vessels. Now was the time when the fitting of emergency flotation gear, as well as the special 'Naval' armament earlier described, came in for particular attention—together with the services of an officer who had already done much for the glory and efficiency of British Naval flying. This officer was Flt Cdr F. J. Rutland (Rutland of Jutland) who was, perhaps, the Navy's strongest advocate and most determined practitioner in the flying of Pups from ships.

Like Harry Hawker after him (see under 'Atlantic') Rutland contended that it would be safer to ditch a buoyant landplane than to trust oneself to an alighting in a seaway with an inherently frail floatplane; and in any case, he argued, the Pup was the only aircraft that could tackle a Zeppelin near its ceiling. (The first Pups actually delivered to a ship may well have been Nos.9910 and 9911, which went to HMS *Vindex* following their acceptance on Boxing Day 1916).

With an air-bag lashed inside the narrow rear fuselage of a Pup—much the same arrangement as Blériot had used to fly across the English Channel in 1909, and much the same also as employed on Hawker fleet fighters of many years later—the aforementioned Rutland took-off from HMS *Manxman* on anti-Zeppelin patrol, but was forced to ditch off the Danish coast, where his Pup remained afloat for twenty minutes only. The date was 29 April, 1917; but on 23 June flotation tests were put in hand on the first Beardmore-built Pup, 9901, moored off the Isle of Grain and having a trial installation of 'Mark I Emergency Flotation Bags'. To confer on the Pup improved flotation qualities these were of inflatable type—their inflation being unconfined, moreover, because they were externally attached (to the undersides of the bottom wings where, while deflated, they lay flat). On this occasion the Pup stayed afloat not for twenty minutes only but for more than six hours.

Another Pup experiment at the Isle of Grain involved the fitting of a jettisonable landing gear, with which, nevertheless, the Pup tended to overturn on ditching. Greater success was achieved when hydrovanes were fitted under the fuselage and on the tailskid.

Early shipboard operation now being our chief concern—with credit being accorded, as due, to the Pup (as to the 1½ Strutter) for its suitability and adaptability—one feels bound to emphasise that careful thought had been given long before the war of 1914 to the use of aeroplanes not only from specialised aircraft-carriers, but from other British Naval vessels also. The following was, in fact, written pseudonymously in 1911:

'Aeroplanes may be carried either in large numbers in a specially built mother-ship, or one or two in every large battleship or cruiser.' (How remarkably accurate

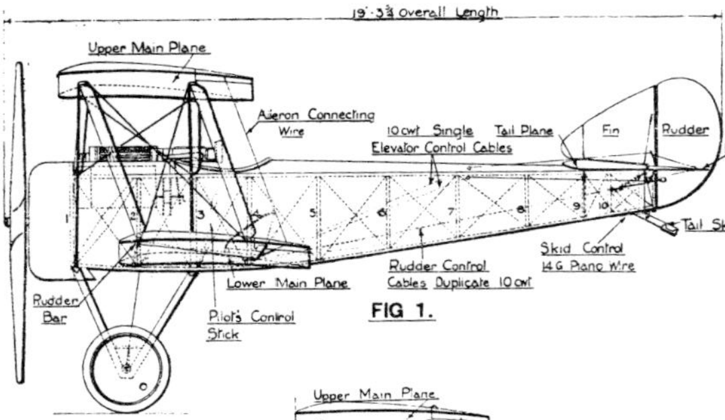

FIG 1.

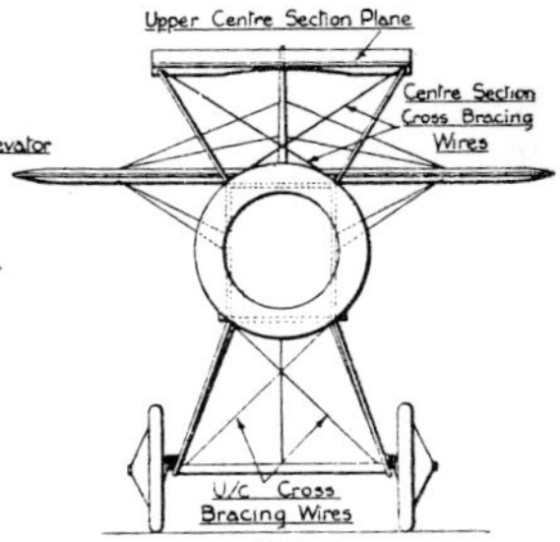

FIG 2.

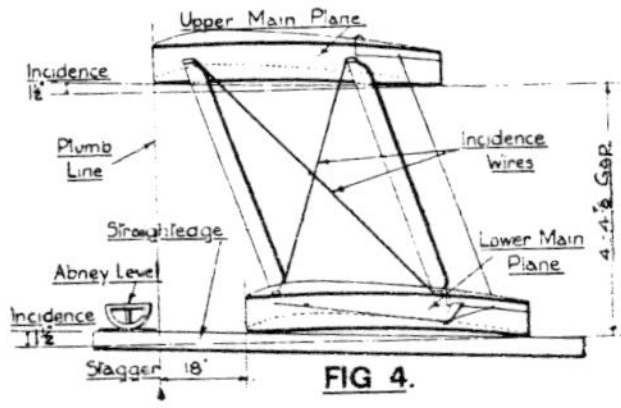

FIG 4.

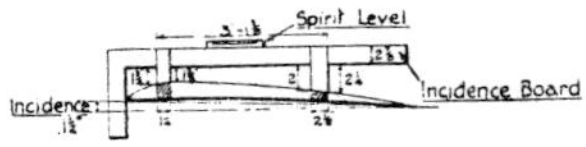

One of the finest sets of official drawings ever prepared, showing not only salient features of the Pup, but details of armament (even the winding-off drum for the ammunition belt)—not to mention the buttons on the seat-cushion!

FIG 5.

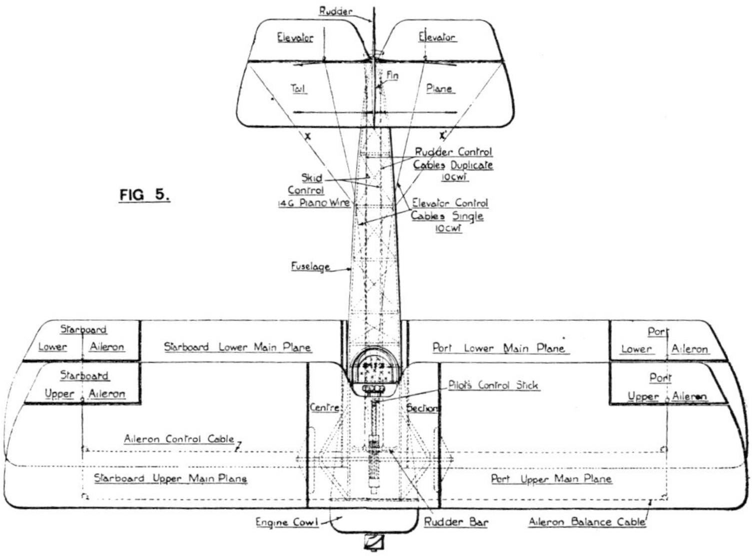

this was to prove is clear from a Naval officer's post-Armistice affirmation, already recorded, that 'My ship carried one Camel and one 1½ Strutter'). The 1911 quotation continued: 'In the case of the mother-ship the stowing space can be made ample. Also, she can be fitted with large decks or any cumbersome but convenient method of starting the aeroplanes. In fact, she can have all the luxuries of an aerodrome. In a mother-ship the aeroplanes would be well looked after, while in a man-of-war they might lack sympathetic treatment. This however, once aeroplanes have established their utility, will be grown out of. In all probability aeroplanes, at first, will be carried one in every big man-of-war. In that case the aeroplane will have to make the best of what it finds there. It will be stowed, mostly in bits, up among the boats. A tarpaulin cover for its engine will be its hangar. No special arrangements will be fitted unless they are small and unobtrusive, unless they in no way detract from the fighting or sea-going efficiency of the ship.'

And—the saltiest and most sagacious touch of all, having special regard to flotation gear and flying-off platforms, which are very much our present concern:

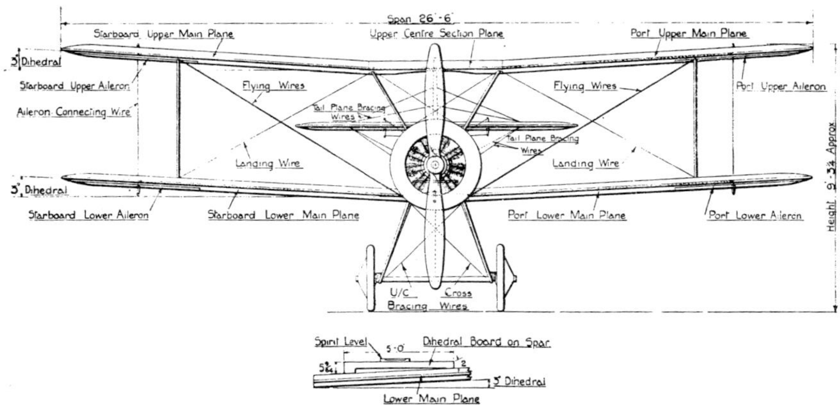

'The deck space will always be too limited to permit a return to it, and so the return will always be made to the water. On his return the aviator will be picked up and his machine hoisted in. If it is not boating weather, well, probably the machine won't be flying . . . '

Against these remarkably prophetic ruminations of 1911 we may now set this backward glance in 1919—noting especially the name 'Deck Pup'. Thus: 'One of the remarkable features of the war has been the way in which all classes of ships (excepting destroyers) have been equipped to carry and fly off small aeroplanes from barbette crowns [*i.e.* turret tops] in big ships, and from small platforms in light cruisers. In the *Carlisle* and "D" types of light cruisers, a high "Arc de Triomphe" hangar has been combined with the fore bridges, the fore portion of which can be closed by wind screens or roller shutters.' The reason for this 'huge' structure was thus explained: 'The small Sopwith "Camels" and "Deck Pups" carried in our warships have to attain their flying speed with a remarkably short run, on account of the limited length of flying-off platform. So anxious are they to get into the air, they tend to rise by themselves when the carrying ship is steaming into a head wind. Lashing down the 'planes resulted in straining. Accordingly, small wind screens have sometimes to be rigged round the aeroplanes. But wind screens are a nuisance to rig or unfurl in a rising wind—hence the permanent structures adopted in new light cruisers.'

Although the design of post-1919 cruisers is not of present concern, the contribution made by the Sopwith Pup to operations from this class of ship assuredly is; and so it must now be recorded that on 28 June, 1917, Flt Cdr Rutland pursued his experiments by flying a fully armed Pup from a 19 ft 3 in (5.8 m) platform fixed in position over the forward 6-in gun of the 5,250-ton light cruiser *Yarmouth* (so that for launching, the ship had to steam into wind) leading to a decision in the following August that one ship in each light-cruiser squadron should be fitted with a flying-off area. Earlier Rutland had used—and this from the slower *Manxman*—a platform of only 15 ft 6 in (4.7 m). Thus this particular platform was shorter even than the tiny Pup itself.

On the warlike side, Flt Sub-Lieut B. A. Smart, flying a Pup from the pioneering *Yarmouth* on 21 August, 1917, shot down Zeppelin L.23. Clearly, this particular Pup had received the 'sympathetic treatment' that the gentleman writing in 1911 thought it 'might lack', though it sank before similar attention could save it.

A few lines earlier in this present account the displacement-tonnage of *Yarmouth* (5,250) was deliberately quoted to emphasise the relatively small size of a light cruiser as compared with the converted 'light battle cruiser' *Furious* (roughly four times as much)—*Furious* being a vessel concerning which there will be far more to say. Meanwhile it may be remarked that, resembling as she did a big destroyer

Airborne over the 15 in guns of a battle cruiser, with the broad bows stretching out ahead, the Pup was given a different nautical scale (sailor-caps give further scale and atmosphere at lower port). The flying-off platform seen here was an experimental downward-sloping one, mounted on 'B' turret of HMS *Repulse*; the pilot Sqn Cdr Rutland; the date 7 October, 1917.

Pup N6453 (which was also tested aboard HMS *Furious*) departs from the aft ('Y') turret of HMS *Repulse*—a battle cruiser having two such twin-gun turrets forward and one aft. Note that the departure in this instance is made over the rear of the turret, and not over the guns.

Yarmouth's success stimulated the notion of providing even destroyers and other small craft with aeroplanes (the 'Kittens') though that particular notion came to naught. The war had not long been over, however, when *Yarmouth* had the then-unrecognised distinction of carrying (as an observer of atmospheric conditions in far lands) a man—Robert Watson-Watt—whose contribution to his nation's defence was in later years to prove certainly no less in value than that of the Sopwith Pup. (HMS *Yarmouth* herself had served at Jutland also, but was sold in 1929).

Whether or not a Pup was ever catapulted is uncertain, though in May 1917 two or more Beardmore-built examples were sent to Hendon for that purpose; but given a cruiser's speed, a modest platform, a headwind (and, of course, a fittingly 'sympathetic' bunch of officers and deck-hands) what would a Pup need with a catapult? One reflects, in fact, that if a Pup had been fitted with rockets as take-off assisters instead of as 'R.P.s' as already related, it would have run well-nigh the entire 'modern' naval operational gamut (VTOL, of course, being taken for granted).

In the matter of launching, the basic difference between flying-off an aeroplane from a fixed platform, as instanced by that on the light cruiser *Yarmouth*, and performing a comparable operation from a turret platform (of the kind already mentioned in connection with the 1½ Strutter) aboard a capital ship having rotating turrets for heavy guns—*i.e.*, a battleship or battle cruiser—was that the latter ships could rotate the turret/platform combination, or 'turntable', into the 'felt' wind, instead of having to steam *into* wind. Thus they could maintain a course. For testing the Pup from a capital ship the chosen vessel was the battle cruiser *Repulse*, and the pilot Flt Cdr F. J. Rutland, whose name has already been acclaimed as 'perhaps the Navy's strongest advocate and most determined practitioner in the flying of Pups from ships'. The first trial was on 1 October, 1917, using a downward-sloping platform on 'B' turret. This trial having proved successful, the platform was transferred to 'Y' turret, and on 9 October Rutland took-off again—on this occasion not over the guns, but over the rear of the turret. By this time, however, the Camel 2F.1 was well advanced in development, and, together with the 1½ Strutter, was standardised for shipboard use.

The problems of *landing* an aeroplane on a ship's deck did not arise like a mountainous sea in the course of naval-flying development as some accounts suggest; nor would it be justice to ascribe to any particular proposal its realisation as a workaday procedure. In token whereof it is needful only to remember that as early as 18 January, 1911, Eugene B. Ely had placed a Curtiss pusher quite firmly down aboard USS *Pennsylvania.* Yet the way ahead was still a rough one, involving in particular the Sopwith Pup and HMS *Furious,* which joined the Grand Fleet in July 1917—the month after the Rutland/*Yarmouth*/Pup experiments already recorded.

To HMS *Furious* now one turns attention—a ship which had the well-nigh incredible distinction of having operated Sopwith Pups in the First World War and Hawker Sea Hurricanes in the Second; so if the Pup was sired by the 1½ Strutter, then it had as its foster-mother-ship a vessel with a history no less curious and distinguished than its own. Although launched in April 1916 as a 'light battle cruiser' designed to have a main armament of two 18-in guns (the Navy's standard then being 15-in) she was completed in March 1918 as an aircraft-carrier—of a kind. The essence of the alteration was in deleting the forward big gun and in building ahead of the ship's superstructure a 228-ft (69.4 m) flying deck, with a hangar below it. In this hybrid form the vessel served from June to November 1917; then the second big gun (aft) was removed and the original flying-provisions more or less duplicated. The ship was re-commissioned on 15 March, 1918, and of her Sopwith links thereafter there will be more to say under the heading of '2F.1 Camel'. Before returning to the Pup, however, let it be recorded that, after yet another rebuilding, chiefly in 1924 (whereby she was given a virtually unobstructed full-length flight deck, *Furious* had aboard at one time or another Hawker Nimrods and Ospreys, Gloster Sea Gladiators—and the aforementioned Sea Hurricanes. In 1948 she was sold to be broken up.

The Pup and the *Furious* established this incomparable Sopwith/Hawker connection in the manner following—already preceded (as earlier intimated) not merely by paper proposals but by the American Ely's example of 1911.

At the Isle of Grain in March 1917 (the very month in which Harry Hawker's compatriot Harry Busteed was posted there to command the Port Victoria Repair Depot) Pups 9912 and 9497—the latter deserving a special place in the history of deck-flying as perhaps the most frequently and extensively adapted experimental machine of 1914–18—had been employed for deck-landing experiments using a dummy deck—a device originally utilised considerably earlier (September 1916) in developing arrester gear involving transverse ropes and hooks, though it could not, of course, contribute to the 'felt' wind like a ship under way. By February 1917 a new and larger deck, circular in outline, was being used at Port Victoria by (in addition to an Avro 504C) Pup 9497, this machine having a rigid hook for engagement with transverse ropes supported on 2-ft posts and weighted with sandbags. The time was now approaching when operational aircraft could be landed back aboard a ship instead of being ditched, though flotation gear was still a 'must'.

With her forward flying-deck installed, HMS *Furious* was made available, as noted, in June 1917, and on 2 August following, Sqn Cdr E. H. Dunning so manoeuvred his Pup by 'crabbing' ahead of the ship's superstructure (or 'Queen Anne's Mansions'), with *Furious* steaming at about 26 kt, that he became

The historic 'Dunning/Pup/*Furious*', or 'crabbing and grabbing', pictures have received different ascriptions from various authorities, though the upper one of this pair—showing no Lewis gun on the tripod mounting—seemingly records the true 'first', on 2 August, 1917. The second picture (wherein a light cruiser is crossing the bows distantly) shows the Lewis gun and rope-toggle hand-holds.

incontestibly the first man in history to land an aircraft on an aircraft-carrier—indeed on a ship of any kind while she was under way. On 7 August Dunning gave a repeat performance (though slightly damaging the Pup); but shortly afterwards, following a third touch-down the engine of the Pup involved (on this occasion known to have been N6452—five Pups in all having been shipped) choked on being opened-up again. The Pup went over the starboard bow and Dunning was drowned.

It is important here to emphasise that Dunning was using no elaborate arrester gear, depending almost solely on the low landing-speed and general handling qualities conferred by the Sopwith company on the Pup—though also on the seizure (by the ready hands of a deck-party) of rope toggles, attached to the fuselage and bottom wings.

In November 1917—within weeks, that is, of Dunning's demonstrations—*Furious* came once more into the hands of the dockyard mateys, and in about three months only had her aft big gun removed and, in its place, an after flying-deck (284 ft or 86 m) with associated hangar fitted prior to her re-commissioning on 15 March, 1918. This fact is restated and amplified here because it accelerated work in hand at Grain (under Busteed's supervision) on special forms of landing gear for the Pup, of deck arrester gears and other equipment. Jointly with modifications to *Furious* herself, such developments showed the way to the 'classic' or 'modern' form of aircraft-carrier. Instead of wheels, skids of several patterns were designed, and tested on Pups, the definitive skid-equipped aeroplane being officially known as the Sopwith Type 9901a. Though the skids on this new standard production-type Pup were of plain wooden construction (the wheel-equipped Sopwith Type 9901, as built for the RNAS by Beardmore, having already been introduced in this account with special reference to armament) experimental skids—sprung and otherwise, and sometimes with adjuncts and variations—were given close attention. On the Type 9901a 'dog-lead' clips were sometimes fitted to engage fore-and-aft arrester cables—athwartship cables (though not in themselves by any means new) being a particular feature of the Armstrong Whitworth arrester gear. One form of this gear (for which L. J. Le Mesurier appears to have been largely responsible, as he also was for catapults designed by the same company) had flexible transverse 'loops' formed on fore-and-aft cables passing over pulleys and working in conjunction with an hydraulic cylinder.

Though athwartships and fore-and-aft cables were schemed, and sometimes tried, in various proprietary, official/proprietary, official, demi-official, unofficial and 'non-attributable' combinations, the frequently-modified Pup N6190 (Sopwith-built, with 15-in instead of 18-in stagger) can be specially mentioned for its part in testing one form of Armstrong Whitworth arrester gear.

Curiously perhaps, the experimental adjuncts to the aircraft did not generally include wheels, though these had certainly been foreseen—in connection, for instance, with an Armstrong Whitworth arrester scheme—and were, of course, later standardised for deck-landing fighters. It fell, in fact, to the Parnall Hamble Baby Convert (with skids instead of floats as on the Sopwith Baby, though with wheels in addition to the skids) to reverse the historic Tabloid landplane transformations—first to the 1914 Schneider racer and then to the Schneider and Baby naval seaplanes.

Experimental Pups used for deck-landing development work included:

Pup (N6190 identified) with forward extensions to skids and with arrester hook, for athwartships cables, attached at about mid fuselage. (A similarly 'hooked' Pup also had clips or horns of V form, to engage fore-and-aft cables).

Having acclaimed the Pup's daintiness, it is needful here to pre-empt the question 'Whatever happened?' by explaining that the specimen uppermost is that described in the text as 'Pup with sprung skids and short, underslung, forwardly located arrester hook'. The aeroplane which, in the lower picture, has suffered 'compressibility' is the first Beardmore W.B.III (modified Pup No. 9950) which is the subject of a note in the text and which led to a new Service type (S.B.3D).

Pup with sprung skids and short, underslung, forwardly located arrester hook. (The springing was achieved by retaining basic components of the standard—wheel—gear, complete with shock-absorber cords).

Pup with fuselage-attached arrester hook and bow-shaped propeller guard on forwardly-extending skid-like members.

Pup with friction attachment on the tailskid.

Pup with wheel-cum-skid (or embryonic-skid) landing gear, nine claws or horns on spreader bar, and combined hydrofoil/propeller guard.

These foregoing are merely instanced as illustrating the adaptability, as well as the tractability, of the Pup as a ship's aeroplane; and though such developments—which stand largely to the credit of the Isle of Grain RNAS station, Port Victoria; Sqn Cdr Harry Busteed personally; and private contractors like Armstrong

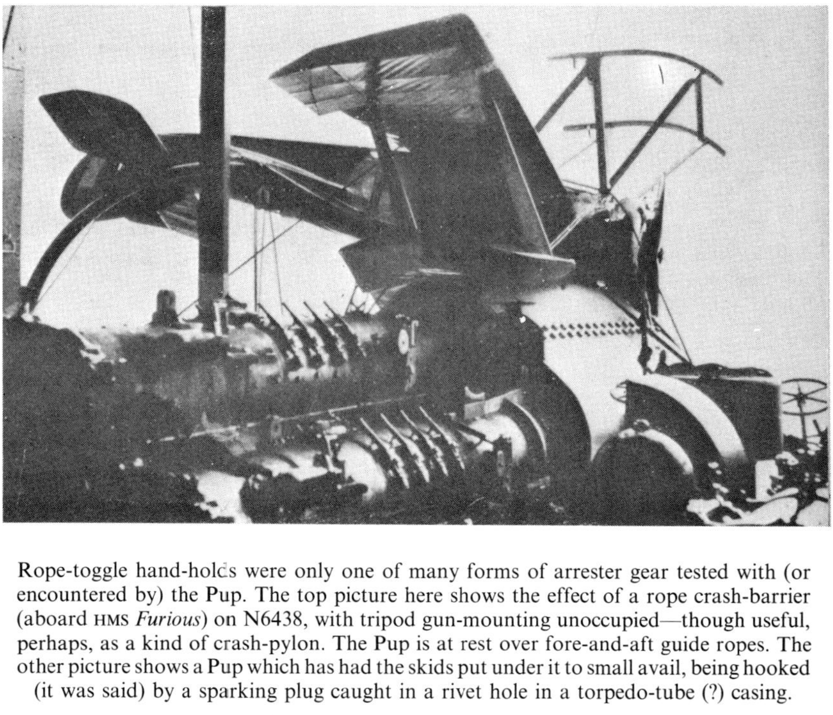

Rope-toggle hand-holds were only one of many forms of arrester gear tested with (or encountered by) the Pup. The top picture here shows the effect of a rope crash-barrier (aboard HMS *Furious*) on N6438, with tripod gun-mounting unoccupied—though useful, perhaps, as a kind of crash-pylon. The Pup is at rest over fore-and-aft guide ropes. The other picture shows a Pup which has had the skids put under it to small avail, being hooked (it was said) by a sparking plug caught in a rivet hole in a torpedo-tube (?) casing.

Whitworth—do not come strictly within the Sopwith compass they must not pass unheeded. Nor can one fail to mention the fitting of a Pup with paired, grooved tandem wheels fixed outboard under each bottom wing to run along parallel wire cables above a ship's deck, thus obviating altogether a flying-off platform. Yet cable-launching—retrieval even, as also tried with a Pup by engaging a loop on an overhead cable—was already old, having been demonstrated by Pégoud on a Blériot before war came in 1914. Wooden troughs to accept the wheels of a Pup and ensure a straight take-off were a less exotic notion, and were, indeed, fitted to several ships; but to conclude our study of the Pup as a Service aeroplane, and not as an experimental vehicle, we must return to our deliberately early mention of the Beardmore company.

Whatever the appendages that grew on Pups for trying-out ideas or meeting special requirements, the type itself never grew in size to any noticeable degree, as emphasised by this distant view of one on the flying-off platform of a light cruiser (HMS *Yarmouth*—or certainly one of the 'Town' class) with only men to give real scale. This ship's length was, in fact, about 450 ft.

The award to Beardmore of the first large Pup contract for the Admiralty and the special armament provisions connected with this early association having been recorded, it remains to note that the Sopwith Pup aeroplanes ordered as such from William Beardmore & Co Ltd, Dalmuir, Dunbartonshire, Scotland, were Nos. 9901—9950 and Nos. N6430—N6459. From these aeroplanes 9950 was selected for a metamorphosis—a transformation, at least, which represents one of the most imaginative (if one of the less successful) Naval-air undertakings on the British technical record, spattered though this record is with 'make-dos', 'mods' and 'variants'.

Stowage-space for Pups in the smaller classes of vessel involved in Naval operations generally and anti-Zeppelin work in particular being clearly at a premium, Beardmore undertook a complete redesign of the Pup accordingly. Not only were the wings (now without stagger, and with less dihedral) adapted to be folded—'Folding Pup' being a popular name for the aircraft—but the landing gear likewise was largely 'retractable' into the fuselage. Later the gear was fixed, but could be jettisoned for emergency alighting at sea. Flotation gear, jury struts and wingtip skids were added in the early stages, the control system was redesigned and the fuselage slightly lengthened—all these features being connoted by the new designation W.B.III. Though some of the novelties were abandoned or mitigated, one hundred W.B.IIIs were ordered; and though not all reached Service units, at one time the carrier *Furious* had fourteen of her own.

From Kingston-on-Thames, through ferocious battles over lands and coasts and narrow seas, the Pup—most affectionately remembered of all fighting aeroplanes, and an object-lesson in design—had played all kinds of tricks at the airman's and sailor's behest. And if Oliver Stewart's long-acknowledged verdict 'The perfect flying machine' was not sustained in every transformation, then this could seldom have been the fault of The Sopwith Aviation Co Ltd, which later sought in vain to perpetuate that acknowledged perfection, as we shall later be seeing under 'Dove'.

In small numbers Pups went to some of Britain's Allies (Australia had eleven or more in 1919); and the following examples passed to the British Civil Register: G-EAVF (scrapped 1921); G-EAVV (scrapped circa 1921); G-EAVW (scrapped 1921); G-EAVX (not repaired after an accident in 1921); G-EAVY (scrapped circa 1921); G-EAVZ (scrapped circa 1921); G-EBAZ (scrapped 1924); G-EBFJ (scrapped 1924). The famous 'Shuttleworth Pup' was G-EBKY, converted from a Dove, as noted under the appropriate heading.

Though precise production is indeterminate (some aircraft, for instance, being delivered as spares) nearly 2,000 Pups were ordered, contractors and numbers being as follows:

Sopwith 3691; 9496—9497; 9898—9900; N5180—N5199; N6160—N6209; N6460—N6479 (N6480—N6529 ordered but cancelled).

Beardmore 9901—9950; N6430—N6459

Standard A626—A675; A7301—7350; B1701—B1850; B5901—B6150; C201—C550

Whitehead A6150—A6249; B2151—B2250; B5251—B5400; B7481—B7580; C1451—C1550; C3707—C3776; D4011—D4210. (The last two Whitehead batches were delivered as spares).

Pup

(80 hp Le Rhône)

Span 26 ft 6 in (8.1 m); length 19 ft 3¾ in (5.9 m); wing area 254 sq ft (23.6 sq m). Empty weight 787 lb (357 kg); maximum weight 1,225 lb (555 kg). Maximum speed at 5,000 ft (1,520 m) 105 mph (169 km/h); maximum speed at 11,000 ft (3,350 m) 101 mph (162 km/h); maximum speed at 15,000 ft (4,570 m) 85 mph (137 km/h); climb to 5,000 ft (1,520 m) 6 min 25 sec; climb to 10,000 ft (3,050 m) 16 min 25 sec; climb to 15,000 ft (4,570 m) 32 min 40 sec; service ceiling 17,500 ft (5,330 m); endurance 3 hr.

Pup

(100 hp Gnome Monosoupape)

Span 26 ft 6 in (8.1 m); length 19 ft 3¾ in (5.9 m); wing area 254 sq ft (23.6 sq m). Empty weight 856 lb (388 kg); maximum weight 1,297 lb (588 kg). Maximum speed at 6,500 ft (1,980 m) 107 mph (172 km/h); maximum speed at 10,000 ft (3,050 m) 104 mph (167 km/h); maximum speed at 15,000 ft (4,570 m) 100 mph (161 km/h); climb to 5,000 ft (1,520 m) 5 min 12 sec; climb to 10,000 ft (3,050 m) 12 min 24 sec; climb to 15,000 ft (4,570 m) 23 min 24 sec; service ceiling 18,500 ft (5,640 m); endurance 1 hr 45 min.

Production Triplane in operational trim.

Triplane

By the above style and title only (though with the inevitable diminutive 'Tripe' or the then-stylish 'Tripehound') was known one of the daintiest and most distinctive fighters ever to leave Kingston-on-Thames. And with the Pup still close in mind the terms 'dainty' and 'distinctive' are not glibly applied. There was, indeed, a very close relationship between the two machines, though in the Triplane—the first aeroplane of its form to be put to practical use—a special effort was made to improve not only fighting view (which in the Pup was deficient) but manoeuvrability also—for whatever its virtues in respect of handling and height-holding the Pup still invited attention to rate of turn and roll. Even so, Harry Hawker himself considered not only the Pup, but the Triplane also, rather too stable—though he himself is said to have recommended for the Triplane the reduction of tailplane area that was to become one further distinction from the Pup.

Design criteria and characteristics were most concisely epitomised by Harald Penrose when he wrote: 'Superimposing the Pup fuselage drawing on that of the Triplane shows, as with the Tabloid, matched profiles, though the spacer-strut disposition varies a little, with particularly clever adaptation at the centre-section struts. The span of the Pup and the Triplane was identical at 26 ft 6 in; the effective stagger from top leading edge to lower wing trailing edge was the same; and to align the three leading edges the wing chord of the Triplane was made 3 ft 3 in instead of 5 ft $1\frac{1}{2}$ in. However, the weight of the Triplane would be at least 200 lb greater than the

Pup; clearly it was desirable to go for the biggest available engine to enhance performance, so a 110 hp Clerget from the 1½ Strutter production line was taken. On 28 May, 1916, the Sopwith Experimental Department passed the machine for flight tests. By that time the Pup was earning tributes everywhere for its impeccable handling.'

To this appraisal the present writer would merely add a note to emphasise the Triplane's quite astounding rate of climb—by virtue of a lower power-loading, which more than offset the higher wing-loading.

One manifest disadvantage inherited by the Triplane from the Pup was the fitting of a single machine-gun only, the installations being more or less identical, with the Vickers gun lying on the centre line ahead of the cockpit and having mechanical synchronising gear (Scarff-Dibovsky or Sopwith-Kauper), an ammunition supply of 500 rounds and the Sopwith padded screen. In some degree the disadvantage of low firepower was offset by the Triplane's stability, enabling it to fly hands-off for clearing gun-jams. Departures from standard armament included the fitting of twin Vickers guns in a few examples; the addition of a Lewis gun at the root of the port middle wing; and provision for elevating the Vickers gun to fire upwards at an angle of about five degrees. This last installation entailed fitting an Aldis optical sight having a special graticule, and was intended for stern or underneath attack at relatively long range.

Structurally, of course, the triplane wing cellule commands our first attention—and this must go beyond a remark by Lord Weir that 'Some of the aerodynamic disabilities of the triplane were overcome by the pronounced forward stagger and the use of a single strut. This single-strut system increased the difficulties of manufacture and repair, particularly as regards truing-up.'

One's first thought here is what was meant by 'the aerodynamic disabilities of the triplane' that could be overcome by merely staggering the wings, and one can only conclude that these 'disabilities' must have been attributed to interference between the juxtaposed wings—though the gap was substantial and the chord quite narrow. Lord Weir's allusion to 'forward stagger' may conceivably have been deliberate, for as we shall later see, the degree of stagger in the later Snark and Cobham triplanes was sharply unequal—conveying almost an impression of backward stagger from some aspects—while, more pointedly, some Sopwith biplanes (the 'fabric' Snail as well as the Dolphin) had actual negative stagger.

As for difficulties of manufacture, repair and rigging, the last of these was, seemingly at least, simple, the cellule being braced much as a biplane structure, though with upper and lower drag and anti-drag wires. Should the terms 'drag' and 'anti-drag' seem too extremely archaic in respect of such external—as distinct from internal—wires, it may be remarked that, well into the 1930s, riggers of RAF Siskins—a type officially declared to have 'unusual' wing-bracing—were advised: 'The drag and anti-drag bracing wires are "threaded" through the plane framework from their anchorages . . . '

Admittedly, there were six ailerons on the Triplane instead of four as on the Sopwith biplane fighters; but the instructions for 'truing-up the main planes' seemed simple and explicit enough—even in this lavishly 'capitalised' verbatim rendering:

'Adjust by Landing Wires, and check by Abney Level and Straightedge, or Dihedral Board and Spirit Level, along the Front Spars of the Upper Main Planes.

'The *Stagger* from Upper Main Plane to Lower Main Plane is 36″, being 18″ between Upper and Intermediate and Intermediate and Lower Main Planes respectively.

'This should be correct at the Centre Section. To ensure that it is constant throughout, place straightedges across the Leading Edges of the Main Planes. Adjust Drag and Anti-Drag Wires until any two of these straightedges are in line.

'Check for Main Planes being square with Machine by taking measurements from Top and Bottom Sockets of Outer Interplane Struts to Rudderpost and Front Drag Wiring Plates. Corresponding measurements should be the same on both sides.

'The *Incidence* is 2° for all Main Planes. Check by Abney Level and Straightedge, placing the latter along the chord of a rib.

'This can only be adjusted if the fittings on Interplane Struts have not been drilled.

'It is important throughout the process of truing the Main Planes to check the Dihedral, Stagger and Incidence to ensure that adjustments for one do not throw the others out.'

These verbatim instructions notwithstanding, it must be acknowledged that there were separate instructions for 'truing-up the centre section', and that Lord Weir's remark may well have had more substance than suggested.

Oliver Stewart recalled that the Triplane was 'reported to be subject to the same trouble as the Nieuport and to have a habit of twisting one of its planes about the front spar so that control and stability were lost.' He quickly added, however, that 'in fact none of these faults was demonstrated to be inherent in the aeroplane, and as pilots got to know it better they got to like it better until, when it was superseded, it was allowed to go with regret.'

In outlining the Triplane's technical development and operational employment —which were very closely linked by the familiar tale of the 'prototype N500' being sent into action a few minutes only after its arrival in France (mid-June 1916)—one would first remark that although a photograph now reproduced supposedly shows 'N500' under construction, the number prominently placarded on the port rear landing-gear strut is, in fact, '490'. Interesting though this observation may be, its significance may be scant, for the airframe depicted appears to match the fine 'taxying-at-Brooklands' study in all discernible respects; and though the transparent panelling in the top centre-section may be absent in the view taken before completion, also absent—and indubitably in *both* views—is the Vickers gun. It is worth noting, nevertheless, that the figures '490' formed no sequence in Admiralty Contract No. C.P. 117520/16, ascribed to the first Triplane order, though they did appear as three digits in N5490, one of the early Sopwith-built production machines.

In both the pictures just mentioned the engine is apparently a 110 hp Clerget (makers' suffix 9Z), and the tailplane—again quite properly, to accord with characteristics usually ascribed to the first, or 'prototype' Triplane—appears to be adjustable. This adjustment, it may here be mentioned, was effected (on production Triplanes at least) by the pilot turning a wheel fixed to the starboard centre-section strut where that strut passed through the cockpit. Externally, at higher level, this same strut served to carry a propeller-driven air pump for petrol delivery.

When the first Triplane joined the first Pup at Furnes (base of 'A' Squadron, RNAS) in mid-1916 it amazed and delighted pilots, especially by its proven ability to reach 12,000 ft in 13 min; and, like the Pup the type was ordered for the RFC as well as the RNAS, the Navy getting the first machines built by Sopwith themselves, and the War Office depending for its initial supplies on Clayton and Shuttleworth Ltd, of Lincoln. As things turned out, however, the Triplane was never used as standard operational equipment by the RFC, the machines originally intended for

This pair of early Triplane photographs has special mention in the text, one point of particular interest being the number '490' attached to a landing gear strut in the uppermost view. That this photograph was taken in the Sopwith Experimental Department is evident not only from the unfinished state of the aircraft (unarmed as well as uncowled) but by the lofty presence of the L.R.T.Tr. in the background—and further by 'Ex D' painted on the part of the tail-trestle seen just below the starboard interplane strut. In the 'Brooklands landscape-with-figures' picture transparent centre-section panelling is seen.

that Service being exchanged for Spads.

Even the excellent rate of climb just noted was surpassed by a slightly later Triplane—apparently the second example, N504—which, by September 1916 was flying with a 130 hp Clerget engine (Type 9B) and averaged a climb-rate of 1,000 ft/min up to a level of 13,000 ft; it may indeed have climbed to 22,000 ft in September 1916—an achievement which, if true, would have been the more remarkable because in that same month Sqn Cdr Harry Busteed recorded its sea-level speed as 116 mph.

Subsequent re-engining was possibly confined to the testing of a 110 hp Le Rhône installation. For the greater part Triplanes in service had the 130 hp Clerget, rather than the alternative 110 hp unit of the same make, occasionally with a small pointed spinner on the propeller, as was sometimes the case with Camels; but probably the most interesting experimental development concerned not the power plant but the airframe. This involved the testing, in December 1916, of N5423 (Sopwith-built) with wings increased in chord by three inches—that is, to 3 ft 6 in. Clearly, an increase in chord may well have been associated with a change in section; but although there is no confirmation that this was in fact the case, it is certain that official laboratory studies were made of triplane wings for which the R.A.F.15 section was substituted for the R.A.F.6. Interestingly enough, there were laboratory tests also of model triplane wings having a gap/chord ratio of 1 (almost exactly that of the standard Sopwith Triplane, which had a gap of 3 ft and a chord of 3 ft 3 in)—and with zero stagger, with +30 deg stagger, and with −30 deg stagger.

Although some aspects of the Sopwith Triplane's performance were apparently improved with the long-chord wings, these wings were never standardised; nor was wing-bracing appreciably interfered with until the Triplane's active-service life was over—though the type was still a fighter classic, even, for instance, with the School of Aerial Fighting at Marske, in Yorkshire. One RAF Technical Order of 1918 required a compression strut to be fitted spanwise above the externally mounted Vickers gun.

In May 1917 N5486 left the RNAS Depot at White City, London, for White Russia, where it was fitted with skis. Jack Alcock's 'Sopwith Mouse' (so-called) embodied some Triplane components and has a note to itself under 'Apocrypha'.

Although Harald Penrose firmly ascribes to Harry Hawker the recommendation that the Triplane's tailplane should be reduced in area (the most important modification made to improve this fighter's combat performance—notably by increasing the rate of turn) the alteration clearly stemmed from operational experience, for it dated from February 1917. The new horizontal surfaces (the elevator as well as the tailplane being involved) spanned only 8 ft instead of about 10 ft, and their area was thus reduced by more than 10 sq ft. Because the leading edge was now shorter than the trailing edge, the familiar inwardly-raked tips which had been so characteristic of the Pup were now reversed in form, and various changes in handling were attributed to this innovation by different pilots.

Short but lustrous was the Triplane's operational career after N500 was sent off on an interception within its first quarter-hour at Furnes in June 1916, though something of an interregnum was implicit in the fact that production Triplanes did not enter service until February 1917. The total number built, in fact, was apparently no more than 150, and from Oakley Ltd, of Ilford, Essex, came only three of their order for twenty-two. Thus in March 1917 the busy-minded and busy-tongued Mr Noel Pemberton Billing asked the Under-Secretary of State for War if he would give the date 'on which the first Sopwith triplane scout was offered to the authorities; the date on which the first order was placed for the same; what proportion of the order has been delivered; and what proportion is now on active service.' As may be imagined, 'P.B.' was told that it would not be in the public interest to give the particulars asked for—and seemingly some mystery remains to this day concerning a substantial reduction in Triplane orders, though the coming of the Camel may have had some influence here. Oakley, moreover, had never built aircraft before, and though their meagre contribution did not begin until the

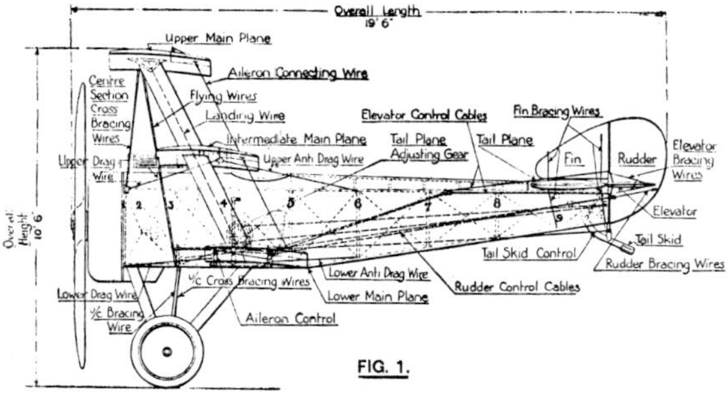

FIG. 1.

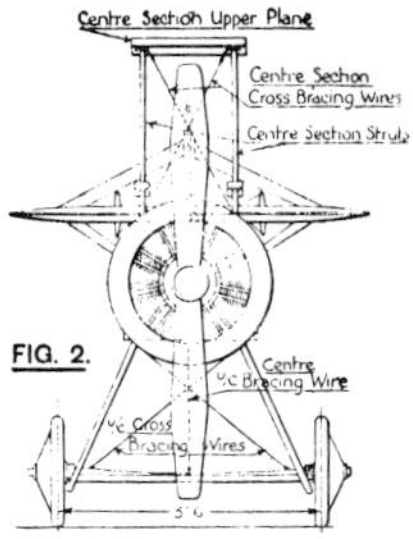

FIG. 2.

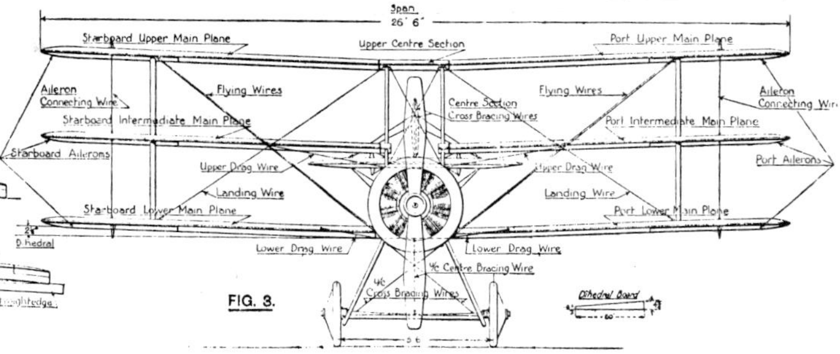

FIG. 3.

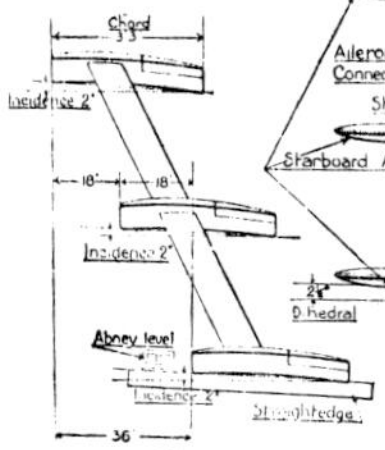

FIG. 4.

FIG. 5.

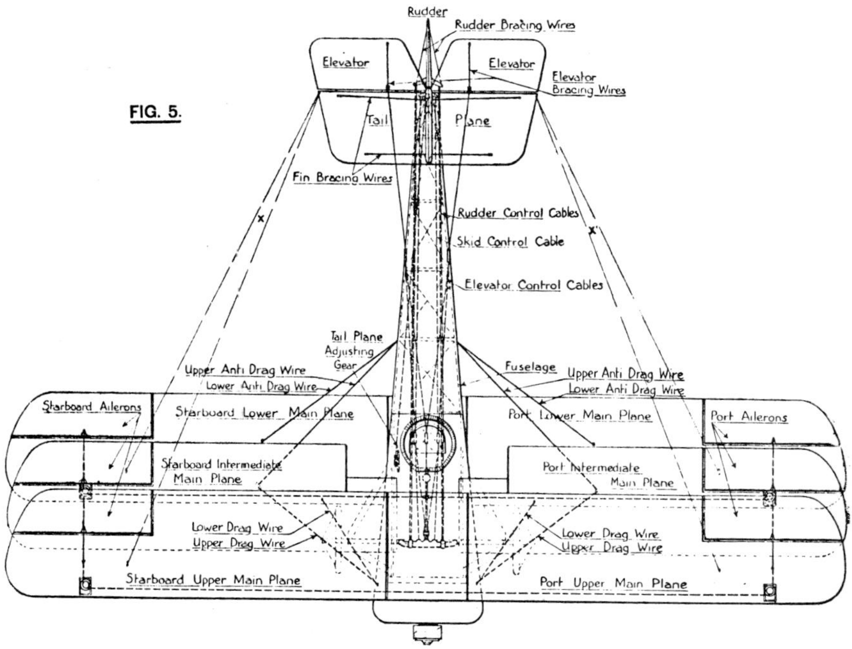

autumn of 1917, they had been asked to fit Camel-style armament (twin Vickers guns).

None of which considerations detracts in any way from an operational career which enabled H. A. Jones to record in *The War in the Air* that 'The sight of a Sopwith Triplane formation, in particular, induced the enemy pilots to dive out of range.' And the Triplane having been, as already shown, very much an RNAS fighter, we take from *British Naval Aircraft since 1912* (Owen Thetford, Putnam) this fittingly brief summation: 'Production Triplanes entered service with No.1 and 8 (Naval) Squadrons in February 1917 and with No.10 (Naval) Squadron in May. Some remarkable engagements were fought by such redoubtable Triplane pilots as Sqn Cdr C. D. Booker, DSC, and F/Sub-Lt R. A. Little, of 'Naval Eight' and F/Sub-Lt Raymond Collishaw of 'Naval Ten'. The Triplanes of Collishaw's 'B' Flight (named *Black Death, Black Maria, Black Roger, Black Prince* and *Black Sheep*) became the terror of the enemy: between May and July 1917 they destroyed 87 enemy aircraft. Collishaw personally accounted for 16 in 27 days and shot down the German ace Allmenroder on 27 June . . . The Triplane's career was glorious but brief. It remained in action for only seven months; in November 1917 the Camel had supplanted it in squadrons.'

To the French Government went Triplanes N5385 and N5388, and, as already noted, N5486 was despatched to Russia. N5458 (after serving its time at the Front) was exhibited in the USA. The Germans and Austrians were clearly influenced by the design, which underwent close scrutiny at Adlershof (for several examples were captured and tested); but one clear advantage possessed by the hardly less famous Fokker Dr. I was the fitting of twin belt-fed guns as standard equipment. That Sopwith Triplanes N533—N538 are known to have been similarly armed (with two Vickers guns) and that this same armament was intended for the Oakley-built machines was small comfort, though obviously an increased warload meant a decreased rate of climb. And rate of climb, perhaps, was the real trump card in the Triplane's symbolic deck of three.

Triplane production was apparently as follows:
Sopwith N500; N504; N524; N5420—N5494; N6290—N6309
Clayton & Shuttleworth N533—N538; N5350—N5389
Oakley N5910—N5912 (N5913—N5934 were not completed).

Triplane

(130 hp Clerget)

Span 26 ft 6 in (8.1 m); length 19ft 10 in (6 m); height 10 ft 6 in (3.2 m); wing area 231 sq ft (21.5 sq m). Empty weight 993 lb (450 kg); maximum weight 1,415 lb (642 kg). Maximum speed at 6,500 ft (1,980 m) 116 mph (187 km/h); maximum speed at 10,000 ft (3,050 m) 114 mph (183 km/h); maximum speed at 15,000 ft (4,570 m) 105 mph (169 km/h); climb to 6,500 ft (1,980 m) 6.3 min; climb to 10,000 ft (3,050 m) 10.6 min; climb to 15,000 ft (4,570 m) 19 min; service ceiling 20,500 ft (6,250 m); endurance 2 hr 45 min.

◀ Superbly informative are these five self-explanatory figures, originally prepared for the guidance of riggers, and presented now for the enlightenment and joy of every aeroplane-lover. Basic Sopwith drawings of this general nature were sometimes used officially, with acknowledgement. Thus, one particular sheet has in one corner the legend 'Sopwith Drg. No. 1720/Copied from Drawings of the Sopwith Aviation Co Ltd, Kingston-on-Thames' and in the opposite corner 'Military Aeronautics Directorate . . . Drg. No. AD 61204'.

Triplanes (Hispano-Suiza)

For heading this chapter as above there are several sound reasons. First, to differentiate the subjects of the chapter from the rotary-engined single-seat Triplane just described (which, for sheer convenience, was sometimes called the 'Clerget Triplane') and to emphasise the basic nature of that distinction. The next reason is that the physical characteristics of the two machines now scrutinised differed so greatly from those of the other Sopwith triplane types—which

Directly comparable views of N509 (top) and N510, showing—apart from differences in engine-cowling—the higher thrust-line of N510, resulting from the geared Hispano-Suiza engine. (The propeller of N509 was directly driven).

numbered no fewer than six quite distinct ones, the present pair being considered as a single type only, for they were essentially alike—that these characteristics alone warrant individual study. A third reason is to dispel an understandably persistent misapprehension that the Hispano-Suiza single-seaters which now concern us were close relations of the 'Clerget Triplane'—which they were not.

Enlarging upon this third point to the verge of over-simplification, the rotary-engined 'Clerget Triplane' can be regarded as a Pup development, while the Hispano-Suiza pair were quite distinctly related to the 1½ Strutter. Like the two aeroplanes themselves, the story of their development has a special interest—even amidst the prolific and interesting Sopwith family of types; and it would hardly be overstating the case to affirm that they represent not merely a distinct type but a distinct class.

These matters being so, one must direct attention to seemingly neglected areas of study, and to note at the start that the two Sopwith 'Hispano Triplanes' (as we may now tidily call them, though one designation used by Sopwith was seemingly 'Sopwith 200 H.P. H.S. Tractor Triplane') were numbered N509 and N510. Thus they were not only RNAS machines, but shared both a numerical sequence and a structural affinity (being triplanes) with N511—N514, which were Armstrong Whitworth quadruplanes. Further, it seems proper to remark, N503 (an apparently unbuilt Clerget-engined Pup development, already mentioned) was numerically preceded by N502—a Blackburn triplane fighter, for the design of which Harris Booth, formerly of the Admiralty's Air Department, was responsible.

Chronologically, the Hispano Triplanes followed shortly after the Clerget Triplane, and thus were flying during 1916—before the Clerget machine was established in service. Indeed, some hiatus in respect of Hispano-Suiza engine development or delivery may well have been responsible for a delay at Sopwith; and that the basic design of the Hispano Triplane (1½ Strutter-related) actually preceded that of the Clerget Triplane (Pup-related) seems not only possible but entirely comprehensible, for the Pup was preceded by the 1½ Strutter.

This particular reference to Hispano-Suiza engines now calls for an explanation of the plural (Triplane*s*) used in the present chapter-heading. In essence this is simple enough—N509 had a 150 hp direct-drive engine whereas N510 had a 200 hp geared engine; but there will be more to say concerning Sopwith's new French affiliation in the engine field, the fate of which could well have been sealed at quite an early date by the pressing demands of the S.E.5 programme—for to the French water-cooled vee-8, or a unit of similar form, this 'Factory' fighter project was wedded. Meanwhile attention must be redirected to the two new Sopwith airframes.

The main components of the tail unit were seemingly identical with those of the 1½ Strutter, and the fuselage being deep, the cockpit coaming was similarly deep—far more so than on the Clerget Triplane and located further aft in relation to the wings. (The pilot's field of view was thus accounted poor). It is the wings themselves that have doubtless been responsible for over-close associations between the Hispano and the Clerget Triplanes; for though visually similar, especially by reason of the 'plank' interplane struts, those of the Hispano machines measured roughly 2 ft more in span and—perhaps more significantly—about 1 ft more in chord. Precise dimensions are lacking, but as the chord of the Clerget Triplane was a mere 3 ft 3 in, the difference is, as suggested, truly significant. Especially true is this because a Clerget machine was tested (as noted in the foregoing chapter) with wings increased in chord by 3 in only.

Though performance figures, like other data, are scanty and possibly suspect, the rate of climb—only 9 min to 10,000 ft—ascribed to one of the two Hispano

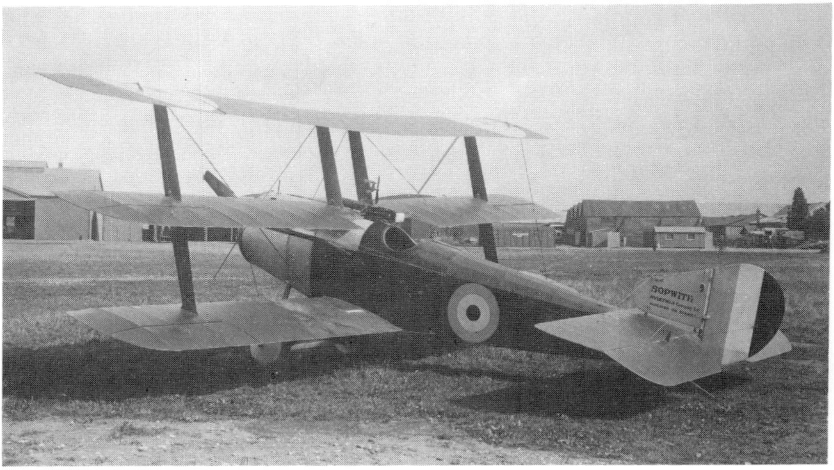

Two more comparative views of N509 (top) and N510. The maker's caption to the top picture reads: 'S.110—Sopwith Triplane. 150 hp Hispano Suiza—1916'.

Triplanes is remarkable. One especially interesting aspect is that the time represents only about twice the corresponding figure for the RAF's Hawker Fury intercepter of the early 1930s. (The Armstrong Whitworth Siskin III, first issued to the RAF in 1924, reached 10,000 ft in about $8\frac{1}{2}$ min). One factor which might conceivably have contributed to the Hispano Triplane's rate of climb would be a carefully chosen wing section, though its pattern is not known. While rigging was similar in principle to that illustrated in the preceding chapter, wing construction must have differed quite considerably, by reason of the greater chord; and though the best available photographs do not confirm that 'the middle wing was on top of the fuselage' as Harald Penrose has it, this may well have been so in one particular form. To this very feature, in fact, Mr Penrose has ascribed a 'decreasing wedge for the air flow, resulting in turbulence which caused the tailplane to break off the more powerful second prototype N.510.'

The catastrophe just mentioned occurred during trials at Eastchurch in December 1916—in which connection it may be remarked that, obscure though N509 and N510 remain in history, they were evidently quite familiar sights not only at Brooklands but at several RNAS establishments also. 'Tail vibrations' were noted on N509 by Harry Busteed when he flew it on 2 January, 1917 (very shortly after the crash of N510), but in the following March this same machine was at Westgate, near Manston, Kent, and it was finally written off (in the clerical sense at least) at Manston on 29 October, 1917.

The installation of the Hispano-Suiza engine in each of the two triplanes calls for special comment, for the greatest pains had clearly been taken to fit Marc Birkigt's masterly creation of 1915 to the best advantage. In each case the frontal radiator was circular in form, but though the cowling generally was neat, the tops of the cylinder-banks were prominently visible, and unfaired either at front or rear. This feature doubtless stemmed from the 90-degree setting of the cylinder-banks themselves, though for 'banks' the term 'blocks' might justifiably be substituted, for the Hispano-Suiza engines were the first in which screwed steel liners were used in conjunction with 'monobloc' cast-aluminium jackets and heads. Tail-pipes, terminating (somewhat curiously, it may be supposed) just forward of the cockpit were fitted to the exhaust manifolds, though the ends of the pipes were toed-out from the fuselage.

On the engine of N510 the reduction gearing accounted for a slightly higher thrust-line and a left-hand propeller.In one form at least N510 had a revised fuel system, with a gravity tank in the top centre section and fuel pipes running down the central interplane struts—the starboard one of which (on both machines) carried a wind-driven pump, as on the Clerget Triplane. Cowling details varied considerably, though a large air-exit louvre just aft of the front landing-gear strut on the port side was seemingly characteristic of N510.

A superbly detailed aspect of N510, captioned by the makers: 'S.113—Sopwith Triplane. 200 hp Hispano Suiza—1916'. A gravity tank is fitted in the top centre section, with a pipe leading out along the starboard mainplane.

Even though the power was higher, and the wing area greater, than on the rotary-engined Triplane, the only recorded armament was a single Vickers gun, with—in one instance at least—Sopwith-Kauper synchronising gear, though one may justifiably quote from *Armament of British Aircraft 1909—1939,* thus: 'The end of the gun was padded and there was a small metal screen forward of the feed block. As on the earlier Triplane, there were drag and anti-drag wires attached to the spars of the lower wing, but on the Hispano-Suiza machines there were two access panels outboard of each of the ribs to which these wires were attached. These panels somewhat resembled those on the later Sopwith triplane fighter, the Snark, and with which auxiliary outboard armament was associated. Definite conclusions, however, would not be warranted, especially so as the panels were at the anchorage points for the lower drag wires.'

To these foregoing remarks one would now add a further reflection: Although a 400-round drum for the Lewis gun was still being considered as a possible provision for fighters in the early 1930s, a drum of this capacity had in fact been designed by early 1917. While recognising that a Lewis gun with a monstrously tall drum holding 400 rounds of S.A.A. might prove too weighty and too drag-producing an item for a (relatively) light and low-powered single-seat triplane to carry in each of its bottom wings, one recognises equally that the Hispano-Suiza engine was sometimes adapted to carry a Lewis gun instead of the more famous Puteaux *canon.* An early S.E.5 scheme had, in fact, involved just such a combination, though not, apparently, with an oversize drum-magazine for the Lewis gun. The Sopwith Hispano Triplane, with its excellent lifting capacity, might well have been adapted for—indeed, intended for—an 'engine gun' combination. Equally (and there is some slight evidence to suggest that positive steps were taken in this direction) the type might have carried a bomb or bombs, though—unlike the single-seat 1½ Strutter bomber and the larger and later B.1—externally, and not in the fuselage behind the pilot.

Finally, respecting weapons, one may have been the merest trifle over-cautious in so deliberately quoting the phrase that 'definite conclusions would not be warranted' respecting outboard guns—remembering that of four Curtiss triplane fighters built in 1916 (and fitted with water-cooled engines) one was tested in March 1917 with two Lewis guns under the top wing. Still, there is no real evidence that either N509 or N510 was ever armed with more than a single synchronised Vickers gun; yet even so, one now restates without equivocation the view of these aeroplanes expressed at the outset—that 'it would hardly be overstating the case to affirm that they represent not merely a distinct type but a distinct class.' Less confidently, alas, one quotes these data:

Triplane (Hispano-Suiza)

(150 hp direct-drive or 200 hp geared Hispano-Suiza)

Span 28 ft 6 in (8.7 m); length 23 ft 2 in (7 m). Maximum speed 120 mph (193 km/h); climb to 10,000 ft (3,050 m) 9 min.

All salient features of L.R.T.Tr., in the form in which it was completed, are shown in this view—notably, nacelle, wing and tail shape; landing gear; and mounting for Lewis gun behind rear cockpit. Apart from the facts that the 's' in 'Rolls' is reversed and that 'Rolls-Royce' is not hyphenated (for this was, perhaps, the first occasion on which the penman was called upon to write the company's name) the Sopwith caption reads: 'S.87—Sopwith Triplane. 250 hp Rolls-Royce—1916.'

L.R.T.Tr.

Appearances notwithstanding, this aeroplane was a fighter, or 'long-reconnaissance' aircraft, having not only three wings but three seats also; and although the question once arose whether the triplane layout was signified by the 'T.' or the 'Tr.' in the designation, 'Tr.' was evidently the operative symbol, the 'T.' connoting 'tractor', as in SL.T.BP.

That the Sopwith triplane now studied was the ungainliest single-engined fighter ever built—with the possible exception of the rival Armstrong Whitworth triplane sometimes called F.K.12—is a point hardly to be questioned, though this dubious distinction can be attributed not only to the triplane wings, with the gunner's nacelle on the topmost one (the rival type mentioned had two such nacelles, both on the middle wing) but even more so to the four-wheeled chassis, which, though embodying the characteristic Sopwith 'divided axle' was more strongly suggestive of agricultural than aeronautical practice. Upon this conception—the forward wheels whereof minimised the risk of nosing over, and hence the peril of the nacelle gunner—the aeroplane normally towered tail-in-air, though there was a tailskid also to allow it to sit tail-down. For lateral protection there was a rearwardly-raked skid beneath each bottom wing, in line with the outermost set of plank-type interplane struts.

Yet another suggestion of the farmyard rather than the flying field was the triplane's sobriquet 'Egg-box'—an appellation that was especially apposite by reason of the neat cockpit recess in the streamlined top-wing nacelle, which smooth receptacle seems never to have been compromised by a gun-mounting. More important, this nacelle may give an essential clue to design philosophy, for its location afforded in theory a well-nigh untrammeled field of fire, and one that certainly precluded any need for deflector or synchronising gear—the latter still being a comparative novelty when the design of the aircraft was under way early in 1916.

Yet the very mention of synchronising gear is suggestive of machine-gun armament; and though it may well be that the nacelle (which was originally designed to be smaller and, in any case, underwent at least one change in shape—the later version being shorter in the nose, blunter, and housing a gravity-feed petrol tank) was intended to have a Lewis gun, one inclines to the view that heavier ordnance may have been in mind. The likelihood is that this would have been a gun of the Davis recoilless type, the installation whereof—whether fixed or free—would have been of relatively small account by reason of the aircraft's low-set vertical tail-surfaces. The pictures clearly show that not only was there an under-fin, but that a substantial part of the rudder was below the fuselage—though it must be recognised that large vertical tail-area must in any case have been demanded by the very deep nose and the jutting-out nacelle and landing gear. As for weight, a 2 pdr Davis gun might have accounted for only about 70 lb, and no great number of shells would have been carried.

Conversely, the fitting of a gravity tank in the nacelle, if this tank was meant to supplement the very large one under the pilot's cockpit (for he was seated bodily above the top longerons) may have signified an alternative emphasis on endurance, at the expense of armament. Even in this case, however, the aircraft would not have been unarmed, for immediately behind the pilot was a cockpit for a second gunner, who had a Lewis gun on a swivelling pillar mounting, characterised by a perforated gun-arm. Additionally, he had a second set of flying controls.

The foregoing facts and surmises are clearly in accord with the prescribed duties of long-range escort work and anti-Zeppelin patrol; and one feature that is obviously compatible with long range, or long endurance, is the very high aspect

A companion picture to the one preceding, showing, in particular, the under-wing skids.

The four-wheeled main landing gear of the L.R.T.Tr. affords, in this instance, a platform of convenience (not to mention apparent pride) for bodies who aid in giving scale. One of these may be Ernest Newman, who occupied the nacelle when it became misaligned after an early test by Harry Hawker.

ratio of the staggered triplane wings, the chord whereof was a mere 4 ft or so, giving an aspect ratio of about 13 to 1. This was a very high figure for 1916—though rivalled it would seem by the Caproni Ca.41 and possibly also by the earlier 80-ft-span Sopwith pusher—but even though the spars were very closely spaced, each of the six wing-panels carried an aileron. Contrarily, these lateral-control surfaces were of *low* aspect ratio.

One other notable feature of the three-bay wing cellule—the struts whereof were Sigrist-patented in form, comprising a wooden nose and tail, held either side of a central H-section metal strut with bolts from nose to trailing edge—was the fitting of upward-hingeing airbrakes at the root-ends of the bottom mainplanes. These braking surfaces were cable-actuated by a handwheel in the pilot's cockpit.

Even so, the interest of the L.R.T.Tr. was not confined to armament and aerodynamics, for the engine was the first Rolls-Royce to be installed in any Sopwith aeroplane. Known contemporaneously as the 250 h.p. Rolls-Royce, or 250 h.p. Rolls-Royce Mk.I, this was an early member of the historic Eagle family (see later under 'Atlantic'), and though there had been a scheme for installing the radiator aft of the engine, this item was, in the event, fitted frontally, with the reduction-gear housing largely exposed.

So, its ungainliness notwithstanding, the L.R.T.Tr. displayed features of uncommon interest, though it was one of those aeroplanes that Sopwith never recalled with pride—which would account for its reported rapid relegation to a Brooklands hangar.

L.R.T.Tr.

(250 hp Rolls-Royce Mk.I)

Span 53 ft (16.1 m); length 38 ft (11.6 m).

In a winter wonderland at Brooklands stands the first of all Camels—with one-piece top wing and 110 hp Clerget engine (immaculately cowled). There is no separate windscreen, but the marked slope of the 'hump' to the front rim of the cockpit is clearly seen. Though frequently copied, this picture is a true original, uncaptioned by Sopwith.

F.1 Camel

To begin what will rightly be expected to comprise one of the most important chapters of this book with what may seem to be a blatant heresy will at least have the merit of originality; and anything original concerning the most famous and successful British fighter of the First World War must inevitably verge on the heretical, for so many facts have been enunciated relative to the Camel by so many authors, and even by authorities. The heresy referred to is one that Rudyard Kipling (who really gave the Sopwith Bat Boat its name) might well have supported, if one gives any credence to the second of his *Just So Stories* (1902). This story Kipling called 'How the Camel got his Hump'—and he himself supplied illustrations in a style which might have got him a job as a Sopwith draughtsman had he failed with his text.

To keep our present tale on strictly Sopwith lines we may restate in typical terms an accepted belief: that the Sopwith Camel—a two-gun development of the Pup—was so named because 'the breeches of the guns were enclosed in a prominent hump-like fairing'.

The heresy now uttered against such statements is in part demonstrable and in part suggested. Demonstrable first is that the 'breeches' (allowing a 0.303 in Vickers gun to have a breech at all, and not a chamber) were in any case invisible, being buried quite deeply among the 'action' contained in the 'action body', 'breech casing' or 'receiver'. So the 'breeches' of the two Vickers guns (which guns were so characteristic of the Camel, for it was the first British machine of its class to have them as standard equipment) were enclosed not by the equally characteristic hump-like fairing or scuttle but by the casings lying under that protuberance. What the 'hump' did in fact cover was not only the rear parts of the guns themselves, but often some quite complicated external fittings associated with firing and loading. Furthermore, the hump functioned partly as a windscreen for the pilot—and contributed directly to the uncommonly poor field of view available to him.

Thus it is plainly suggested that the famous hump on the Camel's fuselage afforded a contributory—and not an absolute—explanation of the aeroplane's name; for in the 1920s, and on into the 1930s, the present writer was repeatedly assured—sometimes by demonstration and usually by men who knew the Camel and its antecedents well—that equally, if not primarily, responsible, was the absence of dihedral on the upper wings and the sharpness of that angle on the lower ones. These features, quite obviously, combined to give a 'round-shouldered' or 'hump-backed' appearance which was (equally obviously) accentuated by the fairing ahead of the cockpit. In some degree this is borne out by an account by Oliver Stewart who, having mentioned the 'hunched shoulders just below the centre section' goes on to refer in the same sentence to 'the arrangement of the planes with the marked dihedral angle on the bottom plane and the straight top plane.' Less known perhaps, but hardly less compelling, is this following—by another Camel pilot of the RFC:

'Those at the C.F.S. in the early months of 1917 will recall when the first rumours of the Camel floated round the mess . . . Someone brought the news "Sopwith has produced a machine with a 130 hp engine. It'll do 130 level and climb 10,000 feet in six minutes. It has *two* guns through the prop."' And later: 'There, in amongst the graceful two-seaters, sat a squat contrivance looking much more like a frog than a camel. I gazed with awe into the cockpit, felt the tail for improvement on that of the $1\frac{1}{2}$ Strutter, and tried to fathom the Kauper gear. In no time I had Mr. Hun well in

With flattened 'hump', this early Sopwith-built Camel is probably N6332. The ribs along the tops of the gun breech-casings are seen; also a loading handle in the cockpit. Maker's caption: 'S.122—Sopwith Camel. 110 hp Clerget—F.1.—2nd. Machine—1917.' There is a separate centre section, with rectangular cut-out, and the ailerons are lengthened.

hand, and if I didn't hit him it was not the fault of the Camel, or the Vickers, or ought else.'

And so we have 'this squat contrivance' presented, as it were, first-hand—with only the performance figures discredited!

Equally, now, we may discredit the 'What's in a name?' viewpoint; for the two synchronised Vickers guns that lay partly beneath its hump constituted, in effect, the Camel's very heart, forming, as they did with the ammunition and the engine beneath them, and with the pilot and petrol close behind them, a concentration of masses that was a primary factor in the Camel's astonishing powers of manoeuvre. 'Mr. Hun', as the quoted pilot called him, was certainly 'well in hand'; and the gentleman thus specifically addressed was generally the pilot of any of Germany's new Albatros single-seaters—armed though these were (like the Camel) with twin fixed guns. Thus, having noted that the Camel's guns constituted, in effect, its very heart, (and were not merely appendages that caused a growth on the top decking) we may consider armament before 'ought else'.

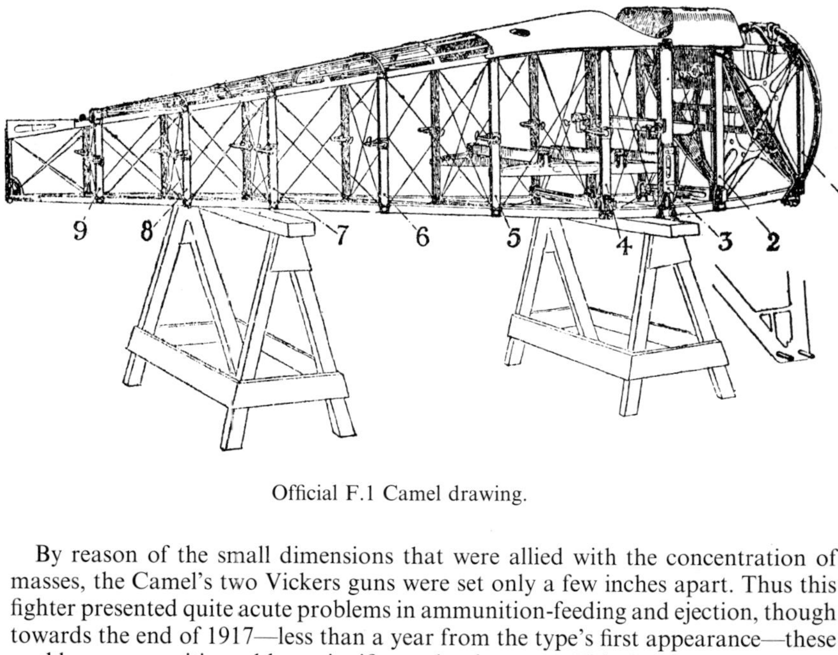

Official F.1 Camel drawing.

By reason of the small dimensions that were allied with the concentration of masses, the Camel's two Vickers guns were set only a few inches apart. Thus this fighter presented quite acute problems in ammunition-feeding and ejection, though towards the end of 1917—less than a year from the type's first appearance—these problems were mitigated by a significant development. This development was the introduction of a 'left-hand feed block' (the standard Vickers gun was fed from the right-hand side) so that 'handed' guns were now generally adopted—on the Camel first of all. It must be emphasised however that the earliest Camels—both experimental and production—had 'R.H. feed blocks' (right-hand) for both guns.

Here one would stress effectiveness of armament as distinct from sheer weight of armament (as commonly expressed by the number of guns installed), having now especially in mind the 'Hazelton attachment' or 'speeding-up set' that was first introduced as standard equipment on the Camel as the war neared its end, and the significance of which has never been generally acknowledged. Lord Balfour of

Inchrye (or Major Harold Balfour as he was at the time) once declared that this fitment increased the chances of a fighter pilot by 100 per cent (that is, the rate of fire of the Vickers gun was increased—very roughly—from 500 rounds per minute to 1,000). Disadvantageous though this development was to the enemy, it came as no unmixed blessing, for wear and tear on the guns themselves could prove quite distressing to an armament officer. Nevertheless, one appends this testimonial from an RFC fighter pilot:

'With two guns firing at the rate of from 400 to 600 rounds a minute it was astonishing what pilots missed. They often got right on the tail of their objective and went on firing, and simply nothing happened at all, and one did not know where the shots were going. Towards the end of the war, when destruction of the enemy increased a great deal, I understood it was due entirely to the speeding up of the firing to 2,000 rounds a minute, and I believe that was the sole cause of pilots coming into the killing class where previously they had been in the missing class.'

What a testimonial indeed, from and for a fighting man—'into the killing class . . . ' Yet what of the exceptional powers of manoeuvre that enabled the Camel's guns to be brought so effectively to bear, and which were attributed (in brief) to the super-sensitive ailerons and elevator—the latter in particular; to the concentrated masses already mentioned; and the gyroscopic, or torque, effect of the relatively powerful rotary engine, which enabled this fighter to turn unusually smartly to the right? 'Take the Sopwith Camel and the S.E.5', invited that master-pilot Sqn Ldr R. M. Hill, 'two conceptions diametrically opposed.' Pronouncing:

'The S.E.5a is stable with elevators free, the Camel unstable with them fixed. The Camel is more lightly loaded and has, with the exception of the rudder, more powerful controls. In a dive the Camel is flicky, due to lighter loading and excessive longitudinal instability; the S.E.5a is very steady, but dull to small intentional movements. In a zoom the Camel improves greatly owing to its lighter loading and instability; the S.E.5a is inclined to become languid, and its stability near stalling

Evidently copied by a Sopwith photographer at the same time as a better-known ¾ front view of a different Camel with a smashed port bottom wing, this is: 'S.481A—Camel F.1 which collided in the air and landed safely.—May 20/18'. Note ring sight on starboard gun and propeller for air-driven petrol pump on rear starboard centre-section strut.

draws down the nose so that a large backward stick movement has to be made. In a Camel the pilot has always to make small movements of controls to pick up steady speed which is difficult to maintain. Going into a dive, compared with the S.E.5a, the elevators of the Camel work the reverse way: the stick, though initially pushed forward, has to be pulled right in again, but the S.E.5a, if the stick is pushed forward, would drop its nose, creep up to trimming speed, and stay there. The S.E.5a is impossible to fly inverted: the Camel may remain so unintentionally. Nevertheless there is no doubt that the conception of the Camel as a fighting aeroplane made an irresistible appeal to a certain class of pilot, but it never could be comparable with the S.E.5a because, even assuming that it possessed certain qualities equal or even superior, the difference in performance—from 7 to 10 mph at 15,000 ft—and the view, gave the S.E.5a a superiority that only fighting pilots really understand.'

'Irresistible to a certain class of pilot'—evidently the 'killing class' . . . Even so, the Camel was not everybody's meat—least of all the enemy's, as witness the 1,200-odd victories that have been credited to it. Which may be reckoned a fair return for just under 5,500 F.1 Camels built.

The F.1 Camel, which owed something to the Baby as well as to the Pup, was designed for land operation, though it was used by shore-based squadrons of the RNAS as well as by the RFC; and that very early (and specialised) Naval interest was manifest in the design is clear from the succeeding chapter on the 2F.1. Unnumbered, though having a sizable blank panel on the fuselage apparently for some identification marking, the first Camel was cleared by the Sopwith Experimental Department on 22 December, 1916. Back at the drawing-board stage, some slight or moderate dihedral had been envisaged for the top wings as well as the bottom ones, the deciding factor in the straight-top-wing decision, it seems, being the possibility of one-piece construction (this demanding coarser dihedral on the lower panels).

On the first Camel, the top line of the 'hump' sloped upwards to the front rim of the cockpit, so that the cockpit looked unusually deep and obstructive to view—the more so as it was flanked by the rear 'centre-section' struts and because there was no cut-out in the top wing other than the little 'bite' out of the trailing edge. The marked trimness of appearance was accentuated by the absence of a windscreen of familiar type; by the cup-shaped fairings over the muzzle attachments and the front ends of the barrel casings of the guns (as on some later Camels); and by the pristine 'engine-turned' and burnished aluminium cowling over and behind the 110 hp Clerget 9Z rotary engine. Beside this aeroplane even a Pup would have looked 'bitty' (though never clumsy), while the snubness of the Camel's nose, with the guns ending in line with it—so short was the fuselage—contributed further to the 'stumpy-humpy' aspect of Sopwith's new creation.

Apart from the one-piece top wing there was little that was startling in the wire-braced wooden structure, with its plywood cockpit-sides. Even so, the whole machine suggested a triumphant conspiracy between an armourer, an aero-dynamicist and a demanding fighter pilot—the last of these being rightly named, even though the four ailerons (later lengthened) finished in line with the interplane struts.

The Camel that was clearly regarded as the first major experimental variant was a very different aeroplane indeed—so much so in fact that the designation conferred was apparently F.1/1, though for reasons clearly obvious it was otherwise called the Taper(ed) Wing Camel. With its single deep-chord interplane struts (recalling those of the Triplane), its blunt-tipped wings sharply tapered both in plan-form and

This altogether superb and unfamiliar view of the 'Taper Wing Camel' is specially referred to in the text. The top wing is possibly in three sections. Over by the Sopwith sheds at Brooklands stands a Camel; between the tapered port wings of the clean experimental machine are glimpsed a D.H.2 single-seat pusher, a windsock and a speed-limit notice. To the left of the picture is a curious automotive vehicle.

thickness, and its 110 hp or 130 hp Clerget engine, this version was otherwise distinguished by a central rectangular cut-out in the top wing, additional to a specially shaped cut-out in the trailing edge, which appreciably reduced the 5-ft (1.52 m) chord above the fuselage. Disappointingly, this 'super-Camel' gave poor results when officially tested in May 1917, landing-speed, for instance, being higher and manoeuvrability inferior, and development was not apparently pursued, though a '4F.1 Taper Wing Camel' was under discussion in December 1917. It is

In this $\frac{3}{4}$ rear view of the F.1/1 'Taper Wing Camel' cut-outs in the apparently one-piece top wing are seen.

worth remarking, nevertheless, that although a one-piece top wing is generally (and respecting one form at least, correctly) attributed to the Taper Wing Camel, the head-on photograph now reproduced (seemingly taken by a Weybridge photographer, and having no lettered-on Sopwith caption) appears to show a top wing in three sections. Also in the context of the Camel variant now discussed, it may be noted that whatever designation may have been formally applied, a side view that has now become familiar, and a detail of which was shown in *Armament of British Aircraft 1909-1939* (page 332), bears the following Sopwith caption: "S.127—Sopwith 'Camel'—110 hp Clerget—Type F.1— Tapered Wings—1917—3rd. Machine".

Suggestive that the description 'super-Camel' earlier applied to the taper-winged form was not over-sensational, and that this development constituted an exercise in aerodynamic cleanness as well as structural refinement, was the fitting of deep-chord fairings on the V-struts of the landing gear (thus more or less matching with the interplane struts). A further point of interest in this regard is that the landing gear fairings may not be present on the aircraft as seen in the head-on view now reproduced—the steel-tube members being wholly exposed.

That there were several experimental Camels is certain, and that some of these were Sopwith private-ventures equally so, though obscurities remain. Nevertheless, it seems definite that the Camel F.1 to which the Admiralty number N517 was allocated was at Brooklands for test as early as 26 February, 1917, and that by the following May N518—with the new Admiralty Rotary A.R.1 engine (first flown in a Camel in April)—was at Martlesham Heath. Not until July was this new British engine redesignated B.R.1 in deference to its designer, W. O. Bentley, who had served at Gwynnes Ltd to keep a Naval eye on Clerget developments.

The significance of the presumably-existing designation F.1/2 is unsure, though it may conceivably have been deliberately 'missed out' (for the practice has not been unknown in the past) to avoid confusion with 2F.1, later defined, or equally conceivably, have connoted the version which retained the one-piece, constant-chord top wing and short ailerons but had a rectangular cut-out at the top-wing centre. Certainly a variant so distinguished existed at a relatively advanced stage of Camel development, as witness the flattened 'hump' ahead of the cockpit (the general form that was standardised, and that may also have been used on the two-gun Sopwith-built Triplane N5445)—with the addition of a transparent windscreen. Even so, Sopwith photograph S.122 shows a form which is declared as '2nd. Machine' and which is characterised by lengthened ailerons, a three-piece top wing, no transparent windscreen and differences of detail (*e.g.* cowling). A four-figure number prefixed by 'N' almost certainly proclaims the subject-aircraft as N6332, thus making this particular '2nd Machine' the second, perhaps, to be completed of the first Sopwith production batch (N6330—N6379). A dead-rear view evidently taken on the same occasion was similarly described, the print being numbered S.121.

The F.1/3 is defined in an official document giving airframe, engine and loading details, and especially a detailed weight breakdown—and affording confirmation that the aircraft thus defined would correspond in all essentials with the first production form of the F.1.

Although the tabulation just mentioned gives the engine as a '130 Clerget' (delivering 127.75 'nominal bhp' at 1,250 rpm) official documentation on the early F.1, in the form wherein the type first entered service, affirms that the '150 h.p. A.R.1' was 'in some cases' fitted. Other engines than the 130 hp Clerget 9B that were installed in the form of the aircraft that was called F.1/3 were the 110 hp Le Rhône

9J (tested in May 1917—the Clerget 9B having been tested in March) and the 140 hp Clerget 9Bf (tested July). As already mentioned, an Admiralty Rotary A.R.1 (Bentley B.R.1) had been flown in April, and RNAS Camels, deliveries of which began on 7 May, 1917, were, as also intimated, fitted with this engine as an alternative to the 130 hp Clerget—and soon giving rise to the term 'Bentley-Camel'. Not until August 1917 were tests made at Martlesham Heath with a 100 hp Gnome Monosoupape (sometimes fitted in Service Camels), this installation being followed in December 1917 by trials with a new type of Gnome Monosoupape, nominally of 150 hp, though never standardised for British Camels. Late in 1918 the 150 hp Gnome Monosoupape installation was revived by the US Air Service, the Americans having been impelled towards the Gnome after dissatisfaction with plans for Clerget manufacture. The most powerful nine-cylinder rotary engine fitted to a Camel however (and all wartime Camel engines were of this general form) was the 170 hp Le Rhône 9R, the first experimental installation of which was made in France in the Sopwith-built F.1 numbered B3891 and which the French Government had acquired.

Repose in a shell-hole—and affording an excellent view of the long Aldis sight between the guns, the air-driven pump on the front starboard strut, etc.

For the RFC, Camel contracts were placed by the War Office in May 1917 (the month in which Naval deliveries began), the first contract recipients being Ruston Proctor & Co who had formerly made 1½ Strutters. 'Rustons' were, in fact, to make a total of over 1,500 Camels, the first of which was finished in June 1917, and peak production by this company was 128 in a month—May 1918, just before phasing-out in favour of the Snipe. Sopwith's own output of F.1 Camels was only about a third of Ruston Proctor's—roughly equalling Clayton & Shuttleworth's though representing only about a tenth of the total of nearly 5,500. Boulton & Paul rivalled even Ruston Proctor, and far surpassed British Caudron, Hooper, and other constructors whose orders are set out at the end of this chapter.

Three with a difference: Hooper-built H826 (top) was one of the H734—H833 batch of F.1 Camels specially equipped for night fighting, though having standard twin-Vickers armament and normal cockpit position. B9287 (centre) was made by Boulton & Paul, was unusually marked, had the rearwardly-moved cockpit and two Lewis guns on Foster mountings. B3801 was a two-seat trainer conversion—Sopwith-built in the first place, but not necessarily the first such transformation.

Production F.1 Camels had a three-piece top wing and lengthened ailerons; but a feature that seems to have caused special comment in Germany was the fixed tailplane, as witness this 1918 report 'from an enemy source': 'The wing spars, which are made of spruce, are spindled out to an I section, with the exception of the bottom rear spar, which is left solid . . . The tail plane, which is deeply cambered on both sides, is rigidly attached to the upper body longerons, with an angle of incidence of 1.5 deg. The tail plane trimming gear hitherto fitted to all Sopwith machines has been abandoned, in spite of the fact that the petrol tanks are placed behind the pilot's seat.'

Engine cowlings varied less than might be supposed, the frontal lip being an unbroken circle, accentuating the snub nose, though sometimes quite extensively slotted round parts of the perimeter for additional cooling in high temperatures or at low altitudes—or, perhaps, in association with a special engine (*e.g.* Le Rhône 9R). Propellers of many patterns were tried, especially as the Camel was noted for climb rather than level speed, though many factors—including engine type, make and state, and sometimes involving such fundamentals as compression ratio and carburation—could easily invalidate investigations. Time to 10,000 ft, for instance, could vary from around a mere 8 minutes to something of the order of 16. 'Rationalised' data are therefore quoted at the end of this chapter for the F.1 Camel in seemingly representative states and with each of the three commonly fitted engines (110 hp Le Rhône, 130 hp Clerget, and 150 hp B.R.1). The prices of these engines were respectively £771 10s, £907 10s and £643 10s, whereas the F.1 airframe without engine, guns or instruments was £874 10s.

Respecting mechanical 'extras' and prices, the item which an earlier-quoted pilot described as the 'Kauper gear' (and which he tried to 'fathom') now commands attention; for, although mention has already been made, in connection with the 1½ Strutter, that 3,950 sets of the Sopwith-Kauper synchronising gear for the Vickers gun were supplied to the flying services, and 2,750 installed, it must now be noted that a good proportion of these sets must have been used on early twin-gun Camels. In any case, the royalty on each is said to have been £1. The nature and recipient of this sum is not known, but the gear was certainly fitted to aircraft other than Sopwiths—the Bristol M.1C Monoplane for instance. Curiously enough, an example of this aeroplane (though without armament) was being operated after the war in Australia by Harry Kauper himself—jointly with Capt Harry Butler.

Though first made in 1916, the Sopwith-Kauper gun-gear was developed in several patterns, but in essence was of 'rod-and-spring' type, the spring actuating the gun's firing mechanism when permitted to do so by a controlling cam rotated by the engine. As for the contention that this was an 'interrupter' gear, as distinct from a 'synchronising' gear (the latter term supposedly, though quite incorrectly, being reserved for the C.C., or Constantinesco, hydraulic type, used on later Camels, though more especially on the Snipe) one would only repeat that the cam was mounted on the engine (Sopwiths' words now, my italics) 'in such a manner that it can oscillate the tappet rod in order to *synchronise* with the passage of the blades of the propeller past the line of fire of the gun, alternatively *holding off or preventing* the fire and then permitting the fire by means of the spring.'

The Constantinesco hydraulic gear was also first made in 1916, and was likewise developed in several patterns. Its teething troubles in service, however, were experienced mainly on the S.E.5a, the first production contracts for which fighter had been placed at the beginning of 1917 with Sopwiths' neighbours Martinsyde and Vickers, Weybridge works. Fortunately, perhaps, the S.E.5a had one Vickers gun only. Early Le Rhône-engined Camels may, in fact, have been the first RFC fighters to have two Vickers guns and the new hydraulic gear; but respecting both the gun and the gear it is needful to recognise that, although it is commonly stated that the 'Vickers automatic gun' was concerned (which is essentially true) the gun's mechanism was so controlled by the hydraulic gear that firing was in fact *semi*-automatic—being initiated for each individual shot. Martinsyde, it may be noted, worked on electrical synchronising gear, though any such system would probably have been heavy and prone to other inherent faults.

As for the Camel, its firing gear, like the rear parts of the guns, was largely, if not wholly, obscured by the 'hump' fairing, though this was sometimes cut away to clear the breech casings almost entirely. Also impeding an onlooker's view of the mechanism (though concentrating the pilot's view of the target) was the Aldis tubular sight, usually mounted on the centre line, with the rear end projecting behind the windscreen—which itself took several forms and situations. Ring-and-bead sights were fitted, to the port or starboard gun (usually starboard, but sometimes to both) the ring being in front, either on a pedestal or attached directly to the barrel casing of the gun. But there were many variations, official or otherwise.

Wholly obscured inside the Camel's cockpit were the Hyland loading handles ('feed-block levers' or 'cocking handles' as they were sometimes called, though they were not to be confused with the crank handle of the Vickers gun as familiar to infantrymen). The loading handles were intended to operate—one-handed on the pilot's part—the loading and cocking mechanisms, without the necessity of having to pull the belt through the feed block.

The weight of the 1914—18 Vickers gun, as adapted for aircraft use, varied considerably—not merely by reason of the fact that it was arranged for air cooling instead of water cooling, but according to which 'land-service' items (*e.g.*, handle block at rear, and sight) were retained, and which 'air-service' ones (*e.g.*, parts of synchronising gear and loading handle) were admitted to the reckoning. A 'typical' figure was 35 lb, which accords precisely with the present writer's statement on another occasion that the weight of the Camel's two guns was 70 lb; and though the quantity of ammunition per gun was then quoted as 250 rounds, it may now be affirmed that this was later increased (*e.g.*, 600 rounds total). Even so, as the F.1 Camel's 'very heart'—as previously noted—was its armament, there was little in the way of major change over the years, though the blank space on an official weight breakdown, for 'bombs & gear', was later filled for Service Camels, as will shortly be explained.

Meanwhile, an experimental fitment of what may have been a large-bore gun firing through, or from, the propeller hub of a night-flying Camel—apparently in US service, though suggesting French influence respecting the Vickers guns—may be mentioned as possibly involving any of a number of British or foreign weapons. These were commonly of smooth-bore type, as was the French 37 mm Puteaux 'engine gun' (in one of its forms, at least) and could fire single or multiple rounds of 'canister', 'grapeshot', 'case shot', high-explosive or incendiary ammunition. On the other hand, in connection with the particular installation mentioned, some engine-starting gear may have been involved, or the intended fitting of a special armoured cowling (as designed for rotary engines), or even one of several forms of target-illumination.

When the F.1 Camel carried bombs (as it quite frequently did) these were generally stowed on a 'Carrier, 4—20 lb., Mk.I', which was attached under the cockpit and could be loaded not only with the familiar 20 lb Cooper, 20 lb Hales or 16 lb H.E.R.L. but with the 40 lb Phosphorous bombs (though in the last-named case several modifications to the carrier were necessary and only two bombs could

Camels were often used for bombing, and a bombed-up machine (4 × 16 lb or 20 lb) is seen in the top picture. But to force-land, as otherwise depicted here, with a bomb still on the carrier (as it obviously is) could surely be unhealthy. Did the horsemen know?

be loaded instead of four). 'Four Coopers' were a usual load (even after the bomb-designer's name was officially dropped itself)—these little bombs, which actually weighed 24 lb and were commonly quoted in bombing returns as 25 lb, having as officially prescribed targets 'personnel and aerodromes'. Release was by a simple toggle, and no bombsight was fitted—not even when a single 112 lb H.E.R.L. bomb was taken by an F.1 Camel (as it sometimes was, using the appropriate pattern of carrier); and although in the context of Camels and relatively heavy bombs one is always inclined to remember the Tondern raid by 2F.1 Camels in July 1918 (2 × 50 lb each) it would be sad to forget that F.1 Camels, diving in line astern to a low height, had hit sheds at Zeebrugge earlier in that year with 112-pounders. (This was not dive-bombing in the later-accepted sense, though for all his modesty, Oliver Stewart claimed to have done some pioneer work on that technique at Orfordness, Suffolk, using a Camel in 1917).

The fitting of various engines, fuel systems and engine-accessories, the occasional addition of bomb carriers, the cutting-away of fuselage fairings and the nature and dispositions of windscreens were, of course, only a few of the alterations made after the F.1 was placed in production. The pilot's view being poor, top-wing cut-outs were enlarged and otherwise re-formed; but (far more drastic) two-seat versions were produced for training, at first in some degree *sub rosa* or 'locally', though official thinking and design work seems to have anticipated such modifications. Whatever the truth behind some mysteries that remain attached to certain exotic two-seat developments, the special (much modified, though single-seat) Home Defence night fighting version commands the greater attention.

Perhaps by reason of its reliability, as much as for its availability and general suitability, the 110 hp Le Rhône engine was favoured for this remodelled F.1, as this same engine was for the single-seat night-flying Avro 504K fighters with faired-over front cockpit; and another point of affinity was the adoption of the Foster gun-mounting. But whereas the Avro had a single mounting of this type (and in consequence one Lewis gun only) the special Home Defence Camel had two. Thus these two Lewis guns could fire their 'special' (incendiary or incendiary/explosive) ammunition—possibly mixed with 'ordinary' or 'ball'—above the propeller without the risk imposed by synchronising gear. Armament of Home Defence Camels, however, varied considerably, and in some degree reproduced techniques applied also to the special Home Defence 1½ Strutter (which, like the Avro just mentioned, had its front cockpit faired over, with the pilot at the rear).

Correspondingly, the Home Defence Camel with Foster mountings and special night-flying equipment, had its cockpit moved aft. The two Vickers guns with their 'hump' were removed, and the rear (bottom) ends of the tracks on which the gun-carrying blocks of the Foster mountings were arranged to slide were fixed to a cross-bar carried on two rearward-sloping members just ahead of the new cockpit. In this region also the petrol tankage was located, with a Neame illuminated sight, having a large-diameter ring, on the decking above. For upward fire (which the Foster mounting was developed to make possible, in addition to the changing of ammunition drums) separate sights were provided, and the head-fairing behind the rearwardly-moved cockpit probably assisted the pilot in aiming.

With a mass of top-hamper the specially modified Home Defence F.1 Camels could hardly have shown a vastly superior performance over the adapted Avro 504K, and in any case the cleaner twin-Vickers Camel was not outstanding for its speed and rate of climb, though this form also was used by Home Defence units—fears that it was too unstable for work at night notwithstanding. As matters transpired, the twin-Vickers armament was ultimately favoured for RAF single-

Two F.1s with a difference: Top, B9268 with a Hythe camera gun to port and a hole in the 'hump' for the starboard Vickers (and with pupil-pilots behind waiting to 'have a go'); bottom, B3891 of Sopwith's second production batch, in which the French installed one of their new 170 hp Le Rhône 9R engines. This was the Camel's most powerful engine, and at least two installations were made.

seat night fighters, in preference to 'fancy' schemes. Thus the standard post-war aircraft of this class, the Hawker Woodcock, had two fixed forward-firing Vickers guns mounted externally on the fuselage sides, in which position, the makers claimed, their muzzle flash at night did not interfere with the pilot's vision. The point now being made is by no means far removed from the Camel, for well-nigh every writer on the subject of this fighter quite correctly observes that Camel pilots at night were momentarily blinded by the flash of their Vickers guns close ahead of them. That the so-called 'flash eliminators' developed accordingly were more accurately described as flash suppressors is suggested by the fact that when the original Gloster Gauntlet was tested at Martlesham Heath in 1933 with Vickers

Mk. V guns (supposedly having such devices fitted, and certainly installed in troughs low in the fuselage sides) it was reported—to quote Derek James' book *Gloster Aircraft since 1917*—'During armament trials the guns each fired some 12,000 rounds . . . producing small flash and no dazzle.'

Be this matter as it may, mere mention of the Gauntlet is a reminder that the F.1 Camel's basic armament of two synchronised Vickers guns remained standard for RAF single-seat fighters far beyond the Woodcock (already named), the Gloster Grebe and Gamecock, the Bristol Bulldog and the Hawker Fury—for the Gauntlet did not enter service until 1935!

Though armament was one differentiating feature between the F.1 and 2F.1 Camels (the latter being specialised Naval machines, as later described) F.1s were nevertheless used for ditching trials with hydrovanes and flotation gear. Even so, more appropriate to this present chapter are the following innovations and experiments:

Special developments put in hand respecting safety belts (lap straps) and full fighting harness (greatly reassuring in some manoeuvres, or antics, by the hyper-sensitive Camel). Here one would add that, although in his word-picture of the Camel in *The Clouds Remember* Oliver Stewart devotes much of his space to Oliver Sutton's 'Sutton harness'—stating that Sutton received no official credit for that invention and that even the name was no longer officially current—an official handbook issued in the 1930s respecting the Bristol Bulldog most certainly has a quite lengthy dissertation on 'Sutton harness release gear'. If the name Pup could prevail over officialdom, then so could that of Sutton.

Likewise respecting pilot-safety in Sopwith's 'firecracker', there were tests with Imber self-sealing petrol tanks and Calthrop Guardian Angel parachute harness.

Further touching the Camel's peculiarities of handling were experiments made with controls and surfaces (larger fins and rudders, for example); likewise the moving of the wings to the rear and the redistribution of weight. Special investigations were made of spinning and inverted flight.

Peculiarities of flight behaviour having been mentioned, engine-handling must also be noted as a factor that was studied with extra care, demanding as it often did skill as well as knowledge. The type of engine could, in fact, affect aircraft-handling quite appreciably, tail-heaviness, for instance, being experienced with the 100 hp Gnome Monosoupape.

All in all, the Camel was a little box of tricks—but how some of those tricks could be turned—no pun intended—to good account is implicit even in a much abbreviated record of its operational career, though to taste the true romance thereof the reader may relish Chaz Bowyer's *Sopwith Camel—King of Combat*.

The first F.1s (as earlier noted) had been delivered to the RNAS in May 1917; N6347 was in action on 4 June, with No.4 Squadron, RNAS, and by the end of June that Service and the RFC together already had a total of 135. By July the Sopwith Triplanes of Nos.8 and 9 (Naval) Squadrons were being replaced by the new biplanes, and on the 4th of that same month Naval F.1s (appropriately of No.4 Squadron) tackled sixteen Gothas that were returning from a raid on Harwich. After the Battle of Ypres began at the end of that July, ground attack (even at night) and bombing were added to the Camel's operational roles, and the Battle of Cambrai (November/December) saw far more low attacks by these 'scouts' or 'fighters', which could dodge round the terrain with the same agility as they could round airborne targets.

These points one mentions early in this 'operational summary' to emphasise that dog-fighting by day (or fighting by night for that matter) were not by any means the

Camel probably from Boulton & Paul, with local colour and local talent.

limit of the F.1's capability, though this particular Sopwith type must surely rank as the supreme dog-fighter of all time. The tales of its battles are here beyond re-telling. 'Legendary', indeed, is hardly too bold a word for its exploits; for have not the 'Biggles' books of W. E. Johns placed them firmly in that class, no less than the works of possibly more meticulous historians? ('The Camels Are Coming', indeed . . . ').

The RFC's first Camel victory (70 Squadron) was on 27 June, 1917; and thenceforth to the end of the war did F.1s turn their pairs of Vickers guns on air and ground targets that varied from other fighters (Fokker Triplanes included!) to big-gun batteries. Italy, the Aegean, Macedonia and Russia (and not forgetting England) were Camel hunting-grounds.

March 1918 (by which time seven Camel squadrons had been brought up to an establishment of 25 aircraft) saw the start of Germany's great offensive, in which connection—in two patrols on one day—Capt J. L. Trollope destroyed six enemy aircraft with C8270. But names (even Ruston, Proctor!) are invidious at this massive turn in the Sopwith story. Not for us here the Brown/Richthofen battle, and other glories of the kind; though we shall do well to note that RNAS F.1 Camels were often attached to the RFC before the RAF was formed on 1 April, 1918.

Though the Armistice, and mounting production of the Snipe, halted further development of the F.1 Camel (with the 150 hp Monosoupape for the USA, for instance) purely as a fighter—as distinct from an ever-challenging research tool for Farnborough—both the United States and Canada had postwar deliveries. Fighting for Poland in 1920 possibly marked the end of the type's military career, and civil use was small. F6302 was registered G-EBER in 1922, but by that time Hubert Broad's G-EAWN (H2700, not flown until December 1919 and registered to Broad in March 1921) already had a good amount of sporting and aerobatic flying to its credit. It was dismantled at Stag Lane in 1922. Grenville Manton's specimen with 45 hp Anzani he spoke of wryly, though it seems to have passed through several hands and forms.

But trivia must not obscure the massive production orders that were placed for F.1 Camels. Apart from N517 and N518, these were:
Sopwith N6330—N6379; B3751—B3950; B6201—B6450.
Boulton & Paul B5151—B5250; B9131—B9330; C1601—C1700; C3281—C3380; D6401—D6700; D9381—D9530; F1301—F1550; F1883—F1957; F6301—F6500; H2646—H2745.
British Caudron C6701—C6800. (Order for H3996—H4045 cancelled).
Clayton & Shuttleworth B5651—B5750; B7181—B7280; D3326—D3425; D9581—D9680; E4374—E4423; F3096—F3145; F4974—F5073. (Apparently only about half of this last order delivered).
Hooper B5401—B5450; C1551—C1600; F2083—F2182; H734—H833; H7343—H7412 (deliveries may have ended with H7363).
March, Jones & Cribb C8301—C8400; F5174—F5248.
Nieuport & General C1—C200; F3196—F3245; F3918—F3967; F8496—F8595.
Portholme Aerodrome B4601—B4650; B7131—B7180; D9531—D9580; E5129—E5178; F1958—F2007; F8646—F8695.
Ruston Proctor B2301—B2550; B5551—B5650; B7281—B7480; C8201—C8300; D1776—D1975; D8101—D8250; E1401—E1600; E7137—E7336; F2008—F2082; F3968—F4067.

F.1 Camel
(130 hp Clerget)

Span 28 ft 0 in (8.5 m); length 18 ft 9 in (5.7 m); height 8 ft 6 in (2.6 m); wing area 231 sq ft (21.5 sq m). Empty weight 962 lb (436 kg); maximum weight 1,482 lb (672 kg). Maximum speed at 10,000 ft (3,050 m) 104.5 mph (168 km/h); maximum speed at 15,000 ft (4,570 m) 97.5 mph (157 km/h); climb to 10,000 ft (3,050 m) 11 min 45 sec; climb to 15,000 ft (4,570 m) 23 min 15 sec; service ceiling 18,000 ft (5,480 m).

F.1 Camel
(150 hp Bentley B.R.1)

Span 28 ft 0 in (8.5 m); length 18 ft 6 in (5.5 m). Empty weight 977 lb (443 kg); maximum weight 1,508 lb (680 kg). Maximum speed at 10,000 ft (3,050 m) 111 mph (179 km/h); maximum speed at 15,000 ft (4,570 m) 103 mph (166 km/h); climb to 10,000 ft (3,050 m) 9 min 50 sec; climb to 15,000 ft (4,570 m) 20 min; service ceiling 18,000 ft (5,480 m).

F.1 Camel
(110 hp Le Rhône)*

Span 28 ft 0 in (8.5 m); length 18 ft 8 in (5.6 m). Empty weight 889 lb (403 kg); maximum weight 1,422 lb (645 kg). Maximum speed at 10,000 ft (3,050 m) 118.5 mph (191 km/h); maximum speed at 15,000 ft (4,570 m) 111.5 mph (179 km/h); climb to 10,000 ft (3,050 m) 9 min 10 sec; climb to 15,000 ft (4,570 m) 16 min 50 sec.
*(Performance with engine actually delivering 137 hp)

F.1 Camel
(170 hp Le Rhône)

Span 28 ft 0 in (8.5 m). Empty weight 1,048 lb (475 kg); maximum weight 1,567 lb (711 kg). Maximum speed at 10,000 ft (3,050 m) 113 mph (182 km/h); maximum speed at 15,000 ft (4,570 m) 108.5 mph (175 km/h); climb to 10,000 ft (3,050 m) 9 min 35 sec; climb to 15,000 ft (4,570 m) 17 min 30 sec; service ceiling 21,500 ft (6,550 m).

T.F.1 Camel

To have appended the above prefix 'T.F.'—signifying 'Trench Fighter'—to the foregoing chapter headed 'F.1' (which constituted one of the first symbolic recognitions of the term 'fighter' instead of the persistent 'scout') would have given the special low-attack version of the Camel now studied unjustified prominence. Yet to grant it this brief chapter to itself incurs no less a risk; though one feels reassurance in that the specially armoured aeroplane which now concerns us was a true sire of the Salamander. There is vindication also here of the emphasis placed in the preceding chapter on the extensive employment of standard F.1 Camels for low-flying attack; and one finds a link here, too, with that other armoured Sopwith aeroplane the two-seat Buffalo. Even if the armour be discounted, furthermore, the analogy between the T.F.1 and the special night fighting version of the F.1, with two Lewis guns which could fire *upwards* from their Foster mountings, is quite a close one, for the T.F.1 had two Lewis guns specially installed for *downward* fire (the two forward-firing Vickers guns being deleted in both instances).

In the joint context of the Camel and the Battle of Cambrai (November/December 1917) one remarked on the F.1's propensity for 'terrain dodging'; and of that same battle it was officially recorded: 'The feature of the battle so far as concerns the air services was the development of low-flying attacks on the infantry, and an extension of this activity, in future battles, was foreshadowed. That the intervention of the low-flying aircraft had an influence on the battle is beyond dispute, but the lessons of that intervention deserve close examination. The casualties to the low-flying aircraft were high, averaging 30 per cent for each day on which aeroplanes were used on this duty. That is to say, a squadron of highly skilled and experienced pilots, flying first-class fighting aircraft, would, so long as it was employed on concentrated low-flying attacks on front-line troops in prepared defensive positions, require to be replaced every four days.'

Here—if ever there was one—was a loud and clear demand for a specially protected aeroplane, though the nature of its armament was apparently less clear, as the evolution of the forward-firing Salamander showed. Thus it came about in January 1918 (by which time the Snipe was already amassing flying hours) Sopwith were invited to design a special armoured single-seater with the 230 hp Bentley B.R.2 engine, this aircraft later emerging as the Salamander. Meanwhile, however, the company built—possibly envisaging the ready conversion of F.1 Camels to the new configuration—the only known T.F.1. This special Camel was itself a conversion of the Boulton & Paul-built B9278, though close scrutiny of the original of Sopwith photograph S.185, now reproduced, discloses on the rudder the Clayton & Shuttleworth trademark—a fact that may be wholly insignificant, beyond, perhaps, exemplifying interchangeability. (B9276 of the same Boulton & Paul batch became the Sopwith Swallow monoplane, and B9275 may have formed the basis for the Scooter); but although B9278 is the only known example of the T.F.1—it had the 110 hp Le Rhône engine, and was photographed, apparently complete, on 19 February, 1918—at least two other conversions were contemplated, if not put in hand.

Having regard to the Royal Navy's special interest in the Camel programme from the start, and also to the low-level activities of Naval Camel squadrons, it was fitting (as well as seemingly judicious, as will be explained) to mount one of the three Lewis

Although the Admiralty Top Plane mounting for a Lewis gun gave the T.F.1 armoured version of the Camel some semblance of the 2F.1, it was, in fact, a direct F.1 development, as reaffirmed by the Sopwith caption to this picture: 'S.185—Sopwith Camel F.1 Armoured Trench Fighter—Feb. 19/18'.

guns which armed the T.F.1 on a fitting of the kind known as the Admiralty Top Plane mounting—described in the following chapter on the 2F.1. The Firth armour plate might also be reckoned as an Admiralty 'speciality', though the RFC had used armour on their aeroplanes from the war's early days.

The Lewis gun on the Admiralty mounting of the T.F.1 could be used for offensive or defensive frontal fire (over the propeller) as well as affording the means—it may be supposed—of moving-along, by the upward fire that this form of mounting also allowed, any multi-seater that might seek to inhibit the T.F.1's activities by taking station above. But offence was the primary task of the 'Armoured Trench Fighter', as the new Camel-development was called, and for this purpose—the raking of trenches in longish runs, for instance—two fixed Lewis guns fired at 45 degrees forwards and downwards, through the cockpit floor. These guns had their spade grips removed (as had the third one, above) and their barrels and gas cylinders were largely exposed between the struts of the landing gear. Vacant ports in the 'hump' proclaimed the absence of the Vickers guns, and no sights of any kind are seen in photographs.

Along the bottom of the fuselage, from the firewall behind the engine aft to the area of the main petrol tank behind the cockpit, and extending slightly laterally under the bottom wing (with cut-outs for the rear landing-gear struts) stretched a sheet of armour plate—a precaution such as had proved its worth as early as 23 August, 1914, when Lieut Joubert avoided thigh-injury from a bullet by virtue of a steel plate under his seat. On 7 March, 1918, the T.F.1 was flown to France, though its visit was apparently brief—possibly because the Salamander was already far advanced.

That Sopwith had equipped B9278 with a direct-reading airspeed indicator on the port interplane struts (see $\frac{3}{4}$-rear view dated 19 February, 1918) suggests that performance was to be accurately measured, though no figures are available.

To conclude this present chapter some allusion must be made to Sopwith-built Camel B6218 which, although apparently having no extensive armour, and not formally designated as being of T.F.1 type (though three T.F.1s were originally intended) served to test a mirror-sight arrangement—using periscopic principles—for the downward-firing guns. This involved a mirror under the top wing and another in the cockpit. The latter feature notwithstanding, it was evidently quite different from the Periscopic Bomb Sight Mk.I, as used on the R.E.7, or the Mk.II (for Martinsyde G.100/G.102)—by means of which sights the ground was viewed, as it were, from below the pilot's seat.

With its wicker seat, and nearby items associated with the engine and flying controls (the joystick is seen tilted slightly forward, close to the point where the two Lewis guns protrude through the floor) the T.F.1 cockpit was neither spacious nor luxurious.

This Camel B6218, with 130 hp Clerget engine, was sent to France, together with the 'pukka' T.F.1 numbered B9278 and having a 110 hp Le Rhône (petrol consumption and weight being prominent considerations in the low-attack business), the two pilots concerned having both been involved in the '45-degree shot' (upward firing) experiments at Orfordness. However, for low-level attack, as well as for the high-level interception of enemy aircraft, the ultimately favoured armament was twin fixed forward-firing Vickers guns. Hence the Salamander.

Familiar—though generally in retouched form, with even the Brooklands huts removed—this view of N6635 remains as the classic study of a 2F.1 Camel. So excellent is the Sopwith original ('S.58—Camel—150 A.R.1 Engine—Type 2.F.1') that even the tiny pillar-mounted bead for the ring-sight on the Vickers gun is distinctly seen above the cowling. Equally so, the aircraft number—on the fuselage side behind the cockpit.

2F.1 Camel

The easy way out in differentiating the 2F.1 Camel (or 'Ship's Camel' as it was commonly known—with the apostrophe variously placed, if placed at all) from the land-based F.1 was by the Lewis gun perched far forward above the wing, and by the single Vickers gun, offset to port and resulting in revised cowling lines, instead of the twin-Vickers installation with the famous 'hump'. But before detailing the finer points of distinction (some of which are hardly less obvious, once pointed out) one must remark how poetically unjust it was that the Navy—which itself had used twin 'bow-chaser' guns in its ships to shoot at an enemy's sails and slow him down—should make such radical changes in a classic form of armament having such a romantic precedent.

The finer points of distinction in the 'definitive' or 'standard' 2F.1 (though many modifications followed, including, for instance, skid landing gear, a steel-tube gear with jettisonable wheels, special armament, slinging gear and lashing-down points) were these:

1. Shorter wing-span, resulting from a shorter centre section.
2. Thinner centre-section struts—these being of streamline-section steel tube instead of wood and being far less splayed-out than on the F.1.
3. A fuselage made in two parts, visibly joined behind the bottom wings, and constructed thus for ready separation and stowage in a ship. (The Ship's Camel was consequently sometimes called the Split Camel).
4. Extra tankage in place of the starboard Vickers gun and its ammunition.

5. External elevator-control cables, running from a lever near the fuselage joint, thus facilitating dismounting and reassembly as well as clearing flotation bags inside the rear fuselage.

6. A tailplane which (*not* in answer to the German comment quoted in the F.1 chapter, and possibly applicable also to some F.1 Camels) was adjustable on the ground. Thus, official instructions read (verbatim and in part): 'Fit the Nut and Screw Tail Plane Adjustment between the Fitting on the rear Spar Tube, and the corresponding Fitting on the Sternpost, not forgetting to lock the Nut to the Eyebolt provided with wire . . . The Incidence of centre line of Tail Plane in normal position is $1\frac{1}{2}°$; in this position the distance from top of Longerons to under-surface of Rear Spar Tube should be $1\frac{15}{16}''$. Any adjustments can be made after tests.'

7. Dihedral angle of the bottom wings increased to $5\frac{1}{2}$ deg. (On the F.1 it was 5 deg).

In the chapter on the F.1 Camel it was declared that it 'owed something to the Baby as well as to the Pup', and that—as now further explained—'very early (and specialised) Naval interest was manifest in the design.' It was added that 'the first Camel was cleared by the Sopwith Experimental Department on 22 December, 1916.' Perhaps of even greater significance here was the affirmation that the two synchronised Vickers guns 'constituted, in effect, the Camel's very heart'; so in now tracing the development of the 2F.1 on the Navy's specialised account we shall do well to focus attention first on armament, but also on the Baby which (it may be recalled) had that uncommon fitment a synchronised Lewis gun, with mechanical gear of Scarff or Hazelton type.

Here it becomes necessary to remark that the Lewis gun, with its 47-round or 97-round magazine, and especially when stripped of its cooling radiator, was a lighter weapon than the belt-fed Vickers (250 rounds or more per gun, plus synchronising gear)—and the Lewis, further, was more amenable to training, or setting, for upward fire, as, for instance, when using incendiary or other 'special' ammunition against Zeppelins (without the danger of hitting one's own propeller). On the other hand, the belt-fed Vickers clearly offered more continuous fire, and was more suitable for synchronising.

It must also be noted here that by 1917 the Navy was feeling a need for a shipborne aeroplane for general duties and anti-Zeppelin work, having wheels or floats or both, and superior in performance to the improved Babies developed and produced by Blackburn and Fairey. Accordingly, Sopwith designed the FS.1, a Camel-type aircraft having two unstepped main floats of pontoon form and a faired tail-float. Thus the formula adopted for Pixton's 1914 racer, and adapted for the post-war Jupiter-engined Schneider racer, was generally followed. Jettisonable wheels to permit deck take-offs were intended (having been pioneered in 1915, as noted in the 'Schneider and Baby' chapter). The link between the two familiar forms of the Camel (F.1 and 2F.1) and the Baby was seemingly the aeroplane numbered N4, which crashed in March 1917 and which was otherwise called 'Improved Baby', 'Floatplane Scout' or 'Camel Seaplane'. The engine was the 130 hp Clerget, as fitted in late Babies by Blackburn, at a time when Sopwith were busy making Camel landplanes.

The aeroplane numbered N4 having now been mentioned, and the names Baby and Blackburn having recurred, it is not unedifying to recall the remark by A. J. Jackson, quoted in the 'Schneider and Baby' chapter and mentioning 'a prototype machine, N300'. Could this one-off Baby 'N300' (one wonders) be in any way equated with 'N3'?

In any case, the next aircraft to be considered was numbered N5, and this was a

Sopwith-built Naval Camel, possibly ordered at the same time as N4. That N5 was essentially a landplane version of the FS.1 design is sure; likewise that an aeroplane numbered N5 (though conceivably incorporating parts of N4) was flown by Harry Busteed at the Isle of Grain on 4 April, 1917, having already been the subject of a Martlesham Heath report in March. One feature that N5 possessed in common with the FS.1, as drawn at Kingston, was a fixed Lewis gun installed *inverted* on the top wing—an installation which, with the pistol grip pointed skyward, was reminiscent of a fitment on a Baby, whereby the grip jutted out to port, the gun itself being fixed to lie on its side, and thus offering its magazine-attachment peg to the pilot. With the upside-down arrangement on N5, it may have been supposed, the changing of the magazine—which projected down through a trailing-edge cut-out—would have been facilitated without recourse to a Foster mounting, or to one of the Admiralty Top Plane type, as eventually adopted. Temporary though it proved to be, the inverted-gun scheme would evidently have demanded a pilot strong in arm, long in reach and high in courage.

As for the Lewis gun's complete inversion, while this extreme position may not have been foreseen in 1914, it was certainly claimed at that time (as a point in the gun's favour for aircraft use) that it could be 'trained freely in any direction from vertically upwards to vertically downwards.' A little later, of course, there was the classic story of Capt Louis Strange and his inverted Martinsyde, with Strange literally hanging-on for dear life to his Lewis gun's magazine—which had obligingly jammed on its peg.

The Admiralty Top Plane mounting, with its Lewis gun, is featured in this photograph. The magazine is of the 'single' (47-round) pattern.

Both guns are clearly installed on the Beardmore-built N7136, seen here at Dalmuir. Note also the external elevator-control cables, running from the lever just behind the fuselage joint.

A single Vickers gun was mounted additionally on N5, to port, as on production aircraft; and though the elevator-control cables on this aeroplane were still internal, the slender centre-section struts (already mentioned as a 2F.1 distinction) were present, and the rear part of the fuselage was seemingly detachable.

During the spring and summer of 1917, N5 was the subject of extensive development, especially respecting its military equipment—though still with the elevator cables snugly inside. A wireless telegraphy set was fitted, the retractable wind-driven generator for which was bracket-mounted on the port side of the fuselage just forward of the cockpit, with the aerial fairlead projecting down between the landing gear struts like one of the downward-firing Lewis guns on the TF.1. Armament, in fact, continued to be of primary concern, and for the Lewis gun on N5 the Admiralty Top Plane mounting was adopted. This fitting was very different from the Foster (track-type) mounting, as used on some night-flying F.1 Camels, though its dual function of permitting upward fire as well as magazine-changing was of similar intent. It is likely, indeed, that the Admiralty Top Plane mounting was developed primarily for the 2F.1 Camel, just as left-hand feed had been developed for the F.1 type.

The Lewis gun on N5's new Admiralty mounting (for the operation of which there was a centre-section aperture forward of the trailing-edge cut-out) was attached to the main, arc-shaped, member of the mounting by lugs and a collar, while at the rear end the gun's pistol grip was slotted-in to a projecting portion. Associated with the gun-carrying member was a system of steel tubes, knuckle-jointed to allow the gun to be raised or lowered, and spring-loaded by elastic cords. Though intermediate positioning of the gun was not possible, as with the Foster mounting, the gun—from which the spade grip was removed—could be readily released by Bowden cable; a second cable of this type was used to fire the gun, the bullets from which passed in forward fire only just clear of the propeller, though the gun sat well above the wing. So close was the gun to the propeller, in fact, that its gas cylinder was a few inches only behind the tips.

By about the same time (early summer 1917) the armament of N5 had been augmented by eight Le Prieur rockets, as described in the chapter on the Pup. The bottom wings and ailerons were then protected from the rocket efflux.

The factors which seem to have deferred inauguration of 2F.1 production until the summer of 1917 were apparently indecision or other difficulties attending the fitting of floats (which were, in fact, never adopted—partly, perhaps, because the improved Babies were performing quite well as water-based maids-of-all-work); the experimental and development work on equipment and armament, already described; and the Navy's obvious attachment to the F.1, with its greater wing area for ultra-short take-offs—and of a type, moreover, which they already possessed. But even with its smaller wing area than the F.1's—roughly 10 sq ft (0.93 sq m) less—and especially when flying below its maximum permissible weight (which varied considerably according to military load) the 2F.1 was a far deadlier 'ship's aeroplane' than was the Pup—and, as Lieut S. D. Culley was to show when using special light armament, as later noted—a mighty hunter of Zeppelins.

Sopwith themselves undertook the first production contract (N6600—N6649) and by the autumn of 1917 examples were leaving the Kingston works (N6603, for instance, was at the Isle of Grain early in November). Sopwith's own fifty were stoutly backed (as might have been expected, having regard to development and production of 'ship's Pups') by a hundred from Beardmore—which company, in fact, received Sopwith-built N6618 very early in 1918, before this aircraft joined the Navy at Rosyth, a short distance away on the other side of Scotland. Beardmore themselves flew their own first 2F.1 very shortly afterwards (N6750, 20 February) and later had a contract for another fifty, though 30 more still were cancelled. By October 1918, in any case, a total of 129 Sopwith 2F.1s were in service. Other constructors were Arrol, Johnston of Dumfries (who did, in fact, erect the last ten of Beardmore's second batch); Hooper; and Clayton & Shuttleworth; but contracts with Fairey (who eventually built the true 2F.1 replacement, the Flycatcher, in 1922), also with Pegler and with Sage the noted shopfitters, were cancelled with the ending of the war.

Certainly it must be noted that the 2F.1 Camel, with Bentley B.R.1 engine, was still regarded as a standard RAF type as late as 1921 (as was the Clerget-engined ship's version of the 1½ Strutter)—which was certainly not the case with the far more widely built and even more famous F.1 Camel.

Although one of the most significant operations by 2F.1 Camels was the strike on the Tondern Zeppelin sheds, using relatively heavy bombs, the 2F.1 was essentially an intercepter and destroyer of aeroplanes and airships in their own element—to which end the type was flown from the rotating turret-platforms of capital ships, from the fixed foredeck platforms of cruisers, from the full-scale flying-off decks of aircraft-carriers, from special destroyer-towed lighters, and even from special rigs on airships.

The three last-named schemes must be considered in particular. Thus, concerning 'real' aircraft-carriers it must be recorded that in one month (June 1918) *Furious* (already familiar in associations with the Pup and Pup-carrying light cruisers) flew-off 2F.1s to attack German seaplanes, as did similar aircraft from the light cruisers *Sydney*, *Melbourne* and *Galatea*. One seaplane was forced down on to the water by a 2F.1 from *Furious*, and (such was Naval/air co-operation) was eventually finished-off by a destroyer. But with *Furious* and 2F.1s, in any case, we most fittingly associate not the interception of enemy aircraft in the air but a classic strike at one of their bases. This operation involved the indispensible elements of sea-and-air collaboration, surprise, special training, swift and telling blows—and audacity and skill of high order. The date was 19 July, 1918; the target was the airship base at Tondern; the Sopwith aircraft involved in the strike were six of seven specially prepared 2F.1s, with pilots specially trained (in two flights) in the techniques of low-

On battleships—especially those with 15 in guns—the 2F.1 was sometimes blasted by the firing of the great guns themselves. Less so in a cruiser—illustrated here by N6779 aboard *Calliope*, with the Camel sitting proud on its platform above the forward 6 in gun. Note that the aeroplane has fully as much extra hamper—in the form of battens and strops, to keep things Sopwith-shape—as the cruiser has herself; and the man in the foreground suggests that the whole affair should be told to the Marines.

bombing with 50 lb bombs—two of which projectiles were carried by each 2F.1; the result was the destruction of Zeppelins L54 and L60 in one shed and damage to another shed. Only two of the Camels managed to return to the Naval force which had conveyed them from Rosyth; three landed in Denmark; one ditched fatally.

For a special note on 50 lb bombs the reader is referred to the chapter 'Folder Seaplane Type 807'. Having regard, however, to the persistent affirmation—apparently based on a remark by Marshal of the RAF Sir William Dickson, who was one of the pilots engaged (then holding the RAF rank of captain) that the bombs were 'specially made 60 lb Coopers', one would reaffirm that the 20 lb Cooper bomb was a common Camel weapon but would add that the Cooper delay-action nose fuse might be applied to bombs of other types. This fuse itself weighed 2 lb, whereas the 20 lb Cooper bomb (which actually weighed nearer 25 lb) contained only 4 lb of Amatol explosive.

'Dickson' (whatever rank we now accord him) was one of the two pilots who, as earlier remarked, 'managed to return to the Naval force which had conveyed them from Rosyth'—this by way of stressing that these stalwarts did *not* land back aboard *Furious* but ditched near a destroyer and were plucked from the deep.

All such considerations aside, the Tondern raid by 2F.1 Camels was the first true carrier-borne strike—and one that was not only successful but historic also. As for details concerning bombs etc, one would only recall Sir William Dickson's virtual apology, after his appointment as Chief of the Air Staff in 1953, for his possibly disappointing personal appearance. To which one's reaction was: did it *really*

Far more than mere 'dignity and impudence' poses, these pictures show (top) N7120 with standard armament, and a Handley Page V/1500 for company and scale; and (lower) twin Vickers guns and a jettisonable steel-tube landing gear, with the Felixstowe Fury flying-boat obliging in a like capacity. Note how distinct are the features mentioned, tiny though the Camels are.

matter—especially with a record like Sir William's, which included service at the Isle of Grain and award to him of the Dunning Cup in 1921 for 'flying from aircraft carriers in experimental machines'. (In any case, concerning the Tondern bombs, we indubitably have the equation 2 × 60 lb—the latter being the 'Dickson' figure—equalling 112 lb. Which is precisely the heaviest bomb-weight quoted in the F.1 Camel chapter!).

Having thus digressed to consider bombs, we must further pursue inquiry into 2F.1 activities aboard what we have earlier termed 'real' aircraft-carriers, including among these not only the big-and-famous *Furious*, *Argus* and *Eagle* but the smaller and less familiar *Vindictive* (a conversion of the light cruiser *Cavendish*, with flying-off deck forward and landing deck aft) and *Pegasus*, which had been laid down as a Great Eastern Railway packet, and is not to be confused with the *Ark Royal* that was renamed *Pegasus* in 1934 and which rendered possibly untold service to the development of Naval flying. For special trials in 1919 with deck-arrester gear *Argus*, with her flush deck, was the chosen vessel, and in the summer of 1920 *Eagle* was operating the 2F.1 Camel type with three large arrester-gear clips on the landing gear spreader bar, a strut-braced propeller guard forward and a quick-release bridle further aft. For Fleet co-operation work with wireless, a few F.1s, as well as 2F.1s, were modified. Indeed, both forms of the production Camel served the Navy afloat—an F.1 with jettisonable steel-tube landing gear being carried on a lighter, and a 2F.1 for similar employment having an F.1's armament of two Vickers guns. Three 2F.1s from *Vindictive* went to Riga in November 1919 to serve

The Royal Marines having now entered the picture, their motto *'Per mare, per terram'* may be borrowed for the 2F.1 Camel—especially as shown in these two views: Top, with skid landing gear for Samson's lighter trial, which ended (see text) very much *per mare*; lower, a 'split' specimen, distinctly *per terram*.

with the Latvian forces, and, as already intimated, the 'Ship's Camel' was still regarded as a standard RAF type as late as 1921, though its lower power and smaller wing area clearly placed it at some disadvantage compared with the 'interim' Nieuport Nightjar (even though that type, which entered service in 1922, was 'derated' from a 320 hp Dragonfly radial to a 230 hp Bentley B.R.2 rotary).

From 'real' aircraft-carriers we turn to destroyer-towed lighters—a form of aircraft-carrier particularly associated with the very large Felixstowe F.2A flying-boat and the very small 2F.1 Camel—though for the Camel not merely as a carrier, but as a flying-off platform also. With the development and operating techniques of towed lighters one name especially is linked—that of Lieut Col C. R. Samson, for that rank 'Sammy' held in August 1918 when he set down some considerations concerning the use of 'Camel Aeroplanes' from lighters. With the 150 hp B.R.1 engine, Samson explained, the Camel could be armed with two synchronised Vickers guns; or one synchronised Vickers gun and one Lewis gun which could be 'moved in elevation' (evidently meaning the standard 2F.1 scheme involving the Admiralty Top Plane mounting), or 'two Lewis guns on the top plane'. Clearly, this last-mentioned scheme was not only light (the guns being fixed, and their magazines unchangeable in flight) but allowed the firing with impunity above the Camel's propeller of 'special' (anti-Zeppelin) ammunition.

It was Samson himself who made the first Camel/lighter test, on 30 May, 1918, using N6623 fitted with skids like those of a Pup, supposedly to run in guide-troughs. The skids got fouled-up; 'Sammy' got a ducking and was run-over by the lighter into the bargain. He survived until 1931.

To Lieut S. D. Culley really fell the honour of vindicating the lighter scheme, and in this wise:

A reminder that the 2F.1 Camel was developed essentially for Naval use with seaborne forces, and epitomising also the glorious victory by Lieut Stuart Culley over Zepplin L53 just before the Armistice, as described in the text. (A reminder also that although the 2F.1 was operated from 'real' aircraft-carriers it was not a true deck-landing aircraft, as were its successors, though like other Sopwiths, it helped to show the way).

A final reminder of the truth of the foregoing caption—and illustrating also salient recognition characteristics of the standard 2F.1 version of the Camel (high-set Lewis gun, and centre section less splayed than on the F.1).

On the evening of 10 August, 1918, the Harwich Light Cruiser Force had sailed for a special operation in the Heligoland Bight. The composition of this force was appropriately 'special', consisting as it did of four light cruisers, eight destroyers, and six attendant F.2A flying-boats (three on towed lighters, the others operating from Great Yarmouth)—the cruisers carrying six motor torpedo-boats for swift and potent strikes, and one of the destroyers—HMS *Redoubt*—towing yet another lighter. Upon this lighter was a 2F.1 Camel (Beardmore-built N6812) for which the pilot was Lieut Culley.

Early on the morning of 11 August the ships transmitted 'spoof' wireless signals to proclaim their presence and lure out Zeppelins—a class of craft which had hitherto been shadowing the Harwich Force, to its discomfiture and embarrassment. Off Terschelling the motor torpedo-boats were lowered into the water and scudded off towards the River Ems. The three lighter-borne flying-boats could not be operated by reason of their overloaded state, a long swell and lack of wind; but the three from Great Yarmouth appeared and went to search for the now-overdue motor torpedo-boats (which had, in fact, met trouble). From one of the flying-boats, however, Maj Robert Leckie saw a Zeppelin approaching at 15,000 ft, and duly warned the surface force by visual means. Leckie was told that the Camel was being prepared for launching, and that he himself was to go home.

At 08.25 the Zeppelin L53 was spotted by the ships; Vice-Admiral Sir Reginald Tyrwhitt, who was commanding the force, ordered the destroyers to lay a smoke-screen (and thus to lure the Zeppelin further); and L53 nosed on from cloud to cloud.

Working up to thirty knots and heading into wind *Redoubt* now had the situation in hand—with her lighter in tow, Culley in the Camel's cockpit, and the sea-drenched handling crew at their perilous posts. The engine having been swung into life (at particular peril) and the quick-release holding-down shackle on the landing gear spreader bar operated, the helpers made themselves scarce by lying flat, and the Camel was up and away in a mere five feet.

Culley climbed, but lost sight of the Zeppelin, though he did see haze and ice. At length through the murk the Camel rose, though slowly, for the height was nearing ceiling—and then the Zeppelin, at 19,000 ft, began to head for home. The Camel was now at just about its ceiling, but from 300 ft below, Culley pulled back on the stick and fired at the airship above. After getting off only seven rounds one of his two fixed Lewis guns (for which there was an Aldis sight, offset to starboard) jammed; but not the second gun—and L53 burst into flame and broke in two. The Camel fell out of the action, and Culley felt the heat from the last German airship to be shot down in the war then soon to finish.

As the sailors of the Harwich Force sent up their cheers and L53 hit the water, Culley began to find himself not only in navigation trouble but in engine trouble too, as the main tank emptied; though with his emergency petrol supply itself running out he was able to ditch near *Redoubt*.

Lighters-and-aircraft engaging in battle with lighter-than-air craft had another symbolism for the RAF, quite aside from Culley's exploit—and one hardly less exciting, though far less known. For this last reason the writer has particular pleasure in acknowledging the researches of Philip Jarrett, who draws attention not only to the antiquity of the idea of lighter-than-air craft launching ones heavier than air, but avers that plans to experiment with the attachment of a Camel to, and its release from, the airship HMA R23 had been formulated before July 1918—referring in particular to photographs taken on or before 25 July and showing 2F.1 Camel N6622 suspended beneath the then-obsolete R23 in a Pulham shed. The Camel was attached by the 'Little-Crook anchoring gear', and one of the co-inventors (Maj I. C. Little, RAF) reported at the time the photographs were taken that the airship's forward engines were run at full speed, the slipstream passing over the Camel, which then ran its own B.R.1 engine. This rig was probably only used for 'dry runs' and not for airborne-dropping trials.

In the September following, Brigadier-General E. M. Maitland, Superintendent of Airships, listed the functions of the airship-carried fighter as attacks against hostile territory (the machines returning to neutral territory) and defence of the airship against hostile attack; while there was still a scheme—propounded by Lieut-Col F. L. M. Boothby—in which the attached aircraft doubled as the principal motive power of the airship.

By 3 October, 1918, R23 had flown for three hours with a Camel (almost certainly the Beardmore-built 2F.1 numbered N6814) in position, and on 3 November, 1918, the F.1 Camel D8250 was air-tested beneath R23, the (unpiloted) Camel's rudder being secured in the neutral position and the elevators lashed about 2 deg down. The airship rose to 500 ft, turned into wind, and the tail support raised clear of the Camel, which was then released (turning over on landing). On 6 November—possibly with 'Ship's Camel' N6814 again—R23 ascended with Lieut R. E. Keys in the fighter's cockpit, and on 15 November Keys reported:

'On Nov 6th 1918 I carried out experiment of dropping a Camel from a Rigid Airship.

'My engine was started before leaving the ground and I kept it running at 500 revs. The Rigid Airship attained the height of 3000 ft. during which time the aeroplane was quite rigid and satisfactory. The saddle fitting over the after end of the fuselage was then removed. I gave the signal to be released and on being released the machine dropped about 10 ft and picked up her glide and was immediately under control. The machine showed no tendency whatever to get out of control. The speed of the Rigid Airship was 30 mph.

'While being released from the Airship I had my controls neutral. I thoroughly tested the machine and found she had in no way suffered from being suspended from the Airship. I consider this method of attachment is entirely satisfactory.'

An element extra to those named in an earlier caption is implicit here, with N6814 snugly stowed beneath the airship R23 for tests connected with air-launching.

It matters little whether the particular fighter concerned was a 2F.1 or an F.1; nor does the inconsistent spelling of Key's name with one or two 'e's—though it was evidently not 'Keyes' as typed on the report. What does, however, matter is that here was one Camel at least which 'showed no tendency whatever to get out of control' in distinctly unusual circumstances (and which, be it noted, remained rigid under its Rigid).

Although 2F.1 Camel N7352 was fitted in February 1921 with an 'Airship and Overhead Wire Landing Gear for Small Craft' (following early-1920 tests with a Camel having an overwing cable-engaging hook, the patterns whereof were varied) no airship trials were involved. N7352, nevertheless, was fitted with a strikingly elaborate propeller-guard, had an entirely stripped centre section for optimum pilot-view, and—with its hook above—presented a singular appearance.

But these and similar schemes were not Sopwith developments, and are noted here in token of that company's wide involvement in curious and stringent trials—if not beyond the call of duty, then beyond a clause of contract. More fittingly, perhaps, this present chapter may conclude as follows:

Quite early in the account of the F.1 Camel it was remarked that the type 'owed something to the Baby as well as to the Pup', while in this present chapter on the 2F.1 the name 'Improved Baby' makes a similarly early appearance. Thus one recalls especially vividly the remarks of a Sopwith Schneider pilot serving in the RNAS (no less applicable to the Baby—and, in fact, even more so) to this effect; 'The duties of the Schneider Flight (at Yarmouth) were extraordinary in their variety. It was the machine for the lover of solitude and independence and a wandering kind of life. The Schneider was a sort of detective, exposing all mysteries, such as whales mistaken for submarines, streaks of oil, and rescuing colleagues in difficulties. Any wild rumour—out went the Schneider to investigate . . .'

How opportune, now, to re-emphasise the relationships between all the Sopwiths by appending this recollection by a Camel pilot (F.1 or 2F.1 seems immaterial) of a certain salty sortie: 'One day in May I was ordered to the Shipwash patrol on a Camel. When almost over this light vessel at a height of approximately 3,000 feet, I saw what appeared to be a submarine on the surface, or rather, going along with only the periscope above water. The visibility was poor, and the submarine, when challenged later, did nothing but proceed in a "suspicious manner".

'After an interval, therefore, two dives were made and several bursts given from the two Vickers guns. Whereupon, the supposed submarine disappeared *downwards*, taking cover supposedly. In great glee I turned inland and landed . . . I was suitably celebrating my success when I was told by the orderly, "You're wanted on the 'phone, Sir".

'My submarine was a whale.'

So, to round off these Camel chapters—from sending down other aeroplanes, and bombs of quite surprising sorts, to being hoisted-up by an airship and taken in by a whale—one would only affirm that if Sopwith's Camel never actually passed through the eye of a needle, then it must have come pretty close even to that attainment. Indeed, as someone once declared, it became so famous that the Arabs named an animal after it.

As for the 2F.1, now under particular scrutiny, production orders were:
Sopwith N6600—N6649.
Arrol, Johnston N7350—N7389 (some not delivered, but company erected last ten of Beardmore's second order).
Beardmore N6750—N6849; N7100—N7149.
Clayton & Shuttleworth N8180—N8229 (deliveries halved).
Hooper N8130—N8179.

N.B. Although the U S Navy may have had some 2F.1 Camels, the six known Camels used by that Service—sometimes from gun-turret platforms on battleships and sometimes also having external flotation bags and a hydrovane—were seemingly of F.1 type. On 9 March, 1919, a Camel made the first turret-platform take-off from a U S battleship (*Texas*). Paradoxically, the pilot's name was McDonnell!

2F.1 Camel

(150 hp B.R.1)

Span 26 ft 11 in (8.2 m); length 18 ft 9 in (5.7 m); wing area 221 sq ft (20.5 sq m). Empty weight 1,036 lb (470 kg); maximum weight 1,530 lb (694 kg). Maximum speed at 10,000 ft (3,050 m) 122 mph (196 km/h); maximum speed at 15,000 ft (4,570 m) 117 mph (188 km/h); climb to 10,000 ft (3,050 m) 11 min 30 sec; climb to 15,000 ft (4,570 m) 25 min; service ceiling 17,300 ft (5,270 m).

Excellent though it is in quality, this Sopwith photograph of the Bee may have been lightly retouched, especially (though by no means drastically) in respect of the landing gear, and in having the Sopwith number 46 inscribed upon it. Quite unretouched is the proud smile on the face of the gentleman—probably Harry Hawker.

Bee

'You could hang it on your watchchain' was the assurance vouchsafed to the present writer by one of the old Sopwith hands (and that was the limit of the information) when seeking some enlightenment concerning the Bee—the smallest, and in some ways one of the most intriguing, of all the company's products. Woven round the little aeroplane by other old hands present on the same occasion were the usual stories (so usual that one became cautious of accepting them all at face value) of its having been 'chalked out on the shop floor by Harry Hawker to his own ideas' etc, etc; though the contention that Hawker was indeed the moving spirit behind the design was so sustained, and so clearly plausible, as to admit but little doubt concerning his personal involvement. Much the same premise (at least that the machine was Hawker's favourite 'runabout' and/or aerobatic-display aircraft) would be eloquently supported by the smile on the face of the airman seen, hand-on-hip/hand-on-wing, in a photograph herewith— if, indeed, the man is Hawker himself. Close scrutiny of the original print suggests that this is so, though the print is lacking one of the customary lettered-on-Sopwith captions, and the inked inscription on the back, beyond repeating the number 46, proclaims the subject as 'Single Seater Biplane 50 h.p.'.

Of the horsepower and the engine more in time. Meanwhile the airframe seems to merit more attention than is usually accorded it: that is, the mere notation that it embodied Pup components (the landing gear especially) and that a curious feature of the tiny wings was that they were arranged to warp for lateral control instead of having ailerons.

The feature, surely, that is even more striking is the form of centre section, which had a cut-out not only for the pilot's head, but another at the trailing edge. Thus it corresponded closely not only with the Dolphin (the tail—with its horn-balanced rudder—was likewise Dolphin-style), but with the Bulldog, Snail and Buffalo. Less close, but nevertheless discernible, was a relationship with the Snipe, Salamander and Dragon, though in these named instances the forward cut-out (or more

In this view of the Bee the very narrow gap and broad chord are accentuated. As in the preceding picture, the characteristic Sopwith 'engine turning' on the cowling for the 50 hp Gnome rotary is in evidence.

descriptively aperture) had the function of improving the pilot's view rather than that of receiving his head. A similar arrangement was a feature of the Westland Wagtail, concerning which fighter there will be more to say.

Though it is difficult to determine a date for the Bee's construction, one would place this fairly confidently in 1917—a year which brings one sharply to consider why such a venerable powerplant as a 50 hp Gnome should have been selected for any new aeroplane of the period. True, though the '50 Gnome' might have been more at home in a pre-1914 'boxkite', it was still the unit that was being offered (unsuccessfully) post-Armistice by The London and Provincial Aviation Co with their Type No. 4 trainer/'runabout' (or tourer); and that the Sopwith Bee was used as, if not solely intended as, a 'runabout' has already been intimated. The fitting of a 50 hp Gnome might, in any case, have been inherited from Hawker's previous runabout the SL.T.B.P.; but whatever the facts of the matter there are indications—other than the form of centre section mentioned—that the Bee was more than a vehicle for transporting Harry Hawker from A to B, whether for test-flying assignments or aerobatic displays. One suggests, rather, that it was regarded as an experimental and development 'hack'—exactly as the Hawker company's famous Fury G-ABSE was to serve in later years.

Failing substantiation, this suggestion commands some further comment—the 'fighter-style' opened-up centre section having already been accorded special notice. Now, how to reconcile this 'advanced' feature with such 'old-fashioned' ones as warping wings and a 50 hp Gnome rotary? One fact is sure: Though Sopwith had seemingly 'gone over' to ailerons from warping wings before war came, it is clear that the company's policy for trying everything at least once (one interpretation of which is 'playing safe') was never abandoned. This very policy, indeed, was carried forward well into the Hawker years, the clearest manifestation being the construction of certain biplane types (Snipe, Bulldog, Woodcock, Hawfinch, Hoopoe) both with single-bay and two-bay wing cellules. As for the use of wing-warping on the Bee, this could well have been for reasons of simplicity—though this latter was a virtue that did not follow automatically. Equally it could have been reasoned (so short was the span) that a high rate of roll was more or less inherent in the design. At the same time, rolling and yawing moments were considerations of high importance to a company like Sopwith, which made a speciality of highly manoeuvrable aeroplanes; and comparative data (as already instanced by the single-bay/two-bay reference) were an ever-present demand in the Kingston design office.

Aside from the ultra-short span and the wing-warping there was yet another feature of the Bee's wings that calls for remark, this being the narrowness of the gap—approximating to the chord of the wings themselves. Curious (especially so having regard to the aperture in the Bee's centre section) is the use of the term 'gap-chord' in the following—possibly corrupt—summary of an official report, issued very early in 1918: '*On the Effect of Cutting a Hole in the Top Plane of a Biplane.*—Results of tests on the loss of lift of an R.A.F.6 biplane without fuselage, and with a gap-chord, due to the cutting of a hole in the top plane, the tests being carried out for analysis of incidence varying from 6 deg. to plus 20 deg., and the results tabulated and plotted.'

Although the wing section, or sections, tried on the Bee are not known, it may be noted that 'R.A.F.6' was a low-camber wing-section, and one that was instanced as such for several years after the Armistice.

As for the Bee's ancient engine, it is permissible to speculate on the possible intention of installing some more modern unit when this became available—a possibility sustained by reasonable suppositions that the Bee may not only have been comparable with the officially-ordered 'Kitten' types (Grain, Eastchurch and James V. Martin) but in some way associated also with the 'Aerial Target' programme (see chapter on 'A.T. and Sparrow'). The nominal power of 50 hp would certainly approximate to that expected from a geared version of the A.B.C. Gnat (45 hp), while higher up the prospective A.B.C. scale were the 60 hp Gadfly and the 120 hp Mosquito. And might there not have been an engine called Bee?—for it must not be overlooked that the A.B.C. Wasp-engined Westland Wagtail was originally called the Wasp, and the Dragonfly-powered Sopwith Snipe development was the Dragon.

Though the seemingly well substantiated experimental fitment of a single Vickers gun would evidently have posed installation problems, the nature of the installation may have been unconventional—otherwise, indeed, the pilot's field of view might have been calamitously compromised. In the armament context, moreover, we may note that the Sopwith-Kauper mechanical synchronising gear outlived all others of its species, and although first made in 1916, continued in development and variation well into 1917, and perhaps even later—the Americans, for instance, having scant regard for the Constantinesco hydraulic system, and favouring instead their own Nelson gear (though Constantinesco also designed a mechanical scheme). In any case, if synchronising gear of any kind (other, perhaps, than electrical) were indeed fitted to the Gnome-powered Bee, some fairly extensive engine modifications would probably have been necessary.

Speculative though much of the foregoing may be, the mere dimensions of the little aeroplane, as given hereafter, are eloquent in themselves. Thus one muses finally on the name 'Tadpole' that was sometimes applied to this same machine.

Bee

(50 hp Gnome)

Span 16 ft 3 in (4.9 m); length 14 ft 3 in (4.3 m).

Certainly one of the earliest and rarest views of the B.1 Bomber is the frontal aspect here shown, bearing the Sopwith caption: 'S.140—Sopwith 'Bomber'. 200 hp Hispano Suiza—1917'.

B.1 and Derivatives

French interest and influence were both apparent by early 1917 in a large single-engined single-seat bomber, intended to succeed the special version of the 1½ Strutter which had been characterised (as already described) by internal bomb stowage behind the single seat. That there should accordingly be a distinct relationship between the wholly new B.1 Bomber (as the type was generally named) and the T.1 Cuckoo torpedo-carrier was not unexpected—the bomber requiring its large wing area to lift a relatively small load to high altitude, with long range and security as prime considerations, and the torpedo-carrier needing it to lift its heavier war-load from an aircraft-carrier's deck and drop it from very low level. (A postwar attempt to combine both the high-level bombing and the low-level torpedo-delivery capabilities in a single type—exemplified by the Hawker Harrier of 1927—proved a notable failure, even though deck-operation was not a requirement).

For the new Sopwith single-seat bomber (or Bomber) the type-designations B.1 and B.2 were both utilised, and as far as possible the significance of these will later be explained. First, however, it must be noted that interest in a two-seat carrier-borne Government-adapted version called the P.V. N50 Grain Griffin was such that this quite widely differing type came to assume a distinct importance and identity of its own; and of this type likewise more later.

The first Sopwith B.1 Bomber was built under Licence No. 6, was initially test-flown at Brooklands early in April 1917 and—such was the future contemplated for the type by both the French and British Governments—that it was ferried to Dunkirk for a joint assessment as early as mid-May. By that time the aircraft had been officially declared to be tail-heavy with full bomb load and nose-heavy when light, in spite of full tailplane adjustment. Further, it was 'tiring to fly'; yet controllability in the air and on the ground were reckoned 'very good'.

In general the airframe was similar to that of the first Cuckoo, the most notable differences being in the cockpit arrangement and landing gear, for in both cases the engine was a 200 hp Hispano-Suiza with a circular frontal radiator. This means that

Possibly taken on the same occasion as the preceding front view (though the propeller is differently angled) this picture of the original B.1 gives the false impression—due to an optical illusion—that the upper-starboard aileron has a balancing surface.

the equal-span non-folding wings were of two-bay construction, with cable-connected ailerons on all four panels, though the pilot sat in the middle of the four 'centre-section' struts (quotation-marks are explained by the top wing being in a single piece) whence he had a good forward and downward view for target-sighting and bomb-aiming; and the landing gear was of the familiar Sopwith V-type. Thus, instead of being of the new and advanced wholly divided form as on the Cuckoo, the gear was 'split' only in the sense that each wheel was on a half-axle of steel tube, this assembly moving vertically in guides against the tension of coiled rubber cord at the apex of each of the two 'Vs'. The tail unit appeared to be identical with that of the Cuckoo.

That the foregoing description applies to a type called B.1 is sure—and that the B.1 designation is used in Sopwith captions to photographs illustrating two forms of a generally similar aircraft (the later one numbered B1496, and otherwise distinguished by external elevator-cables running from rocking-levers near the front of the bomb bay) is equally sure. No less certain, however, is that the designation B.2 was also applied to the second form mentioned—B1496—and neither letter/number combination can be identified with or linked with the Grain Griffin two-seater.

B1496, probably at Martlesham Heath, with external elevator-cables.

Clearly the B.1/B.2 Bomber (as it may here be conveniently styled) was something quite new in its class, especially because of its relatively large size—roughly that of a Bristol Fighter—its single seat and its armament, for in this last regard the interest lay not only in the bomb stowage and in the bombs themselves, but in respect of gunnery also.

The bomb bay was close behind the cockpit, its location being proclaimed by access panels that are clearly seen in photographs. Though the bay appears to have had provision for nine British 50 lb bombs, carried nose-up as in the D.H.9, bombs of French type were provided for, and—as witness an official report dated April 1917—were actually carried as a test-load. This last-mentioned load was contemporarily described as comprising twenty 28 lb 'Analyte' bombs, whereas a later report (May 1918)—quoting a bomb load of 560 lb without reference to the bombs themselves—doubtless involved the same projectile load. That the rearranged elevator controls earlier mentioned could well have had some connection with bomb-stowage may have been gathered from the remark, earlier in

Three superb studies of B1496 in the snow at Brooklands, the views being numbered by Sopwith S.151 ($\frac{3}{4}$ front), S153 (starboard side) and S.154 (opposite). The original captions read additionally: 'Sopwith Bomber—200 hp Hispano Suiza—Type B.1—Jan. 1918.'

this present account, that the external rocking-levers were 'near the front of the bomb bay.'

Two distinct types of bomb must now be considered in the order of their mention; thus, first, the 50 lb British pattern, or 'Bomb, H.E.R.L., 50 lb'. Designed primarily for vertical (nose-up) stowage, this amatol-filled bomb actually weighed, when fused, $49\frac{1}{8}$ lb. Its length was $28\frac{1}{4}$ in—well within the depth of the Sopwith bomber's fuselage—and its diameter was 7 in.

The second type of bomb was not only French, but was entirely different in conception. Alternatively (and apparently more correctly) called the anilite bomb—or liquid-anilite bomb—it required very careful handling on account of its sensitivity, for the explosive content was (according to one account) 80 per cent nitrogen peroxide and 20 per cent hydrocarbon, the mixing of these two substances taking place after the bomb had been released. Professor A. M. Low, who may well have possessed first-hand knowledge by virtue of his wartime activities, described this 'ingenious French bomb' (or one form of it) as follows: 'The explosive was actually manufactured during the flight of the bomb. The bomb had two separate compartments, one containing petrol and the other liquid nitric oxide. Pressure of the air on a small propeller in the bomb opened the two compartments immediately the missile was released and the two liquids flowed together. Nitric oxide is a violent oxidizing agent and the mixture formed an exceedingly sensitive explosive which detonated as the bomb struck. No detonator was required, the impact itself being sufficient.'

Certainly, at least one official British list of bombs includes the entries '10 kg anilite' and '20 kg experimental model'—the latter, by implication, likewise connoting 'anilite', while another British document alludes to tests with an anilite bomb of 'about 20 kg' weight. This last figure—representing about 44 lb—would approximate to the 'actual' weight of the 'Bomb, H.E.R.L., 50 lb', already quoted at 49⅛ lb.

But the armament interest of the new type of Sopwith bomber extended also to gunnery; for though it originally had no gun at all, the first example, as service-tested in bombing raids together with D.H.4s of the RNAS Fifth Wing at Dunkirk, was fitted with a synchronised Lewis gun on the centre line of the fuselage ahead of the cockpit. The type of synchronising gear employed is not known, though for the Lewis gun in particular the French Alkan system was devised in 1916.

Finally, a brief note on the closely-related two-seat reconnaissance aircraft, the P.V. N50 Grain Griffin, the development of which followed the delivery to Port Victoria of the Sopwith Bomber which had been flown to Dunkirk for assessment in its designated role. After close deliberations in October 1917, folding wings and wireless were installed in a modified example, numbered N50, and the addition of a hydrovane landing gear and a pillar-mounted swivelling bracket for a free Lewis gun behind the rear cockpit further proclaimed the new-found application. Drastic redesign of the whole aircraft was quickly found to be necessary, and the seven aircraft formally named Grain Griffin (N100–N106) were built accordingly. These were somewhat larger aeroplanes, powered by the Sunbeam Arab or Bentley B.R.2 engine; and though they still owed much to the basic Sopwith design, they were not true inmates of the 'zoo'. Certainly they would have done it little credit respecting handling, though during 1919 Griffins, together with Camels, 1½ Strutters and Short 184s were aboard HMS *Vindictive* (formerly *Cavendish*) in the Baltic on anti-Bolshevik operations.

B.1 Bomber

(200 hp Hispano-Suiza)

Span 38 ft 6 in (11.7 m); length 27 ft (8.2 m); wing area 460 sq ft (42.7 sq m). Empty weight 1,700 lb (770 kg); maximum weight 3,050 lb (1,380 kg). Maximum speed at 10,000 ft (3,050 m) 110 mph (177 km/h); maximum speed at 15,000 ft (4,570 m) 98.5 mph (159 km/h); climb to 10,000 ft (3,050 m) 16 min 25 sec; climb to 15,000 ft (4,570 m) 34 min 10 sec; service ceiling 17,000 ft (5,180 m).

N.B. Weight and performance data relate to the aircraft with Lang 5150 propeller. Tests were also made with a Lang 3280 propeller, the aircraft's maximum weight—with the same bomb load of 560 lb (254 kg)—then being given as 2,945 lb (1,335 kg), the service ceiling as 19,000 ft (5,790 m), the climb to 15,000 ft (4,570 m) as 29 min 36 sec, and the endurance as 3¾ hr. At least two different Hispano-Suiza engines were installed, and the greatest altitude attained (possibly the absolute ceiling without bombs) was 22,000 ft (6,700 m).

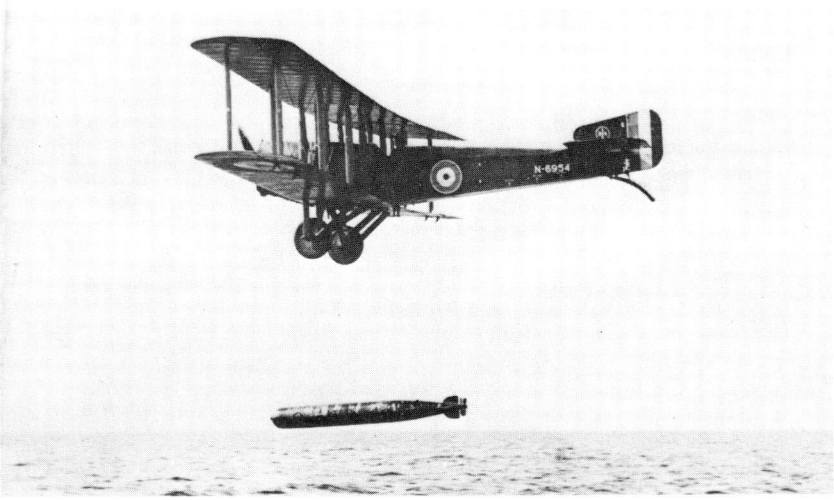

Cuckoo N6954 drops a Mk.IX torpedo, the contra-rotating propellers of which are clearly seen. Two torpedo-crutches are fitted, but the massive pistol-stop structure for the torpedo's nose, as seen in the later close-up study, is absent.

T.1. Cuckoo

To the first Sopwith torpedo-dropper (the Type C seaplane reviewed much earlier in this volume) the Cuckoo—the first torpedo-dropper in the world built to operate from the flying-deck of an aircraft-carrier—owed little or nothing. To the B.1 Bomber, however, it stood, if not precisely in debt, then in close relationship—though less so than first appearances suggest. To the Blackburn Aeroplane & Motor Co Ltd it owed a great deal, not only for the development of its specialised torpedo gear, but for its production in quantity. (Blackburn, indeed, did much the same job with the Cuckoo in the First World War as they were to do with the Fairey Swordfish in the Second, though in the latter instance they were much concerned with applications other than torpedo-dropping).

The relationship between the B.1 Bomber and the Sopwith T.1, as the Cuckoo was styled until given its name after the Armistice—one official rendering in 1918 being doubly parenthetical, viz 'Sopwith Torpedo Plane (200 h.p. Sunbeam Arab) (T.1)'— is of the kind which may confidently be classed as a 'chicken or egg?' affair. And here the 'egg' analogy is especially apt, for the very name 'Cuckoo' has sometimes been associated with the propensity of this aeroplane for 'laying its eggs in other people's nests'. More properly, perhaps, it could be said that this particular Sopwith 'bird' found its true incubation in the sometimes chilly Blackburn nests up in Yorkshire—a phenomenon thus expressed in 1919:

'Their [Blackburn's] first connection with the torpedo arose over the design of the G.P. (general purpose) seaplane in 1915–16. Although not purely a torpedo machine this twin-engined machine was arranged to carry a light torpedo as one of the numerous forms of armament to which it was susceptible, and with it the firm gained some experience of the fitting, carrying and dropping gear necessary to this class of work.

'When the Sopwith "Cuckoo" was produced, and had been tested and found satisfactory as far as the aeroplane part was concerned, it was decided that the work of putting it onto a production basis should be put into the hands of a firm conversant with the essential torpedo problems, and the work was therefore entrusted to the Blackburn firm, who with the assistance of experts from the R.N.A.S. carried out much work in connection with the torpedo-dropping and aiming gear, the silencing of the engine exhaust, and the fitting of warming gear to prevent freezing-up of the air passages in the torpedo itself—a phenomenon which occurred frequently at high altitudes and which effectively demoralised the torpedo by preventing the torpedo engine functioning.

'The problems of the "Cuckoo" being satisfactorily solved and the final pattern in production, Blackburns were asked to design a machine to carry a still heavier torpedo and at a higher speed and with a greater climb . . .'

The account then proceeded with reference to the Blackburn Blackburd and Kingfisher designs—though not, having regard to the date (1919) to their line of successors in the service of the Royal Navy after the Cuckoo, namely the Dart, Ripon, Baffin, Shark, Firebrand and Buccaneer.

Yet the Cuckoo was very much a *Sopwith* aeroplane, and its full technical and operational significance, rather than minutiae of development and deployment, must be our first concern, in order that to Sopwith may rightly go the credit for having designed the aeroplane described at the outset as 'the first torpedo-dropper in the world built to operate from the flying-deck of an aircraft-carrier'. The concept of such an aeroplane had, in fact, far earlier origins—possibly in the ideas of the Frenchman Clément Ader, thus expressed in 1909:

'Air is everywhere. We know how aeroplanes have to land on the ground. And on the sea? The ever-increasing power of the navy, the possibility of having to fight an ironclad, make the problem apparently impossible to solve. However, if we do not hope to succeed in finishing-off an ironclad straight away, we think it will be possible to damage it considerably at the first hit and even to sink it if attacked by a sufficient number of aeroplanes. We foresee . . . the use of the big torpedo of 100 to 200 kilos; but now we must work out how to use it against warships. If we had to attack an enemy squadron in French or allied waters within proximity of land, the operation would be easy, aeroplanes could land and load their torpedoes on areas near the coast. It would be different in the middle of the sea. Therefore, an aeroplane-carrying ship becomes indispensible.'

Having described a 'modern' form of aircraft-carrier, Ader declared that the 'torpedoes' used (which clearly, from his quoted weight, would not correspond to any form of locomotive torpedo current at the time, though the weight might conceivably have represented the explosive content) would have a device 'which will make it possible to have them explode under water at various depths . . .'

Irrespective of the precise form of weapon proposed, however, Ader had the notion of the aircraft-carrier much in the form that the Sopwith Cuckoo was to use. That a torpedo-dropping aircraft was itself a difficult proposition will have been gathered from the note that introduced the chapter on the Type C floatplane of 1914—to the effect that the Italian Guidoni in 1912 had been led, in his pioneering torpedo-dropping work, to resort to twin engines and hydrofoil floats. But if a floatplane was 'difficult', then how much more so was a wheeled aeroplane, capable not only of operating from the confines of a ship's deck but of being borne in numbers in its hull.

SOPWITH TORPEDO AEROPLANE.

Bearing as it does its original caption, this page of views from a 1918 handbook, depicting the first (Sopwith-built) aeroplane of its class, calls for no other—except to remark that the torpedo is a dummy and that the front view emphasises the high thrust line given by the geared Hispano-Suiza engine.

Against this background, then, must be viewed the familiar (though still historic) letter to T. O. M. Sopwith from Commodore Murray Sueter. Sent in October 1916, this document requested inquiry into the feasibility of torpedo-carrying aeroplanes having specified performance and lifting ability and 'probably' catapult-launched. This last-named technique was already old in concept, though new to Britain, for the US Navy had made experiments in 1912—these having stemmed, it seems, from interest expressed in 1911 by the Bureau of Ordnance in a catapult for launching aeroplanes 'somewhat in the manner of launching torpedoes' (!)

The British Naval officer who (with the possible exceptions of Sueter and Longmore) has been most closely associated with the Cuckoo is Sir David Beatty, who had opened the Battle of Jutland in 1916 and later succeeded Lord Jellicoe—'the victor of Jutland'— as commander of the Grand Fleet, which was responsible for guarding the shores of Britain. Nevertheless, one other officer quite intimately concerned was none other than 'Rutland of Jutland', whose name will be remembered from the chapter on the Pup and now forms yet another link between the Cuckoo and the greatest British Naval battle since Trafalgar. (One point of contention after Jutland, incidentally, was whether a fleet should turn away from or towards a massed torpedo-attack). Flight Commander Rutland had faith in ship-borne torpedo-dropping aircraft, and this faith was given expression in proposals jointly prepared by Rutland and Admiral Sir Herbert Richmond. Towards the end of 1917 Beatty brought to the Admiralty's attention the Richmond/Rutland considerations relating to an attack by torpedo aircraft on the German High Seas Fleet in the Wilhelmshaven area, the date then foreseen being the Spring of 1918, and eight specially adapted merchant ships being involved—though the state of torpedo-aircraft development was not considered by the Admiralty to warrant the provision and the preparation of the ships.

In the ultimate, as we shall see, the aircraft-carrier HMS *Argus* became the chosen instrument; but meanwhile what of the development of the 'Sopwith Torpedo Plane' (T.1), as it began to germinate from the Sueter/Sopwith letter of October 1916, already mentioned? Sueter himself was removed from the immediate scene by a posting to Italy only a few weeks later (January 1917) though the testimony of Wing Cdr (later Air Chief Marshal Sir Arthur) Longmore survives to this effect: that in February 1917 he visited the Sopwith works; that there he saw the T.1 fuselage suspended from the beams of one of the shops; and that the airframe was completed at his instigation—these facts according well with the recorded clearance of the T.1 by the makers' Experimental Department at Kingston-on-Thames on 6 June, 1917. By this time, let it be remembered, the closely related B.1 Bomber had already been delivered to Dunkirk for Service trials—a point of particular interest as the complementary use of bombs and torpedoes in attacking an enemy fleet (especially at base) was much in the Navy's mind. It must immediately be emphasised, however, that the T.1, or Cuckoo, was *not* a deck-landing 'torpedo-bomber', though it has sometimes been so styled; for the term torpedo bomber connoted an aircraft capable of carrying either a torpedo or a bomb-load, and was first introduced by the Blackburn Dart which, though still a single-seater like the Cuckoo, remained with the RAF from 1922 to 1933.

But although it could not carry bombs as an alternative load to its torpedo, the Cuckoo had one other characteristic in common with the Dart: it carried no guns, either for offence or defence. In the latter regard its faculty lay more in its good manoeuvrability (after release of torpedo) and in its structural strength for taking evasive action—perhaps at near sea-level—against fighters or anti-aircraft fire.

The first T.1 airframe having been cleared for flight-testing on 6 June, 1917, as

Seen with folded wings, the first Sopwith T.1 shows the 'split' front inner interplane struts; the slack fabric at the wing-fold joints; and a torpedo crutch and sling. Although all ailerons are lowered, this was not always the case with folded production-type Cuckoos.

earlier noted, it was quickly sent for official trials at the Isle of Grain, the engine then being a 200 hp water-cooled eight-cylinder vee-type Hispano-Suiza, installed as in the B.1 Bomber, with circular frontal radiator. In both cases the high-set propeller boss signified that the engine had reduction gearing (as was especially desirable in the T.1 for the lifting of a torpedo) and this may have been a factor in the allocation of Licence No. 6 to cover the construction of both types of aircraft.

Tests of the T.1 at the Isle of Grain during July proved successful, and an order for 100 machines of the type was placed (conceivably 'confirmed' might be more apt) on 16 August, 1917, with the Fairfield Shipbuilding and Engineering Co Ltd of Glasgow. Though inexperienced in aircraft construction this company had long experience with fast ships, often with special characteristics and sometimes having special armament. Interestingly enough (the Cuckoo's torpedo, like other British Naval patterns, being basically of Whitehead type) Sir Hiram Maxim recorded the following facts not long before his death in 1916:

'In the winter of 1884–5 there was a good deal of discussion among naval officers and others regarding the efficiency of the Whitehead torpedoes. It was claimed by many that it would be very difficult to hit a ship even at short range if the ship were in motion at the time. While this discussion was at its height, Bryce Douglas, a very clever and well-known Scotch engineer, came to see my gun at Hatton Garden. The very fact that I had made a gun that would load and fire itself more than ten times in a second seemed to make him believe that I might be of some use in other directions. He told me that he did not believe the Whitehead torpedo would be of any use in the Navy . . . He was in favour of increasing the size of the torpedo and of propelling it through the air instead of through the water. He believed that if a large torpedo were exploded within a few feet of the hull of a ship it would open a large hole which would let in more water than could be dealt with; and he asked me if I could produce a gun of very large bore for throwing aerial torpedoes. I told him that I could . . .

'Having designed the gun, I took the drawings up to Glasgow, where Bryce Douglas was employed as Chief Engineer at the Fairfield Shipbuilding Works. My drawings being approved of, Bryce Douglas made a model of the gun . . .'

Novel methods of launching torpedoes (of whatever denomination, for the term 'aerial torpedo' was confusingly applied in many contexts) were thus nothing new to Fairfield; and, in particular, they had been concerned with the highly secret, amazingly fast, but sadly doomed K-class submarines—one of which ultimately became the aircraft-carrying M.2!

It has already been emphasised, however, that Blackburn, not Fairfield, was the name pre-eminent in Cuckoo development and production; but in any case, back now from the north country to Kingston, Surrey (or establishments likewise in the south), with the Cuckoo-to-be still more or less in its original form, with Hispano-Suiza engine, and bearing the Service number N74, or simply the letter T.

The main differences between this aeroplane and the B.1 Bomber were in the fitting of three-bay wings (instead of two-bay) thus giving notably greater span and area; provision of wing-folding arrangements (the fold occurring at the innermost set of the three-a-side pairs of interplane struts, the foremost strut being 'split' accordingly); provision also of a special divided type of landing gear to permit operation with a torpedo—the outermost attachment-points of the landing gear being at the lower centre-section extremities, near the wing-fold point; and the placing of the pilot's cockpit further aft, precisely in line with the trailing edges of the unstaggered wings. Thus, compared with the Bomber, the pilot and projectile load changed places, though the torpedo could not be stowed internally, as were the companion-type's bombs.

The pilot's view, of course, was by no means ideal for torpedo work, having regard to the length of the nose (the same consideration applying to operation from a carrier's deck). On the other hand, the absence of upward view from the Bomber (the pilot being under the top wing) had led to the criticism that attack from above would be hard to avoid; whereas with the new cockpit position this handicap was mitigated, while avoiding action—or even straight flying—near sea-level would itself preclude the danger of fighter attack from below.

However unorthodox its purpose, N74 embodied no new structural features, the wooden framework being braced by steel wire and covered with fabric, except for the decking round the cockpit—which, in official notes later prepared, was quite understandably described as the 'top deck'.

After its initial trials at the Isle of Grain, N74 was returned to Sopwith for further work before being re-erected at Grain for further tests, especially at high all-up weights, beginning in January 1918. At about this time Hispano-Suiza engines were in heavy demand for the S.E.5a and the corps reconnaissance version of the Bristol Fighter, though in the event the latter had the Sunbeam Arab instead. Now, once again, the Sunbeam associations with Naval torpedo-dropping aircraft (already established by the Short '225' and '320') were renewed by the adoption of the Arab for the Sopwith T.1. This was not a very dramatic departure, for the Arab resembled the Hispano-Suiza not only in general outline, but even in bore and stroke (120 mm × 130 mm). It was made in both geared and direct-drive forms, and fitted nicely in the Sopwith torpedo-dropper, having a frontal radiator as formerly (for the torpedo pistol stop precluded an underslung arrangement as in the Arab-engined Bristol Scout F) though the radiator was now of inverted-horseshoe form, instead of round.

In February 1918 (at about the time when N74 was having an Arab fitted by Blackburn) two hundred and thirty T.1s were ordered from Blackburn as well as fifty from Pegler & Co Ltd of Doncaster (again, up in Yorkshire) though Pegler, like Fairfield, were inexperienced in aircraft construction, and part of their contract was taken over by Blackburn. Problems, numerous and intricate, were encountered, and, hardly surprisingly it was Blackburn—backed with a certain amount of indigenous construction, as well as 100-odd B.E.2cs and series-production of the Sopwith Baby as already recounted—who were first to deliver the new Sopwith torpedo-droppers in quantity. These deliveries began in May 1918, whereas the first Fairfield-built example emerged only in September and the first from Pegler in October. When the Armistice came on 11 November, 1918, orders totalled 350, of

Top, Arab-powered Cuckoo I with enlarged rudder; lower, Viper-powered Cuckoo II with extensive postwar modifications, notably long exhaust tail-pipes to warm the torpedo.

which just over ninety had been delivered. By August, in fact, Blackburn alone had finished eighty; so our opening laudation of that company was by no means out of order. Deliveries, however, were cut back well short of orders—for the Cuckoo, alas, was numbered among those aeroplanes that were 'just too late for the war'.

When, eventually, the type was given publicity, one typically effusive claim—perhaps officially inspired—was thus advanced: 'It was a type designed to replace destroyers and submarines in attacking enemy surface craft and would undoubtedly, but for the Armistice, have rendered brilliant service. Its career has only just begun.'

Effusive though the foregoing may have been, it was partly true at least, especially at the end—and more especially if the final 'Its career' be interpreted as the instigation and development of a long line of carrier-borne torpedo-dropping aircraft of which the T.1, or Cuckoo, was the archetype. The career of the Sopwith type itself was in fact brief, though its design had originated in late-1916 and it was not operational until some two years later—and even then largely by reason of sub-contracting and official experiment and development, with Sopwith playing a part by supplying drawings and having some hand in modifications.

After the initial Arab installation by Blackburn, already recorded (and not overlooking the fact that this company's G.P. seaplane, with torpedo potential, as mentioned at the outset, was Sunbeam-powered) N74 was delivered back to the Isle of Grain, where it used for further trials a special dummy deck and apparatus for measuring speed against distance. The most serious mishap seems to have been the shearing of a propeller shaft in the air (while the aircraft was carrying ballast in an under-fuselage container, in place of the torpedo) some damage thus resulting to the port lower wing.

Modifications to the Sunbeam Arab engine (which was especially prone to vibration) had been, and continued to be, frequent, and production of this engine fell far behind schedule. Although it has been described—especially in the Cuckoo context—as a heavier engine than the corresponding Hispano-Suiza, thus requiring tail-adjustment on the aircraft to compensate, the actual weight difference was not, apparently, very great; for though engine-weights differed widely in those times by reason of modification standard, materials, accessories and other factors, it can be said with some confidence that a typical Hispano-Suiza would weigh about 500 lb and an Arab about 530 lb. Special engine-bearers for the Arab, however, could have made an added contribution; radiator and propeller were other variables— and even torpedo gear could make a difference.

In any case, the airframe had an adjustable tailplane as a standard fitting; in which connection it was notified: 'The Tail Plane is braced on the under surface by two Bracing Wires, which extend from the Bottom Longerons to Front and Rear Spars respectively, and on the upper surface by two Bracing Wires, which extend from the Fin to the Front and Rear Spars respectively . . . The Tail Plane, which is adjustable, is set at the zero Angle of Incidence for trial flight. The Rear Spar is supported at the centre by a vertical telescopic fitting, which is fixed in the Fuselage just in front of the Sternpost. The telescopic fitting has an inner screwed spindle, which may be raised or lowered by means of a sprocket wheel in the Pilot's Cockpit, a cable from which passes round a drum on the spindle. To allow for movement of the Tail Plane, when the Rear Spar is raised or lowered by means of the telescopic fitting, the Front Spar is hinged by means of a cylindrical casting in bronze bearings, which are bolted to Top Transverse Strut No. 15.'

The four ailerons were operated by a wheel on the control column, though control-cable arrangements differed, those to the tail of production Cuckoos being wholly external. The tailskid was taller on production Cuckoos—and taller still when (later) fixed to a massive inverted pyramidal structure. Sometimes associated with this last-named feature were a larger rudder and an offset fin, though on the Arab-engined aircraft as first taken into service riggers were instructed to see that the fin and rudder were 'set straight and square' with the machine.

Having regard to the torpedo-dropping and deck-landing requirements it was the landing gear which called for special study—especially so as its steel-tube components were at first prone to fracture. An authentic description of the production-type landing gear follows:

'The Chassis consists of two Undercarriages which are identical. Each Undercarriage is formed of steel tubes welded and pinned together, and consists of two portions. One portion, which is in the shape of a V, is placed parallel to the fuselage, the End Lugs being bolted to the front and rear Spars of the Centre Section Lower Plane. The other portion has one End Lug bolted to the Bottom Longeron, and is bent so as to form the Axle for one wheel, the other end resting in the apex of the V. Shock Absorbers are formed by 28 feet of 15 mms. diameter elastic, which is given ten complete turns round the Axle and the apex of the V. The Steel Struts are stream-lined by means of wood fairing attached by metal clips, the whole being wrapped with fabric and then doped.'

By the same tokens (torpedo-dropping and deck operation) the wide-span, three-bay, folding wing cellule was no less a basic feature, and the nature of its assembly was thus described for the edification of technical personnel: 'To assemble the Centre Section, first attach the four Outer Centre Section Struts by their Bottom Sockets to the Centre Section Lower Planes. Before fitting the Centre Section Upper Plane into position place it upside down on trestles, and fix and split-pin all Centre Section Bracing Wires . . . The Hinges on the Centre Section Upper Plane should be

Early Blackburn installation of a Mk. IX torpedo on a Cuckoo—Arab-powered, as proclaimed by an engine-maker's plate reading in part 'Sunbeam Coatalen Aero Engine Arab I', with spaces for rpm and consumption figures. The torpedo sling is glimpsed between the landing gear struts. The little handle near the wicker seat controlled the torpedo depth-setting gear.

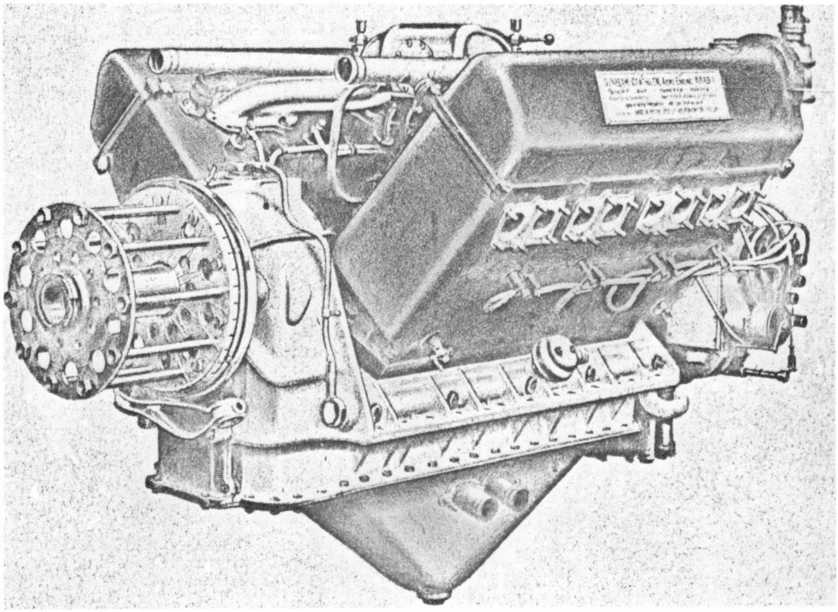

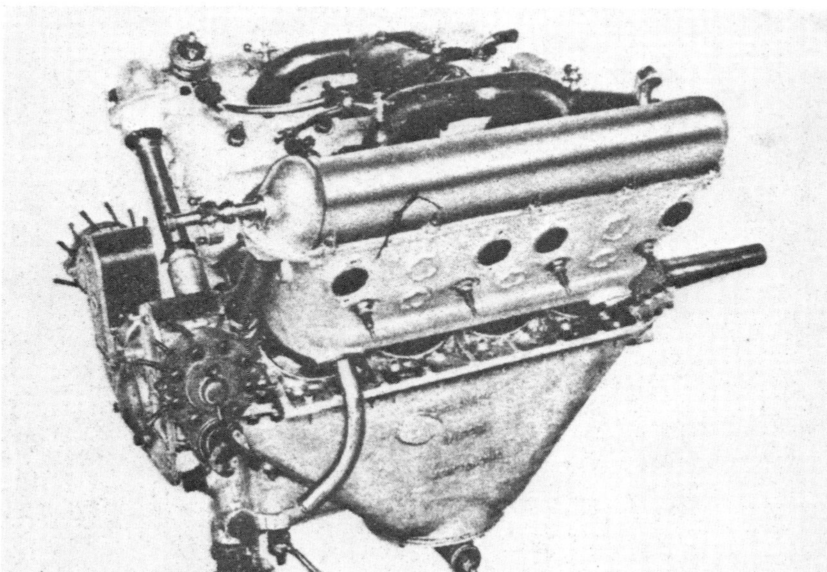

After the 200 hp Hispano-Suiza in the original Sopwith T.1, engines for the Cuckoo were the Sunbeam Arab ($\frac{3}{4}$ front) and the Wolseley Viper ($\frac{3}{4}$ rear).

vertically over the Hinges on the Centre Section Lower Planes. Adjust by the Side Bracing Wires and Incidence Wires, and check by dropping plumb lines from the Hinges on the Upper Plane . . . The Incidence is 3° throughout both Upper and Lower Centre Section Planes . . .

'The Main Planes are assembled with their Leading Edges on the ground. All Interplane Struts are fitted and the Incidence and Outer Flying Wires are loosely connected . . . The Main Planes are hinged at the Root of the Rear Spars. Fit the male hinge pieces at the ends of the Rear Spars into the forked hinge pieces at the ends of the Rear Spar of the Centre Section Lower Planes, and insert the hinge pins. The fork and eye attachments, by which the roots of the Front Spars are connected by the Front Spars of the Centre Section Lower Planes, are secured by inserting a locking pin through the Leading Edge. The locking pin is fixed in position by means of a wood screw . . . The Dihedral is $2\frac{1}{2}°$ for both Upper and Lower Main Planes . . . The Incidence is 3° . . . There is no "Wash in" or "Wash out" . . .'

On themes more mechanical, it must first be remarked that, as an alternative to the Sunbeam Arab engine, some aircraft had the Wolseley Viper—another unit of Hispano-Suiza vee-8 character. Post-Armistice Cuckoos so powered were designated Cuckoo Mk. II, the Arab version being Cuckoo Mk. I. Flotation bags, larger rudder, extra-large tailskid, and torpedo-warming exhaust tailpipes running beneath the fuselage were likewise characteristic of the Mk.II—also a folding pistol-stop for the torpedo. (Had this last not been the case, and a massive pylon structure been fitted under the nose, as seen in the accompanying close-up study of a Blackburn-built aircraft with torpedo slung, then the Cuckoo might have been seen in flight with an inverted pyramid at each end—the prominent tailskid-attachment having already been mentioned. A veritable mirage . . . or was it a case of 'hence the pyramids'?

In describing the Wolseley Viper as an engine of Hispano-Suiza character one has understated the matter, for more precisely it was a development—so much so in fact that on at least one occasion in 1921 the Cuckoo was officially listed as having an 'Hispano-Suiza Viper I'. (The engine series-number was at least correct).

Well before the Cuckoo was declared obsolete in April 1923, the biggest of all its engines had been tried experimentally—the twelve-cylinder Rolls-Royce Falcon, fitted in N7990 during 1919. Having four extra cylinders compared with its predecessors, and weighing the best part of 700 lb (317 kg) the Falcon must have posed some pretty installation problems, and was never standardised.

The standard postwar engine was, in fact, the Viper which, having no reduction gear, could be identified in a Cuckoo by a lower thrust-line. Postwar modifications and additions (including sometimes wireless) were numerous and possibly dangerous, and in 1920 RAF pilots were warned to dive 'modified' and Viper-engined Cuckoos only with half-empty petrol tanks.

The Cuckoo's 18-in Mk.IX torpedo, nominally weighing 1,000 lb, though typically nearer 1,100 lb, was a special lightweight (and short) aircraft pattern of the Whitehead type. In one early form of installation, on Blackburn-built aircraft at least, the massive inverted-pyramid pylon structure already mentioned apparently served the dual function of helping to steady the torpedo and of acting as a pistol-stop, though latterly the standard pistol-stop was of simple rearward-folding type. In any case, a steel crutch, or pair of crutches, semi-circular in form, were the principal means of steadying the torpedo, jointly with the flexible sling.

The torpedo sight was a small ring near the pilot's eye (fitted port and starboard, as shown in the instructional drawing) used in conjunction with a transverse row of beads laterally displaced by distances corresponding to the speed of a ship (for

The method of sighting the Cuckoo's torpedo (allowing for a ship's speed) is apparent here and is further referred to in the text. The basic method was retained in the RAF for many years.

example, 5, 10, 15 knots and up). Training aids such as the one shown ('Aerial Diagrams' as they were officially called) should certainly have proved helpful to the new torpedo-dropping pilots needed for the Cuckoos—pilots, incidentally, who were far more concerned with taking-off from a carrier's deck than landing back on it, for a land base was foreseen as 'journey's end' whenever possible (though the Cuckoo enjoyed a good 'ditching' reputation).

The 'working-up' and operational career of the Cuckoo can be summarised as follows:

Summer 1918, Blackburn-built aircraft to the Torpedo Aeroplane School, East Fortune (near Dunbar and North Berwick, Scotland); pilots thus trained were posted to an operational squadron which joined the Fleet on 7 October, 1918, and embarked (19 October) in HMS *Argus*—14,450 tons, flush-decked, and formerly—before conversion by Beardmore—the Italian liner *Conte Rosso* (at one period Cuckoos were aboard *Argus* together with Camel 2F.1s and Short 184 floatplanes); after the war Cuckoos served also in the carriers *Furious* and *Eagle*; the type briefly equipped Nos. 185, 186 and 210 Squadrons, and was used for development work in torpedo-carrying and dropping at Gosport, Hants (near HMS *Vernon*, the Royal Navy's torpedo 'school'); No.210 Squadron disbanded at Gosport in April 1923, when the Cuckoo was declared obsolete—even for coast defence from shore bases, though the RAF continued to develop aircraft for this function (*e.g.*, a version of the Hawker Horsley).

After the Armistice many Cuckoos on order were cancelled; but in 1921 six Viper-engined examples were taken to Japan by the British Air Mission to the Imperial Japanese Navy, the Mission being led by Col the Master of Sempill, who had joined the RFC in 1914, transferred to the RNAS and was at one time assisting in the testing of Robey-built Sopwith Gun Buses. In Japan, the former Sopwith designer Herbert Smith was responsible, at about this time, for the Mitsubishi Navy Type 10 (1MT1), adopted by that country as a standard carrier-borne torpedo-dropper. This, however, was a triplane, and was first flown (by former Camel-pilot Capt W. L. Jordan) in August 1922. Twenty examples were built with the British Napier Lion engine. More closely resembling the Cuckoo in external form—being a three-bay biplane—was Smith's Navy Type 13 (B1M), completed in 1923 and very extensively developed and operated by the Japanese. This type, however, was used for duties other than torpedo-dropping when operated from shipboard—a technique pioneered by the historic Sopwith Cuckoo.

Apart from N74, Sopwith themselves built no other Cuckoos, production orders being:

Blackburn N6900—N6929; N6950—N6999; N7150—N7199; N7980—N8079 (production of final batch apparently ended with N8011).
Fairfield N7000—N7099 (production ended with N7049).
Pegler N6900—N6949 (first part of contract taken over by Blackburn—see above). N6930 completed September 1918.

T.1 Torpedo Plane (N74)

(200 hp Hispano-Suiza)

Span 46 ft 9 in (14.2 m); length 28 ft 6 in (8.7 m); wing area 566 sq ft (52.6 sq m). Empty weight 1,840 lb (835 kg); maximum weight 3,370 lb (1,529 kg). Maximum speed at 10,000 ft (3,050 m) 100 mph (160 km/h); climb to 6,500 ft (1,980 m) 14.5 min; climb to 10,000 ft (3,050 m) 26 min; range 420 miles (676 km).

Cuckoo Mk.I

(Sunbeam Arab)

Span 45 ft 9 in (13.9 m); length 28 ft 6 in (8.7 m). Empty weight 2,199 lb (993 kg); maximum weight 3,883 lb (1,761 kg). Maximum speed at 2,000 ft (610 m) 103.5 mph (166 km/h); maximum speed at 10,000 ft (3,050 m) 98 mph (157 km/h); climb to 2,000 ft (610 m) 4 min; climb to 10,000 ft (3,050 m) 31 min; service ceiling 12,100 ft (3,960 m); endurance 4 hr.

N.B. Performance of the Cuckoo Mk.II (Wolseley Viper) with typical 'extras' was generally poor, and figures would not surpass those given above for the Mk.I. Tests with the Rolls-Royce Falcon, made in 1919, were disappointing (possibly in part because the propeller used was suited to a Bristol Fighter). Maximum weight was increased to 4,350 lb (1,970 kg), which, nevertheless, was still over 2,000 lb (910 kg) less than that of the Cuckoo's successor, the Blackburn Dart (Napier Lion engine). Speed with the Falcon was much the same as given above for the Cuckoo Mk.I.

The unmistakable first form of the Dolphin (with frontal radiator and deep fuselage to match). The original aircraft at Brooklands.

5F.1 Dolphin

In several respects the Dolphin was the most remarkable (though by no means the most renowned) of all the Sopwith fighters. The backward-staggered two-bay high-aspect-ratio wings, by which it was chiefly distinguished, conferred upon it (having an area roughly the same as the Snipe's, and appreciably more than the Camel's) an especially good high-altitude potential, or alternatively the ability to carry an exceptionally heavy armament. The pilot's view for combat was the prime consideration in providing backward stagger, while the invariable use of an efficient Hispano-Suiza water-cooled engine conferred additional advantages—not least for further development, with French encouragement for the 'Dauphin' being shown in the more exotic variants. (In describing Hispano-Suiza engines as 'efficient', which in a purely technical sense they surely were, one is not oblivious to the faults of some when fitted with reduction gear).

To Sopwith the significance of the Dolphin was not, however, merely technical,

View of the revised Dolphin, again with emphasis on the radiator.

for they themselves were given orders for well over a thousand of the 1,500 built before the Armistice, with production at Kingston succeeding that of Camels and preceding work on Snipes at Ham. Even so, as production went ahead the parent firm continued to set the pace for airframe and engine development. That only four squadrons were Dolphin-equipped reflects little diminution in the merits of the type, already emphasised. Though no Naval version is known to have existed, night-flying for Home Defence was an area of specialised application; and so successful and adaptable did the Dolphin prove as an 'all-round' fighter—and so full of promise did it remain in spite of its defects (real or imagined)—that American interest ran high, and French, perhaps, even higher. Curiously, the last Dolphins on active service equipped some Polish units in the fighting with the Russians during 1920, when Polish forces penetrated deep into the Ukraine. Less surprisingly, a single Dolphin only came upon the British Civil Register, though even this (G-EATC) was a demonstrator for Handley Page. The few two-seat trainer Dolphins were Service conversions.

Like so many other aeroplanes, the Dolphin suffered badly at the mouths of rumour-mongers, their tales of woe and terror being aggravated by recollections of the similarly back-staggered D.H.5; by the Dolphin's unusual spinning characteristics; the vulnerability of the pilot's head in a landing accident; engine difficulties; and prejudice in any case against the unconventional. (Even the sloping nose of the Service type was sometimes regarded not so much as an aid to the pilot's outlook as a feature detrimental to his flying ability—in that he was unable to 'keep the nose on the horizon'). The fact that advantages were rarely set against these strictures is understandable; likewise that the Dolphin was never acclaimed in its time—as it has been by later commentators—as 'the world's first multi-gun fighter', for not only was there secrecy to be observed, but the full complement of guns was seldom mounted.

The Dolphin's back-stagger having been mentioned as a salient feature, with concomitant advantages to pilot-view, a word on the fuller significance of this feature is in order. Although it is well known that the D.H.5 of 1916 was the first operational aeroplane of any note to have a negative, or backward, stagger, and that the dubious reputation of that fighter, especially respecting the stall, was attributed to this same feature, one is none the less left wondering why, in his autobiography *Sky Fever*, Sir Geoffrey de Havilland made no allusion whatsoever to the '5', though types of lesser fame are there. Perhaps the unusual wing-arrangement was officially inspired (though Harald Penrose attributes it to 'D.H.'s' own 'daring', and Oliver Stewart to his 'genius'). But whatever the facts of the matter, the staggering of wings in either direction was not entirely novel, for even in 1902 there was a mention in Wright/Chanute correspondence of 'staggering surfaces back'—or 'arranging the surfaces in steps' as Wilbur put it. Nevertheless, the D.H.5/Dolphin wing-arrangement in a *fighting* aeroplane (with the upper wing supposedly blanketed in a spin by the lower one) was something to which initially high accident rates for both types were sometimes ascribed. Little did pilots know that America's 'Staggerwing Beech' was to remain in production from 1933 to 1948! To Sopwith moreover back-stagger meant not only 'Dolphin' alone, but Hippo, Snail and Cobham also—while even the D.H.6 trainer received a slight degree of negative stagger, jointly with other 'improvements'.

Whatever the arguments for and against back-stagger, it was officially affirmed that 'A negative stagger increases the interference between the planes and is therefore only employed when, for overwhelming reasons, some property such as unobstructed upward view is required.'

The first Sopwith 5F.1, later named Dolphin, was apparently designed more or less in parallel with the Hippo, and was cleared by Sopwith's Experimental Department on 23 May, 1917. It carried no identifying number, and differed from the production form in notable respects. Most notable of all, it had a frontal radiator (instead of flanking surfaces) though the particular radiator fitted, which was to be officially criticised as 'inefficient', was far too deep and narrow for the common description 'car-type'—so deep, in fact, that even the high-set thrust-line of the geared 200 hp Hispano-Suiza engine was well below the upper lip. Of such a depth, moreover, was the whole nose ahead of the top wing that the two Vickers guns were contained wholly within it, while behind the wing the fuselage was similarly deep, meeting as it did the rear steel-tube cross-member of the open 'centre-section'. The tail resembled a Camel's though its small size was largely disguised by the lengthy lever-arm of the sharply tapered fuselage.

Certainly *not* emphasising the radiator, but nevertheless bearing witness to the very deep fuselage—with the cockpit coaming actually slightly above the front 'centre-section' steel-tube spar, is this unfamiliar Sopwith photograph, captioned: 'S.129—Sopwith Dolphin—200 hp Hispano Suizo[*sic*]—Type 5.F.1—1917—1st Machine'.

This first Dolphin was flown by Harry Hawker at Brooklands before May 1917 was out; was officially tested—with plywood decking extended aft—at Martlesham Heath in the following month, when it was deemed to be nose-heavy and was ballasted accordingly; and on the 13th of that same month (June 1917) was flown to St Omer for Service trials. The ferry pilot was H. T. (later Sir Henry) Tizard, who had joined the RFC from the army in 1915 and who did so much not only for 'Martlesham methods' but in founding the station itself.

At the end of June (confirming how well the Dolphin was regarded—being faster than the Camel and more manoeuvrable than the S.E.5) five hundred Dolphins were ordered from Sopwith themselves, with contracts quickly following to the Darracq Motor Engineering Co Ltd, and Hooper & Co Ltd, both of which firms were London-based. The name Darracq must be especially remembered here for at least two reasons. First, this same company had earlier received an order for two hundred examples of the D.H.5 back-staggered fighter; second, as the Dolphin was built to take a vee-8 engine of 200 nominal horsepower, it may be recalled that Darracq had made a racing-car engine of these same characteristics—though far heavier—as early as 1905. (That a young man named Moore-Brabazon had been apprenticed to the company is possibly less relevant). Hooper, of course, were famed for their fine coachwork, and already had Sopwith associations through the 1½ Strutter and Camel. Thus was Dolphin production centred round Surrey and South London.

Three aspects of the Dolphin in its exceptionally interesting second form—with radiators let in to the roots of the upper wings, cut-outs in the bottom wings, and with horn-balanced rudder. The radiators themselves are barely visible, even in a revised, larger, and more forward form (front view), though the $\frac{3}{4}$ rear view shows associated fairings having vents projecting from their peaks. In the front view (Sopwith No. S.132, captioned '2nd Machine') the rear cut-outs in the bottom wings may be transparent, with the horizontal tail surfaces showing dimly through them. Nevertheless, they mark a definite 'kink' in the trailing edge. (In this view also there are vibration-preventers for the inner main bracing wires).

The first Dolphin, with its unmistakable deep frontal radiator, having earlier received attention in this chapter, it can now be recorded that the second machine of the type embodied new features which went some way to meet Service desires, notably respecting pilot-view, but met new problems of its own. Most notable on this second specimen, the seemingly simple frontal radiator was abolished, the cooling for the engine being now provided by two small triangular surfaces, set one near the root of each upper wing, and revised at least once. However inefficient, this new system (apart from being in the top wing rather than the bottom one) did at least presage the twin-radiator installation tried on the Hornbill—Hawker's very first 'water-cooled' fighter, though as the Hornbill's Condor gave nearer 700 hp than 200 the radiators were in greater prominence.

Though the Dolphin's new wing-mounted radiators were not themselves successful, they did allow incorporation of a downward-sloping nose, which was in essence to become so characteristic of the type in service and which left the two Vickers guns partly exposed, though in the second form of Dolphin, now discussed, the top of the cowling retained its great depth at the rear. The enhanced field of view conferred by the sloping nose was supplemented by large cut-outs in the bottom wings; and had these spaces served to accommodate radiators (as they might well have done in the ultimate) then the Hornbill analogy would be all the more apt.

Also to be seen on the second form of the Dolphin were a new fin and a horn-balanced rudder, this last-named feature having been officially proposed because the first Dolphin had been tiring to fly by reason of the coarse left-rudder required at full throttle.

The third form of the Dolphin had flank radiators, a revised fin and rudder and two Lewis guns, as seen in these two views (not to mention a coy little spinner). The front and port-side pictures respectively bear the Sopwith numbers S.136 and S.139, though only the frontal one carries the legend '3rd Machine'.

Though not instantly apparent (perhaps because of the distracting Lewis guns) the fourth form of the Dolphin had a shallower fuselage fore and aft of the cockpit. This form set the pattern for production.

Clearly, however, the radiator system was the Dolphin's real *bête noir*, and the third form of the aircraft had an altogether new arrangement, for although this was again based on the use of two surfaces, these were now of deep block form, mounted not in the wings but on the fuselage sides, well aft, and in line with the rear of the cockpit. Just forward of each block was a shutter for varying the cooling area exposed. The fin was now enlarged, so that the horn-balanced rudder was matched to its contour. Ahead of the cockpit the decking was at one stage somewhat lowered; but interest lay very largely in the armament, which, although it was not exactly 'doubled' as is sometimes averred, comprised two Lewis guns (drum-fed, and mounted on the tubular, front, centre-section spar) in addition to the two fuselage-mounted synchronised Vickers guns (belt-fed, with Prideaux disintegrating links). Here then, we have the inception of what has sometimes been termed, as earlier noted, the 'world's first multi-gun fighter'; though of Dolphin armament there will be more to say.

On the fourth pre-production form of the Dolphin (circa October 1917) the fuselage behind, as well as forward of, the cockpit was shallower, and the cockpit rims cut deeper into the fuselage sides, giving a generally 'leaner' look; though the landing gear struts were still of sturdy ash, and not of thinner steel-tube as later, on production machines. The backward stagger, too, remained unaltered at 13 in (330 mm), though on production Dolphins this was reduced to 12 in (305 mm).

As Dolphin production was to run in parallel with that of the similarly engined S.E.5a of the Royal Aircraft Factory one of the pre-production Dolphins was tried (though unsuccessfully) with a four-blade S.E.5a propeller; but a two-blade Lang pattern was standardised—one advantage of such a form being not merely in respect of propulsive efficiency, but in the synchronising of the two Vickers guns.

By the end of 1917, production of the Dolphin was so well advanced that 121 had been delivered, the first unit equipped being No.19 Squadron (January 1918). Together with Nos.79, 23 and 87, No.19 remained Dolphin-equipped until the war was over, and perpetuated thereafter in its unit badge was the image of a Dolphin. (The name and symbol of the Elephant, of course, was stolen from the Sopwith 'Zoo' by No.27 Squadron, while the Fox, Hind, Hart and Gamecock passed in due time to the custody of Nos.12, 15, 33 and 43 respectively).

C3786
S.28.
DOLPHIN — 200.HP. HISPANO SUIZA.
TYPE 5.F.1.

Dolphin C3786. This machine has special mention in the text by reason of its armament. In the left-hand views it is without Lewis guns. The pictures above show how the Lewis guns could be trained.

From the armorial, however, to the harder facts of armament; for in the Dolphin special problems had been presented by the closely concentrated masses of two fixed Vickers guns and two movable Lewis guns. The solution of these problems had, in fact, been discussed at a meeting between Sopwith and Service personnel held as early as June 1917—shortly after completion of the very first Dolphin, which had two Vickers guns only, as had the standard Camel. Soon after the meeting just mentioned there was another, and on this occasion the RNAS was represented as well as the RFC—the first-named Service by Engineer Lieutenant F. W. Scarff (note promotion to commissioned rank since the early days of the 1½ Strutter). Details of how the Lewis guns were to be installed were apparently the primary concern of Sopwith's Mr Allman, and to limit the training of these guns a three-position ratchet was the fitting approved. The extent to which two Lewis guns were actually fitted as well as the two 'built-in' Vickers—either at the manufacture or the service stage—remains unclear; for although a single Lewis gun was far more normal in the field, a familiar photograph of Sopwith production shows C3786 at least, prominently in the foreground with both Lewis guns fitted, while C3787 and others far beyond along the lines have their Vickers guns only, complete with C.C. hydraulic synchronising gear.

Special mountings for six Home Defence Dolphins were apparently the responsibility of the Royal Aircraft Factory; but to Lieut 'Guns' Knight of No.87 Squadron credit is evidently due for the design of the fixed installation of two Lewis guns on the bottom wings—each gun about 18 in (460 mm) inboard of the inner pair of interplane struts, though with the lines of fire outboard of the propeller arc, so that synchronising gear was not required. Though ground attack may well have been the primary object of the outboard wing-mounting scheme referred to, which certainly preceded that on the Snark—though we must not too readily dismiss the American scheme of 1917 referred to under 'Triplanes (Hispano-Suiza)'—the Dolphin weapon-load for low attack could be augmented by the usual 'four twenty-pound Cooper' bombs. These little anti-personnel bombs (the targets officially prescribed were, in fact, 'personnel and aerodromes') were crutched in a carrier under the fuselage.

For work at the higher altitudes the Dolphin's inherent attributes showed clearly at their best—a warm cockpit being not the least among them; thus for Home Defence duties at night the type was much to be desired, especially so with the German raiders coming over at great heights, and with defending fighters having difficulty not only in merely intercepting them and keeping them clearly in view, but in reaching their level at all (at least, with sufficient time in hand for effective attack).

Militating against the Dolphin's safe employment at night, however, were not only engine difficulties (relative slowness in warming-up, persistent unreliability by reason of reduction-gear troubles and other factors) but the pilot's obvious vulnerability in the event of an accident. ('This would be an unpleasant machine in which to turn over on the ground' was Oliver Stewart's first remark on entering the cockpit, later explaining: 'The pilot's head came above the top plane, and he was completely surrounded by longerons, spars, cross-bracing wires and tie rods, and the feeling of being boxed in with the head exposed in a vulnerable position was experienced at once. With the engine in his lap and the petrol tanks in the small of his back, it seemed to the pilot that he had little chance of escaping injury in the event of a bad landing').

Not for nothing was 'Blockbuster' one vulgar name conferred on Sopwith's fine new fighter, and—much as on the Bristol M.1 monoplane—special pylons or

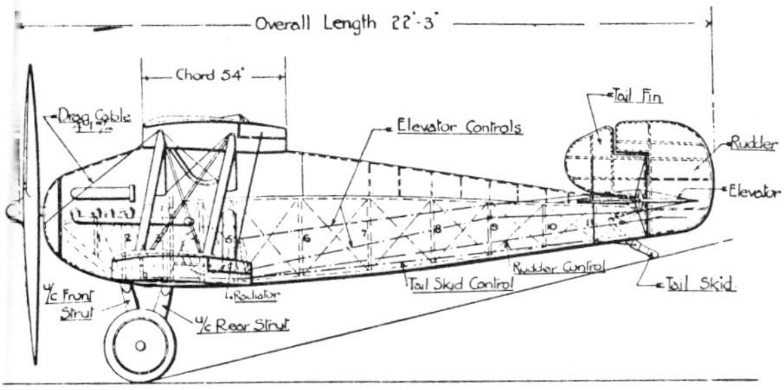

FIG 1.

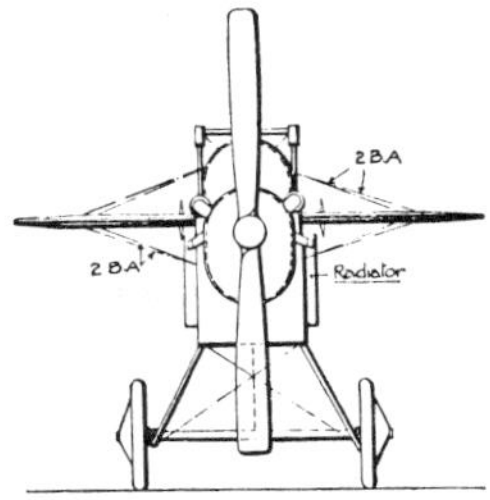

FIG 2.

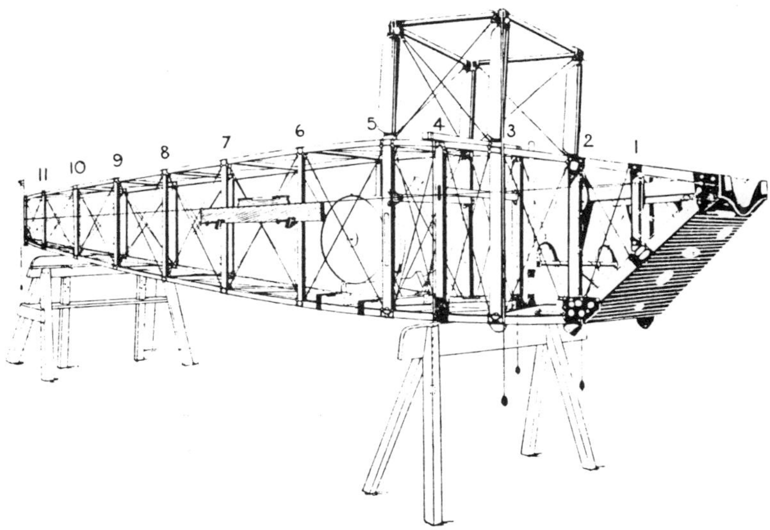

Official Dolphin drawings.

'cabanes' (or even a so-called 'rolling hoop') were in requisition, though whether the mounting of a single Lewis gun on one particular form of crash pylon above the cockpit was primarily anti-German or pro-British remains conjectural. Half-hoops of steel above the attachments for the inner pairs of interplane struts were a feature of Sopwith's own 'Dolphin Night Flyer', shown in a photograph. Although no aircraft-number is visible on the fuselage, the fin is stencilled C3858, and, of more technical interest, this surface is associated with a variable-incidence tailplane. (Rigging instructions for the standard Dolphin remarked that normal incidence was zero, adding that 'any adjustments can be made after tests').

Although this night fighter Dolphin (C3858) has protective half-hoops of steel above the wings and appears to be armed with a single Lewis gun only, provision for the Vickers guns is denoted by the case and link chutes behind the engine. Maker's caption: 'S.189—Sopwith Dolphin Night Flyer—Type 5.F.1—Feb. 19/18'.

In the happy event of a Dolphin pilot surviving a bad upset, he might, with luck, find his escape facilitated by specially modified centre-section bracing or even by a removable cockpit-side panel, with quick-release. As used by No.141 Squadron at Biggin Hill (a unit not already numbered among the Dolphin-equipped squadrons, for it used only a few of the type) flare-brackets were fitted under the lower wings, though there is no evidence of flame-damping exhaust tailpipes—possibly because these last might impede the pilot's exit.

Operational requirements aside, the Dolphin's development was very closely linked with powerplant vicissitudes, and it was, in fact, a shortage of Hispano-Suiza engines which precluded the operational fulfilment of at least one other Dolphin squadron (No.90) early in 1918.

The engine for which the Dolphin was designed was the compact vee-8 200 hp Hispano-Suiza, which, although it had reduction gearing, nevertheless possessed an excellent power/weight ratio. In France this engine was built by many companies, and eventually in other countries also, the wartime total of engines of this general pattern reaching nearly 30,000. In Britain the 200 hp Hispano-Suiza was made by Wolseley Motors Ltd, who called it (in developed form) the Adder; but from French production—notably Mayen—came other units of the type in big numbers. As is well known, crankshaft failures plagued the Wolseley-built geared engines; but there was far more to the story than this, and the following official notes, prepared just after the war of 1914—18, are relevant.

Wolseley had received a British contract for 100 direct-drive 150 hp Hispano-Suiza engines, but also (now to quote the official notes) 'for a much larger number of engines of similar type but provided with a reduction gear and adapted to run at 2000 r.p.m. developing over 200 h.p. at this speed. This particular geared model was based upon drawings supplied by the French, but no engines of this kind had at that time been built. The reduction gear consisted of a pair of large diameter spur gears (with helical teeth of an angle of 4° 50′) which raised the propeller shaft above the crankcase, and the shaft was hollow so that a machine gun could fire through it. The Wolseley Co. obtained permission to modify the drawings in the direction of fitting a scavenger oil pump and employing a different method of securing the reduction gear on the propeller shaft.

'The French at the same time were working on a 200 hp geared engine without the scavenger oil pump and with the propeller shaft gear wheel keyed on a taper on the propeller shaft. This method of fixing gave considerable trouble and the reduction gears on the French engine were the cause of frequent failure due partly to the high tooth pressure and partly to the use of air hardening steel. The Wolseley Co. used a 5% nickel case hardening steel for both gears and had little or no trouble . . . The feature of firing through the propeller shaft was not used however.

'The 150 hp engines of both French and English manufacture gave practically no trouble. In a short time the compression was raised from 4.7 to 5.3 to 1 and the speed raised to 1750 r.p.m. resulting in about 200 hp under these conditions.

'The first difficulty to arise in the manufacture of the Hispano engines in this country concerned the propeller hub fixing . . . The remedy finally adopted was to use a different taper for the hub from that of the shaft . . . The tests are still (Dec. 1918) continuing at the Isle of Grain as this trouble was found most serious on Seaplanes.

'The cylinder holding down studs frequently broke and to meet this difficulty the studs at each end of each block were lengthened and either a long nut or nuts and deep collars were used.

'When about ten 200 hp engines had been delivered from the Wolseley Works, an epidemic of crankshaft breakage was experienced.' (Then followed a lengthy account of measures taken) and it was later recorded:

'In view of the crankshaft failures, and the trouble with the propeller hubs, and the serious failures of reduction gears on French engines, it was considered necessary to reduce the number of geared English built Hispanos, and turn out an ungeared engine capable of a normal speed of 2000 r.p.m. This was the "Viper", which without doubt proved the most satisfactory of the Hispano series . . .

'The French 200 hp engine evidently did not receive the required care in manufacture as is evidenced by the fact that individual engines gave exceedingly good results, particularly from the point of view of weight/power ratio, but the majority require continual overhaul, chiefly owing to the difficulty in maintaining the required oil pressure . . . It should be recognised that the 200 hp Hispano engine was a development of a very satisfactory 140 hp engine, but that certain features, including the oil pump, were pushed beyond their capacity in the higher powered engine.'

In February 1918 (at which time Dolphins were arriving in France in some numbers) the Ministry of Munitions Department of Aircraft Production issued a *Report upon Troubles with 200 H.P. French Hispano in Service*, touching especially on excessive vibration and defective lubrication and largely relating to the 'SPAD Bitrailleuse' (or 'Bi-mitrailleuse' as it was otherwise called in the same document) and the 'SPAD two-seater'. Like these French fighters and the British S.E.5 series, the Dolphin was wedded to the Hispano-Suiza form of engine (an installation of the Sunbeam Arab was schemed, if not tried) and the end of the war found the following British Marks of the Dolphin in being—all distinguished by their engine: Dolphin I with 200 hp geared Hispano-Suiza engine; Dolphin II with the new 300 hp Hispano-Suiza, of which more later; Dolphin III with engine essentially as Dolphin I though with reduction gear removed. (*N.B.* An officially styled '200 h.p. Mayen Hispano Engine Converted to Direct Drive', when tested at the Royal Aircraft Establishment in the summer of 1918, 'ran satisfactorily throughout' and gave 201 hp at 1,800 rpm and 220.5 hp at 2,000 rpm. Some geared engines also delivered about 220 hp).

Of the three Marks of Dolphin listed the Mk. II is now of greatest interest—and equally it interested the French, who made the first installation in D3615 and placed the new combination in production. The presence of the new engine was distinguished by the greater bulk of the cowling (with the two Vickers guns completely submerged beneath it); by exhaust tailpipes extended further aft than usual, and thus having to be cranked to clear the tops of the radiator blocks on the fuselage flanks; and by a telescopic carburetter-air intake (intended to reduce the risk of a carburetter fire) the front whereof was prominent between the cylinder banks.

The number of 300 hp Hispano-Suiza installations made is uncertain; but what is sure is that Harry Hawker went over to fly the first conversion. Though snags were suspected or known, Hawker did rolls and spins. Later the structure was strengthened for production. As for the engine itself, a Mayen-built example was tested at Farnborough, and in July 1918 was the subject of the following interim report on calibration tests:

'The engine was dismantled for examination, then reassembled and placed on a test bed for calibration tests. The maximum power obtainable after tuning up was 280 h.p. at 1800 r.p.m. although on previous tests on a similar engine (150 m/m stroke in both cases) 316 h.p. was obtained at 1800 r.p.m. On trying the compression three cylinders were found to be O.K., three had poor compression, and on the remaining two there was no evidence of compression, when the engine was turned by hand. The cylinder blocks were therefore removed and were found to be leaking badly round several of the spark plug adaptors . . .'

As for the supercharged Hispano-Suiza engine much earlier mentioned in passing ('Other Men's Aeroplanes') this was nominally of 220 hp, but had a Rateau 'turbocompresseur', which gave the Dolphin thus powered its best performance at considerable heights. The maximum speed of 130.5 mph (210 km/h) was, in fact, attained at 8,700 ft (2,650 m), but at low level the speed was reduced to only 119 mph (192 km/h).

For the 300 hp Hispano-engined Dolphin, as intended for French and American use, a variable-incidence tailplane was standardised, partly owing to the large petrol capacity demanded and the distribution of its changing weight. (Petrol and oil systems for the Dolphin varied widely, but the 200 hp versions typically called for 27 gal (123 litres) of petrol and 4 gal (18 litres) of oil). Like the British, the French experimented with various propellers (typical for a 200 hp British Dolphin was a Lang of over 9 ft diameter) though for 'Sopwith Dolphin C1 No.3618. 300 h.p.' designs by Lumière, Gallia, Ratmanoff and Levasseur were tried. The best speed measured in one series of trials was apparently with a Gallia—221 km/h (137 mph) at 4,000 m (13,120 ft) though the best climb was afforded by a Ratmanoff—4,000 m

(13,120 ft) in 12 min 13 sec. Ceiling with several patterns of propeller was quoted as 7,500 m (24,600 ft).

Thus we perceive the Dolphin positively entering the 'postwar' performance bracket; and in furtherance of this perception it can be noted that one Dolphin was experimentally fitted (and flown to France) with a Calthrop parachute, stowed in the top decking, and that another (Sopwith-built D3747) had a jettisonable petrol tank. Even so, one is left with a feeling that, like its namesake in nature, this fighter may not, even yet, have yielded up all its secrets; though in partial proof of our contention at the outset that the Dolphin was one of the most remarkable of all the Sopwith fighters there may be instanced the victories of No.79 Squadron's aircraft alone—64 enemy aircraft and eight kite-balloons destroyed.

Nevertheless, ferocious though it was in combat the Dolphin was in its way tame, Martlesham Heath, for instance, crediting one of the first experimental models (with the first flank radiators) with an unstick run of 60 yd (55 m) and the ability to 'pull up with engine stopped' in 90 yd (82 m).

Production orders for the Dolphin were as follows:

Sopwith C3777—C4276; D3576—D3775; E4424—E4623; E4629—E5128.
Darracq C8001—C8200; F7034—F7133 (J151—J250 were cancelled).
Hooper D5201—D5400; J1—J150 (order not completed).

Dolphin I

(200 hp geared Hispano-Suiza)

Span 32 ft 6 in (9.9 m); length 22 ft 3 in (6.7 m); wing area 263.25 sq ft (24.7 sq m). Maximum weight (with two Vickers guns and one Lewis gun) 1,959 lb (889 kg). Maximum speed at 10,000 ft (3,050 m) 121.5 mph (195 km/h); maximum speed at 15,000 ft (4,570 m) 114 mph (183 km/h); climb to 10,000 ft (3,050 m) 12 min 5 sec; climb to 15,000 ft (4,570 m) 23 min; service ceiling 20,000 ft (6,095 m).

Dolphin II

(300 hp direct-drive Hispano-Suiza)

Span 32 ft 6 in (9.9 m); wing area 263.25 sq ft (24.7 sq m). Empty weight 1,566 lb (710 kg); maximum weight (two Vickers guns only) 2,358 lb (1,068 kg). Maximum speed at 10,000 ft (3,050 m) 140 mph (225 km/h); maximum speed at 16,400 ft (5,380 m) 133 mph (214 km/h); climb to 10,000 ft (3,050 m) 8 min 20 sec; climb to 16,400 ft (5,380 m) 12 min 10 sec; service ceiling 24,600 ft (8,050 m).

Dolphin III

(200 hp direct-drive Hispano-Suiza)

Span 32 ft 6 in (9.9 m); wing area 263.25 sq ft (24.7 sq m). Empty weight 1,466 lb (655 kg); maximum weight (two Vickers guns only) 2,000 lb (907 kg). Maximum speed at 10,000 ft (3,050 m) 117 mph (188 km/h); maximum speed at 15,000 ft (4,570 m) 110 mph (177 km/h); climb to 10,000 ft (3,050 m) 11 min 20 sec; climb to 15,000 ft (4,570 m) 21 min 50 sec; service ceiling 19,000 ft (5,790 m).

This particularly fine study shows Hippo X11 in its later form (with new tail and landing gear and Scarff ring-mounting) and is one of a set of maker's photographs, some of which are later reproduced—bearing, except for their different numbers, the same Sopwith caption as that here applicable, viz: 'S.277—Sopwith Hippo 3.F.2. 260 hp Clerget Blin Engine. 2nd. M/c—April 6/18.'

3F.2 Hippo

The Hippo was a two-seat fighter, very closely comparable with, and related to, its contemporary the Bulldog, and resembling that aircraft not only in having a new type of Clerget rotary engine with eleven cylinders (the Hippo having been designed with a view to replacing the 1½ Strutter in French production) but also in the disposition of the crew. The Hippo differed essentially from the Bulldog, however, in having a backward, instead of positive, stagger; and this was a key feature in the company drawings that were approved on 30 April, 1917. As shown in those drawings the Hippo nevertheless differed from the first example seen in photographs in having plain, instead of balanced ailerons, and vertical tail surfaces with a full, rounded, typically 'Sopwith' shape. A 200 hp Clerget 11Eb engine having arrived from France in September 1917, the first Hippo was completed (with French production in mind, as a private venture—by virtue of Licence No. 16—and not to an official British Air Board contract) quickly enough to make its first flight on the 13th of the same month.

The wing span being nearly 40 ft, and the ailerons of unbalanced type, lateral control was heavy; thus balanced ailerons were indicated (may, indeed, already have been schemed) and it could well have been the incorporation of these ailerons—though alternatively some structural weakness—which led to the fitting of a new set of wings in December 1917, prior to trials at Martlesham Heath in January 1918.

During early November 1917 the actual Clerget eleven-cylinder engine fitted for makers' trials at Brooklands had been taken from a Bulldog, and so far as is known no alternative installation of a Bentley B.R.2 (as once foreseen) was ever made. Between the Hippo and Bulldog airframes an interchange of Clerget engines appears to have occurred more than once, which is understandable because the 11E series was new and in demand, especially so as an output of about 250 hp was in prospect. Aircraft-performance was to be further improved by the fitting of an extra-large spinner, as designed by Clerget.

That the number X11 was borne by the Hippo in the form wherein it was tested at Martlesham Heath during January 1918 is certain, and photographs show clearly that this same number was painted on the fuselage when a Hippo was photographed at Brooklands on 6 April, 1918. In the context of these same photographs the machine depicted is described as '2nd. m/c.', although it obviously has plain ailerons and other differences which strongly suggest that '2nd form' might have been a more precise description, the number X11 having been allocated to the Hippo in two quite different states. At the same time, the possibility is recognised

Whether or not the very first (or 'first-form') Hippo was numbered X10, the two forms of X11 are shown here for comparison—the distinctive features of the later form being clearly shown also in other photographs. Here the Scarff ring-mounting of the later-form aircraft (lower view) is the most prominent feature, partly by reason of the Lewis gun. Nevertheless, in the upper view—of the earlier form—the balanced ailerons and small, angular fin are clearly seen.

Sopwith photograph S.278, showing—apart from more obvious features—a small part of the Lewis gun (without magazine) on the Scarff ring-mounting.

that the Hippo as first built may have been numbered X10—a new set of wings (for example) being deemed to constitute a new machine.

In any case, the first form of the Hippo set the general pattern for the type in having the pilot seated ahead of the top wing and the gunner stationed within (not behind) the wing, his field of view and of fire being enhanced by a trailing-edge cut-out—a double cut-out, in fact as illustrations show. But although the gunner was 'within' the top wing, he was nevertheless behind its rear spar, and was thus far removed from the pilot—an arrangement which clearly led to difficulty in crew-communication, though it was adopted (jointly with the negative stagger) in the interests of view. Nevertheless, the pilot's forward field of vision was compromised by the bulky cowling which largely enclosed his two fixed synchronised Vickers guns—a compromise that was almost inevitable as the fixed armament was double that of the Bristol Fighter, and the forward cowling was shorter, though fatter.

The wide separation of the pilot and gunner clearly invited criticism—especially so with D.H.4 experience in mind; and this criticism was indeed forthcoming in a Martlesham Heath report which declared, in part: 'The machine is very slow and heavy on lateral control, also the pilot and passenger are too far apart for easy communication, these points being disadvantages to a fighting machine. The rudder and elevator controls are fairly light. The pilot's view could be improved by cutting away more cowling, and better lighting of the instruments would be obtained by the insertion of a window in the cowling.'

Further concerning the first form of the Hippo, this had a particularly heavy back-stagger of 2 ft 3⅜ in (0.69 m) on its otherwise conventional wing cellule, the cut-away upper centre section whereof (though mounted directly on the fuselage) was braced to the top longerons by extremely short, and barely visible, vertical struts. Dihedral angle was a noticeable 3 degrees. Between the cockpits were a 30-gal (136 litre) main petrol tank, with pressure feed, and above it an 11-gal (50 litre) tank having gravity feed. The landing gear V struts were of wood.

Sopwith photograph S.274, showing how the Scarff ring-mounting overhung the sides of the narrow fuselage.

On the 'first-form' Hippo the gunner, as well as the pilot, had two guns, though these of course, were of Lewis type, each on a rocking-pillar mounting—these separate pillars demanding less manual effort than paired guns on a Scarff ring-mounting, though, as will be seen, a mounting of the latter type was eventually installed. Thus the armament was exceptionally heavy, the provisions for the pilot commanding this special comment in *Armament of British Aircraft 1909–1939:*

'The pilot had two Vickers guns in a remarkably neat installation and one which imperilled his frontal features less than in some other Sopwith types, the breech casings being located lower and further ahead. But although the familiar Sopwith padded windscreen was thus rendered unnecessary, the leading edge of the top centre-section was padded in the interests of head protection. There were separate case and link chutes low in the cowling and a small fitting, possibly for a sight, ahead of the windscreen. The gun gear was of Sopwith-Kauper type, and 500 rounds per gun were provided. The total ammunition weight of 260 lb which has been recorded for the first [*i.e.* 'first-form'] Hippo, seems somewhat excessive, even if the four guns were included, for the guns themselves would weigh no more than 100 lb and the ammunition not much over 130 lb.'

It may now be added that, whatever the facts of the matter, the Hippo now discussed was 290 lb (131 kg) overweight.

Although official British interest in the Hippo evaporated in February 1918, Sopwith themselves sustained development, and after X11 had returned from Martlesham Heath it was modified extensively. A landing gear of streamline-section steel tubing, and having larger wheels, was substituted for the former wooden gear; a Scarff ring-mounting and provision for eight 97-round ammunition drums for a single Lewis gun were installed for the gunner (the ring of the mounting being considerably greater in diameter than the fuselage width); the back-stagger

Sopwith photograph S.276, showing—even better than the close-up study—the installation of the Vickers guns, with their chutes.

was reduced to 1 ft 9$\frac{3}{8}$ in (0.54 m); and—perhaps most interesting of all—dihedral was not decreased, as might have been expected, but actually increased to 5 deg, though jointly with the fitting of new long-span plain ailerons. Fin area was increased by a fully-rounded outline, in continuance of the rudder top-line.

More than this, there was yet another Hippo (or another 'experimental' number at least, associated with the airframe)—X18, the characteristics of which are not known but which was flying in June 1918. By that time, however, greater power than the Clerget 'E' could offer was not only in prospect but was clearly demanded, though the output of the nominally 200 hp Clerget (or Clerget-Blin) rotary was already being quoted as 225 hp, and even 260 hp had been mentioned by Sopwith themselves.

But here the rotary type of engine was at a terminal point in its development. The short day of the radial Dragonfly was dawning; and the Bulldog was in any case the preferred new Sopwith two-seat fighter of a generally unsuccessful pair.

Having drawn this present chapter towards a conclusion with particular reference to rotary engines, and especially those of Clerget type, a final note is called for concerning Clerget and Clerget-Blin. That the latter conjunction had early origins is attested by the fact that the Imperial War Museum once possessed a 200 hp water-cooled Clerget engine made by the Etablissements Malicet et Blin in 1911, and that Gwynns Ltd, of Hammersmith, London, held a licence from Clerget, Blin et Cie, of Levallois-Perret, for the air-cooled Type 7Z rotary.

3F.2 Hippo

(200 hp Clerget 11Eb)

Span 38 ft 9 in (11.8 m); length 24 ft (7.3 m); wing area 340 sq ft (31.6 sq m). Empty weight 1,481 lb (671 kg); maximum weight 2,590 lb (1,175 kg). Maximum speed at 10,000 ft (3,050 m) 115.5 mph (186 km/h); maximum speed at 15,000 ft (4,570 m) 101 mph (163 km/h); climb to 10,000 ft (3,050 m) 13 min 25 sec; ceiling 18,000 ft (5,480 m).

N.B. The weight and performance figures quoted relate to the Hippo in its first form. It may be noted, however—with the utmost caution—that late in 1918 figures were issued for a Hippo to which a shorter span was imputed, this version having a reduced ceiling (as would be expected), a speed at sea level of 119 mph (192 km/h), and an endurance at 10,000 ft (3,050 m)—including climb—of 3 hr.

2FR.2 Bulldog

Its close relationship to the Hippo notwithstanding, the Bulldog—especially in its very compact single-bay form—was a fighter (or fighter-reconnaissance machine) of great individuality, the reconnaissance function implicit in the 'R' after the 'F' having been envisaged in a possibly earlier, though more or less contemporary, design. This design was likewise Hippo-related, but was Hispano-Suiza powered, and was called Sopwith FR.2; it had two-bay wings with positive stagger and a slightly greater span than the Hippo, the measurement being 40 ft 3 in (12.2 m). Thus it would have been amenable to high-altitude operation with the camera and wireless that were specified for it, especially so as armament was to be lighter than that of the Bulldog in the form wherein this type materialised, the pilot having only one Vickers gun instead of two as actually fitted. A point of special interest is that, as in the earlier Armstrong Whitworth F.K.3 reconnaissance aircraft, the pilot (seated near the top-wing trailing edge) shared the same cockpit as the observer; and hardly less noteworthy is the fact that during the period 1917/18 new Sopwith two-seaters were built both with pillar-mounted rear armament (as on the F.K.3) and with the Scarff ring-mounting (as on the F.K.8).

In the form wherein it materialised, the Bulldog was designed during August/September 1917—confusingly then called Buffalo, and allotted the construction-licence number 14, for like the Hippo, which had Licence No. 16, it was a private venture.

The most striking point about this new two-seat fighter (for as a fighter it was primarily regarded, reconnaissance and contact patrol being subsidiary roles) was that it looked for all the world like a variant of the original Snipe, in particular by reason of its single-bay wings, but also in having twin Vickers guns. Certainly, with its span of only 26 ft 6 in (8 m) and length of 23 ft (7 m) it was about the most compact machine that could have borne its four-gun armament; yet even so, its vertical tail surfaces proclaimed descent (however indirect) from its pioneer forebear in the two-seat fighter field, the 1½ Strutter—which, though considerably larger, had mounted only half the weight of armament.

That the stubby little wings were, in fact, too small to be true detracts but little from this first appraisal; and their inadequate surface may, in any case, be partly attributed to leading-edge and trailing-edge centre-section cut-outs (somewhat as in the Hippo) and even more so to the large gap between the spars through which the pilot's head protruded. The depth of the fuselage in this area was accentuated by a 'hump' fairing for the breech casings of the two synchronised Vickers guns; and thus the installation accorded closely with that of the Snipe. For each of the two guns there was a 600-round belt, and—so near the pilot's face were the ends of the breech casings that these were stoutly padded. Separate case and link chutes were let in to the cowling flanks and the windscreen was perforated for an Aldis sight. The apparent absence of trigger motors on the guns is not necessarily indicative of an intention to fit Sopwith-Kauper gear, or some other mechanical or electrical type, as standard; for had the Bulldog been produced in quantity, as was the Snipe, the C.C. hydraulic gear would doubtless have been specified. The matter may in any case have been an academic one, depending on the gun-timing provisions made, or allowed for, on the new Clerget 11Eb engine—a powerplant unfamiliar in Britain and one that only the Hippo had in common.

These three pictures have a special interest, not only in showing how remarkably clean and compact the first Bulldog appeared in its single-bay form and without its Lewis guns fitted, but in marking the apparent inauguration of the maker's system of numbering photographs, or negatives. The Sopwith caption to all three views reads: 'Sopwith Bulldog—1 Bay—1st. Machine—Type 2.F.R.2', and the prefixed numbers are S.6 ($\frac{3}{4}$ rear), S.7 (front) and S.8 ($\frac{3}{4}$ front).

These three views of the first Bulldog in its single-bay form but with Lewis guns fitted have the same maker's caption as those preceding. Their Sopwith numbers are S.12 ($\frac{3}{4}$ front), S.14 ($\frac{3}{4}$ rear), S.16 (front).

Of greater interest was the gunner's armament, comprising two Lewis guns, one on a pillar mounting at each end of the elongated, forwardly-tapering cockpit. Having quoted from one's own writings in regard to the Hippo's armament, one now accords like treatment to the Bulldog. Thus:

'To extend the field of fire in the forward upper hemisphere, the front pillar was extensible, and, as it was projected upwards, it raised, by means of a connecting tube, a rectangular screen hinged to the rear spar of the upper centre-section. Jointly with a small windscreen, this afforded the gunner a measure of protection. The rear pillar could be traversed from side to side. Possibly in the interests of gun handling, the ammunition drums were of single (47-round) type. In the original armament scheme, which did not materialise, two Lewis guns were to be mounted between the cockpits, capable of firing, like the single gun actually fitted, above the airscrew. An attachment noted on the forward Lewis gun may indicate an intention that this gun could be fired also by the pilot, but this is conjectural.'

Rear and frontal aspects of the first Bulldog in two-bay form with horn-balanced ailerons. Sopwith caption to rear view reads: 'S.3—Sopwith Bulldog—2 Bay—1st. Machine—Type 2.F.R.2'. Front view caption is similar, but number is S.4.

The first Bulldog as tested at Martlesham Heath—numbered X3 and with plain ailerons.

Comparative views of the first Bulldog (Bulldog I) with horn-balanced ailerons on its two-bay wings, and —lower—the Bulldog II numbered X4 and with Dragonfly engine.

Clerget and Dragonfly engine installations compared. Sopwith caption (top) reads: 'S.5—Sopwith Bulldog—2 Bay—1st. Machine—Type 2.F.R.2'. Bottom, 'S.505' (engine given as '360 hp A.B.C.' and dated 'June 24/18.')

Lacking armament, and with Dragonfly engine, the Bulldog II in this picture was captioned by Sopwith: 'S.503—Sopwith Bulldog 2.F.R.2—360 hp A.B.C.—June 24/18'.

The Bulldog's heavy armament, it must be re-emphasised, was distributed over a very large portion of the original tiny single-bay version of the aircraft, and may well have accounted for heaviness both in weight and in control-forces that were clearly manifest by the end of 1917—air-testing having started in November of that year.

It is known that a new set of wings had been fitted as early as 13 November, 1917, but though these were of greater span and area and of two-bay rig, they were not of definitive form—incorporating, for instance, horn-balanced ailerons and probably having been intended as an alternative installation from an early stage. Strength was evidently to be compatible with the greater span, for, although the outer bays were braced by streamline-section 'Rafwires', the inner bays were sustained by stout wire cables. As tested on the Bulldog at Martlesham Heath during May 1918, however, the two-bay wings had no horn balances; nor was this the sole distinction, for the number X3 was now painted on the fuselage, as shown on page 223.

At the time of the Martlesham Heath tests Bulldog X3 still had its two pillar-mounted Lewis guns, though a Scarff ring-mounting with a twin-gun attachment (or 'carrier' as the fitting was called) was in prospect for a second specimen of the aircraft then in the Sopwith works. Another intended installation was that of an A.B.C. Dragonfly radial engine, of higher power than the Clerget 'Type 11 No. 3' to which a much-overgenerous output of 260 hp was credited for the official tests just mentioned.

As a potential replacement for the Bristol Fighter, and as such competitive with the Avro 530, the Bulldog was unleashed in mock combats against a 'Brisfit' and was found to be just about as manoeuvrable. Clearly, however, any worthwhile improvement in speed and climb must await the fitting of the Dragonfly, and a Clyno-built example of that engine was duly installed in Bulldog X4, the new combination being ready for testing in the second half of June 1918. By this time production of the Bulldog as a fighter was no longer contemplated, and X4—or the Bulldog II as it was otherwise styled, to distinguish it from the Clerget-engined X3 Bulldog I—was manifestly a flying test-bed. Not only was all armament deleted, but the coaming for the rear cockpit was carried high to make the occupant snug if not safe—for the Dragonfly-Bulldog was deplorably unreliable. Even so, this aircraft was flying at the Royal Aircraft Establishment, Farnborough (after undergoing various powerplant 'fixes' and at least one complete engine-change) as late as March 1919, having first arrived there on 25 June, 1918.

Although a third Bulldog had once been in prospect, work on this had long been stopped.

Bulldog I

(200 hp Clerget 11Eb)

Span 33 ft 9 in (10.23 m); length 23 ft (7 m); wing area 335 sq ft (31 sq m). Empty weight 1,441 lb (653 kg); maximum weight 2,495 lb (1,132 kg). Maximum speed at 10,000 ft (3,050 m) 109 mph (175 km/h); maximum speed at 13,000 ft (3,960 m) 101.5 mph (163 km/h); climb to 5,000 ft (1,525 m) 6 min 5 sec; climb to 10,000 ft (3,050 m) 15 min 35 sec; climb to 15,000 ft (4,570 m) 38 min 55 sec; service ceiling 15,000 ft (4,570 m); endurance 2 hr.

N.B. All performance data for the Bulldog II Dragonfly test-bed would obviously vary greatly according to engine state, but the much-improved climb is implicit in the recorded time to 15,000 ft (4,570 m) of only 16 min 28 sec.

2B.2 Rhino

The fact that the Rhino, or Sopwith 2B.2, was a triplane tended to screen from view, and thus from full appreciation, aspects of design and equipment that merit careful study, and command far more respect than derision. Certainly it was a worthier Sopwith product than—and in some ways very sharply in contrast with—that other large single-engined multi-seat triplane the L.R.T.Tr. of 1916, a fact that must be attributed at least in part to its later design, for work on this was not in hand until the latter part of 1917. The Rhino was, in any case, produced to meet wholly different requirements (being a bomber, and not a fighter); and even though it shared the distinction of having a water-cooled engine, the unit concerned was of the tall-and-slender B.H.P. six-in-line type, as fitted in the comparable D.H.9. As triplanes go (or went) the Rhino was a notable example.

To the form of the 230 hp B.H.P. engine must be attributed in part the very deep fuselage, though another influential factor was the internal bomb-stowage beneath the pilot's seat. The matter of bomb-stowage is one that renders this Sopwith private-venture bomber (for its construction was authorised under Licence No. 14) an especially valuable object-lesson when compared with the D.H.9—a bomber strongly stamped by heredity, expediency and official dictation. True, the D.H.9 itself possessed internal bomb-stowage; but this was a secondary, as distinct from a primary, feature, and was forward of, and not below, the pilot.

The first of the two Rhinos built (X7) was air-tested at Brooklands late in October 1917, was delivered to Martlesham Heath for official trials on 4 January, 1918, and was followed to Brooklands by a second specimen (X8) in February 1918. The differences between these two machines were interesting (as will be recounted) though fairly minor ones, whereas those which distinguished the Rhino from the L.R.T.Tr. were fundamental. First, whereas the big fighter had been a three-bay 'three-winger' with surfaces of high aspect ratio, and further characterised by a huge landing gear, the new bomber was not even of two-bay, but of single-bay, form, with low-aspect wings and a landing gear that looked minute. Lacking on the Rhino, of course, was the gun-nacelle perched on the topmost wing, though height

Most conspicuous of the features which differentiate these two versions of the Rhino—the (presumed) X7, top, and the positively identifiable X8, in lower view—are the suppression of the aileron horn-balances in X8 and the fitting in this airframe of a Scarff ring-mounting, the base-ring whereof is very clearly seen, at the revised station for the gunner.

The Sopwith caption to this undated view of Rhino X7 (presumed), with horn-balanced ailerons, and pillar mounting for the Lewis gun hard astarboard, reads: 'S.64—Rhino Triplane 220 hp B.H.P.—1st. Machine—Type 2.B.2'.

was emphasised by the very broad centre section which, notwithstanding the widely splayed struts supporting it, overhung those struts by a noticeable margin. Chord of the top wing was constant, but at the roots of the middle and bottom wings were trailing-edge cut-outs, while between the spars of the middle wing were oblong apertures to improve the pilot's view (for his seat was above and a little behind). More prominent still were the large horn-balances for the ailerons on all three wings and the strut between the middle and bottom ailerons on each side (the middle and upper ailerons being connected by cable only). Later the horn-balanced ailerons of X7 were replaced by plain surfaces, as fitted also on X8; and when plain ailerons were fitted the mainplane tips were reshaped.

That the tail surfaces resembled those of the Bulldog is not surprising, for the two aircraft types were more or less contemporary; thus, taken in sum, the foregoing facts may help to emphasise that—its physical appearance and unenviable reputation notwithstanding—the Rhino was a relatively late-comer to the Sopwith menage or menagerie, and by no means as quaint or 'old-fashioned' as sometimes suggested. This point is further stressed by the recognition that almost the last of the military Sopwiths (the Snark single-seat fighter and the Cobham twin-engined bomber) were themselves triplanes. Nor must we forget here Herbert Smith's Mitsubishi triplane for the Japanese Navy (1922).

A particular point was made early in this account of the influence exercised by the fitting of a 230 hp B.H.P. engine—a powerplant chosen for large-scale production (especially by the Siddeley-Deasy Car Co) and precursor of the Siddeley Puma as installed in the Cobham, the Sopwith 'twin' just named. The cooling of the Rhino's

Vickers gun and uncowled B.H.P. of Rhino X8.

An especially fine study of the engine and Vickers gun installations on Rhino X8. The photograph bears the Sopwith number S.182 and is captioned '2nd. Machine', and dated 'Feb. 15/1918'.

engine was achieved in part by admitting air through a deep nose-intake, into which the front end of the crankcase projected, but more particularly by two low-set radiators in the sides of the cowling and flanking the bottom half of the crankcase, below the engine's eight supporting 'feet'. The radiators were of the general type used on production Dolphins: that is, each block was fronted by an adjustable flap to regulate the exposure of cooling surface. By reason of the cowling shape, and because each of the two cylinder blocks of the B.H.P engine (for two it had, even though it was a 'six-in-line') comprised the cylinder heads, water jackets, valve passages and inlet manifold for three cylinders, these cylinder blocks were largely

Rhino X7 (presumed) without horn-balanced ailerons, the photograph bearing the maker's caption: 'S.146—Sopwith Rhino Triplane. 220 hp B.H.P.—Type 2.B.2—1st. Machine—Dec. 1917'.

exposed. This necessitated provision of a fairing at the forward end, though even so, the cast-on legend '230 BHP' was clearly visible on the front cylinder block. Though changes in coolant, fuel and oil systems were made in the course of development none appears to have been basic; so attention may now be transferred to armament.

Although—as on the $1\frac{1}{2}$ Strutter—bomb-sighting presented major problems, the stowage of the bombs themselves was exemplary. As already noted, this stowage was beneath the pilot's seat; but, although this situation was favourable to c.g. location, it inhibited—jointly with a petrol tank—the fitting of a Negative Lens sight for the pilot's use (as, for example, on the D.H.9 and comparable types) while the employment of an external sight, of C.F.S. or other pattern, was rendered difficult, if not impossible, by the fuselage shape, with its tumblehome decking. The bombs—four 112 lb or nine 50 lb or twenty 20 lb—were in a cellular 'crate' which was winched into place, complete with closely associated bomb-release gear, by a system of pulleys attached to the middle-wing spar, or spars, inside the fuselage.

Between Rhinos X7 and X8 (Nos. 1 and 2 as they were otherwise known) the differences were largely in respect of gun-armament. Whereas on X7 the pilot's Vickers gun was mounted on the fuselage centre line immediately ahead of the cockpit (with the feed block faired over, the fairing also affording some protection to the pilot) on X8 the gun was wholly forward of the pilot's normal-type windscreen—a fitting absent on X7—and there was a fairing ahead of the feed block.

Rear armament on X7 was a Lewis gun on a rocking-pillar mounting at the rear of the second cockpit (as on the Bulldog) but X8 had a redesigned gunner's position, with a Scarff ring-mounting fitted on the top longerons, the gunner thus having a deep protective coaming ahead of him. Provision for a downward-firing 'belly gun' has been mentioned in connection with the Rhino, and would not seem incompatible, for such an installation was not unknown on the D.H.9.

Certainly, X8 was tested at Martlesham Heath during February and March 1918 not only with a revised armament installation and plain ailerons but with a Lang 4020 propeller instead of other patterns tried. This being so—and also having regard to the Sopwith type-number used jointly with the Rhino's name to head this present chapter—it may be remarked that a magnifying glass proclaims the following stamping on the propeller hub seen in the Sopwith 'nose close-up' picture S.182 on page 229: 'DRG. L.4020. 230 H.P. B.H.P. SOPWITH. 2.B.2.'

Its unimpressive showing and seemingly derisory name notwithstanding, the Rhino, if developed with a later engine (see under 'Cobham') might have made a distinctly useful addition to the final rhino-like bombing 'charge' by the RAF. And should this prospect, jointly with the pictures shown, occasion shock, then one would only remark that this merely shows how deceptive (as well as instructive) photographs can be; for the Rhino was little bigger than a Hawker Hart!

Rhino

(230 hp B.H.P.)

Span 41 ft (12.5 m)*; length 30 ft 3 in (9.2 m); height 10 ft (3 m); wing area 612 sq ft (56.8 sq m). Empty weight 2,185 lb (990 kg); maximum weight 3,590 lb (1,628 kg). Maximum speed at 10,000 ft (3,050 m) 103 mph (166 km/h); climb to 10,000 ft (3,050 m) 24 min 50 sec; service ceiling 12,000 ft (3,658 m); endurance 3¾ hr.

N.B. Rhino X8 without bombs and weighing 3,061 lb (1,388 kg) is known to have reached a service ceiling of 14,500 ft (4,420 m). In this instance the fuel and oil load was 465 lb (211 kg), though in another case, with a military load of 538 lb (244 kg) the fuel and oil load was 507 lb (230 kg).

* Although the span would clearly differ according to the type of ailerons fitted, the generally quoted figure of 33 ft (10 m) is apparently incorrect whatever allowances are made.

A.T. and Sparrow

The wireless control of aircraft, boats, torpedoes and fighting vehicles was a topic which, by 1918, had engaged the attention of many inventors for several years. By early 1919, indeed, one authority gave this reminder: 'Experiments upon the wireless control of self-propelled bodies such as boats, torpedoes and airships have been in progress in various countries during the past twenty years, and when the war broke out some considerable success had been achieved with models, and in some cases with craft too large to be called models. In America Tesla and Edison were early workers in this field, the former in 1897 having succeeded in devising a small model boat which could, it is said, be made to follow a short prescribed course by the action of Hertzian waves. Hammond, also in America, although he met with much to discourage him at first, carried his investigations and experiments so far that by 1912 he was able to claim that he could control a 25-knot motor launch with precision over a range of three miles or so. In Germany the most noteworthy results were those obtained by Wirth, of Nuremberg, who, round about 1911, exhibited a 50 ft. motor boat which was under radio-dynamic control and which could be made at will to discharge guns, lights and fireworks. In France Deveaux, in 1906, demonstrated to the French naval authorities the working of a wirelessly [*sic*] directed boat carrying an 18 in. Whitehead torpedo in an ejecting tube. In Gt. Britain Gardner constructed and successfully tested a fair-sized craft of a similar nature, while Roberts, an Australian, in 1912 exhibited an airship about 20 ft. long which flew under wireless control round lecture halls and theatres. In all these pre-war instances the distance over which the control could be exercised was short, varying from a few yards to a few miles. Taking account of the developments which other branches of wireless communication have recently undergone there is, however, no difficulty in believing that means are now available for extending the control over ranges of a hundred miles and upwards.'

Evidently descended from the SL.T.B.P.—the dainty little Sparrow, with its A.B.C. Gnat flat-twin (or horizontally opposed) engine, of a general type first designed and built in 1917.

The possibilities of jamming were mentioned by the same authority in the context of 'a method of neutralising signals coming from any particular direction' which had been used 'during the last days of the war to prevent the Germans at the Nanen station from "jamming" messages passing between Allied wireless stations.' Thus (it was added): 'The dependence of the ground-controlled aeroplane or torpedo upon a system of wireless communication which it is within the power of someone else besides the operator to transmit, and the possibility of that someone issuing contrary commands or nullifying those sent out by the operator, have long presented one of the chief problems to the inventors of wirelessly directed bodies. Many attempts have been made to devise means which would render interference impossible, but very little success attended these efforts . . . It would seem that for military purposes the high-flying wirelessly controlled aeroplane or airship provided with an automatic course recorder enabling its position to be laid down from instant to instant on a map at the control station is the only development of this latest aeronautical application which is likely to prove of military value.'

At the time of the foregoing—historically relevant if not, perhaps, scrupulously accurate—pronouncement (early 1919) nothing could be said of the work which had been pressed forward by Capt Archibald M. Low, RFC, (better known as Prof A. M. Low) at Brooklands, Feltham and Farnborough; this despite the fact that Capt Low had applied for a patent on 9 January, 1918, which patent was accepted on 10 December of that same year, though publication was witheld until late in 1925. Nevertheless, even during the preceding year (1924) Prof Low—whose work on radio control had begun in 1916—was looking yet further ahead, for as he himself then declared:

'The pilot of an aeroplane can talk to his base: he will soon be able to write and transmit vision from a plane which could be controlled by wireless. The time will come when low-flying wireless planes will explore, and render visible at many miles distant, places where no human pilot could remain for any length of time in safety.'

These preceding declarations have been set out here to emphasise that during 1914—18 radio control was not a complete novelty; that the British authorities were by no means as backward or as stupid with regard to this technique as might be supposed; that Capt Low was encouraged in—and by—his work; and that the radio control of motor boats (a form of craft in which T. O. M. Sopwith took a personal interest) was very much in mind—special 'floats' or 'sea-sleds' having been proposed.

One must now proceed to show how Sopwith's involvement extended to the aircraft side, and how the A.B.C. engine company contributed to Britain's programme with a series of horizontally opposed two-cylinder air-cooled units, latterly, at least, given the name Gnat.

To conceal the true purpose (bombing, or possibly the attack of airships) of several small pilotless radio-controlled aircraft made (or intended to be made) by the Royal Aircraft Factory and the de Havilland (Airco) and Sopwith concerns, the cover-name 'Aerial Target' (A.T.) was adopted; and to stress that Britain's programme of the First World War was not the mere superficial—even light-hearted—affair that it has sometimes been made out to have been, one would add that on the engine-development side not A.B.C alone, but Armstrong Siddeley also, with their tiny Ounce, were active into the 1920s. (Later still the Armstrong Siddeley Lynx engine gave a name to the Larynx flying-bomb developed by the Royal Aircraft Establishment).

Sopwith's first contribution to the 'A.T.' programme was never flown, apparently because of damage to the bottom starboard wing and upper port wing, sustained during erection at Feltham. An existing photograph shows this damage and two other features of special interest: first, the sharp dihedral angle for lateral stability; second, the four-wheel landing gear. Both these features, of course, might naturally be expected on an experimental flying-bomb or aerial target (now using the latter term in its 'open' sense), though one is nevertheless tempted to relate them—with the utmost circumspection—to the following facts:

By early 1918 tests had been undertaken in the William Froude National Tank at Teddington to find a 'suitable float for an aerial target'; so, while giving due observance to 'National', it is also permissible to note that Teddington was almost literally 'next door' to Feltham (as well as Sopwith's Kingston works) and thus to suggest that the tests may have been related to a known request by the Royal Navy for an air-launched 'glider target'. To minimise damage on striking the sea this was designed as a floatplane, was made longitudinally stable, and was steered by obsolete spring-driven torpedo gyroscopes.

Clearly, the Sopwith A.T., with its massive four-wheel landing gear might fairly readily have been adapted as a floatplane and used either as a glider target or as a wireless-controlled pilotless aircraft or flying-bomb.

Conjecture aside, the Sopwith A.T. had a rudimentary structure, respecting both the square-section fuselage and the single-bay wings—these last having very few ribs and being attached to inverted-V struts on the centre-line. For wireless control, the aerials were wound round the rear part of the fuselage and the outboard sections of the upper wings. Light-painted discs on the fuselage might be associated with tracking—possibly photographic.

The Sopwith Sparrow—for such was the undoubted name of the A.T.'s successor—was greatly refined in design, and quite clearly related to the SL.T.BP. Certainly it had a cockpit for a pilot (as had the comparable de Havilland Queen Bee of later years) and it was somewhat larger than the earlier A.T., the span apparently being just over 25 ft (7.6 m). The refinement mentioned largely concerned the nose, which was aerodynamically cleaner than that of other A.B.C.-powered Sopwith types (the Rainbow racer included)—largely by reason of the fact that the 35 hp Gnat engine had two cylinders only, and these horizontally opposed. The Sparrow differed also from the original A.T. in having a conventional centre-section, with trailing-edge cut-out; and lateral control was positively by wing-warping, whereas the A.T. seems to have lacked even this 'refinement', being dependent for lateral stability on its sharp dihedral angle. Vastly different also was the clean, typically Sopwith, V-strut landing gear.

These points notwithstanding, if the Sparrow airframe underwent appreciable development at all, then this was not in the direction of a wireless-controlled flying-bomb, the design possibly having been superseded by the Ounce-engined monoplane designed at the Royal Aircraft Establishment, Farnborough—a design which itself had a precursor in the 50 hp Gnome-engined RFC Experimental Works 'A.T.' That the 50 hp Gnome was apparently the designated powerplant for four 'Sopwith small scouts' with warping wings, which aircraft were seemingly at Brooklands in May 1917, may well have signified some association with the Bee (see earlier) rather than with the British 'A.T.' project—wherein, as shown, Sopwith nevertheless played their part. Certainly the Sparrow appears to have dated from a later period.

As for the little A.B.C engine which represented such an important element in the 'A.T.' project, and for which a power of 35 hp has already been mentioned (this figure evidently being a maximum—output, of course, depending much upon tuning, life expectancy and state of development) it may be noted in this regard that during 1917 a small Lang propeller was designed 'primarily for use' (as it was later stated) 'on an A.B.C. Gnat engine of 28–30 hp, mounted in an R.E. [*sic*] experimental machine for control by wireless waves.' It was added: 'This same design of airscrew was used for the small "Kitten" type aeroplanes, designed at Eastchurch Air Station and the Isle of Grain Experimental Station during 1917.' (The Lang propeller works, it may be noted, were at Weybridge, near Brooklands). Thus, in the matter of powerplant, as well as in other connections, Sopwith may have had some particular association with the Eastchurch Kitten. (See under 'Apocrypha' at end of book).

7F.1 Snipe

First constructed—specifically as a successor to the Camel—late in 1917, the Snipe was destined to remain in service with the RAF until 1926, even though the Sopwith company had gone into voluntary liquidation in 1920. The operational history of the Snipe, therefore, is largely 'post-Sopwith', and somewhat obscured by the fact that the first Hawker Woodcocks were replacing Snipes in May 1925 (the Woodcock having first been flown in 1923). Harry Hawker, whose surname the new company bore, had been killed in the Nieuport Goshawk in July 1921. That same aeroplane (the Goshawk) was the precursor of the famous Gloster racer nicknamed

Bearing the Sopwith numbers S.9 and S.10 respectively, these two views of the single-bay Snipe with flat-sided fuselage and open centre-section are both captioned by the makers 'Sopwith Snipe. Type 7.F.1. 1st. Machine'.

'Bamel', though, unlike the Snipe, the Bamel was not related to the Camel—its hump containing nothing more lethal than fuel!

There is one further point to make here: Both the Goshawk and the Bamel belonged to the Nieuport Nighthawk family—and so, also, did the Nightjar, which saw RAF service as a Naval fighter and was a direct counterpart of the Snipe, having the B.R.2 rotary engine; in being rigged as a two-bay biplane; and in being fitted in one form—called Sparrowhawk III—with a hydrovane landing gear (plus the normal wheels) similar to that which was tried on the Snipe when that Sopwith fighter was itself envisaged as a replacement for the 2F.1 (Ship's) Camel.

While the Snipe's record of RAF service as generally published has rightly surveyed developments after Maj. W. G. Barker's monumental Fokker-fight of 27 October, 1918, by naming Home Defence duties jointly with commitments in occupied Germany, Russia, Egypt, Turkey, India and Iraq, and has given due credit to the Snipe as a trainer as well as a Hendon highlight (formation aerobatics in 1921, for instance) one aspect of this Sopwith fighter's use seems nevertheless neglected. This particular aspect was not, perhaps, of great importance; but—having gloomy regard to circumstances that prevail in Ireland at the time of writing (as of yore)—it has indubitable interest, and helps to place the Snipe in true perspective.

Here one has in mind the employment of RAF Snipes in 1921, when 'Snipes for the Snipers' was an item in the news—though 'snipe' instead of 'for' might have been more explicit, having regard to the explanation: 'A patrol of this type of aeroplane recently located [in Ireland] an ambush, and the result was disastrous for the ambushers, five of whom were killed by our airmen.'

This superb Sopwith study has the maker's caption 'S.20—Sopwith Snipe—200 hp—Type 7.F.1—1st. Machine'.

A perfect comparison with Sopwith photograph S.9 (on page 235) this one has the maker's caption 'S.48—Snipe—200 hp Bentley Rotary Engine—Type 7.F.1. 2nd. Machine.' Note especially the splayed-out centre-section struts.

Another point concerning the Snipe's lengthy retention in RAF service is that this was not solely due to economic considerations, but also to difficulties with the A.B.C. Dragonfly radial engine (See later chapter headed 'Dragon').

Thus, while Woodcocks, Siskins, Grebes and Gamecocks took up their stations, it was in large degree the Snipe—both by its physical presence around the RAF and by artless representations in official publications on fighting tactics and airmanship—which was to nurture many leaders for the '40s and the '50s and beyond. (Quite often—very long after the Armistice—was the Snipe depicted in various attitudes and in varying company, some members of the latter having a distinct resemblance to L.V.G.s or Rumplers; and such was the colouring of the Snipe and its consorts in those 'tween-war years that pilots were counselled how, in stalking an enemy, they should not bank so steeply that sunlight might flash 'from planes or struts'. They should watch, moreover, an adversary's rudder, because its movement could foretell the direction of a turn). In such an environment—linking

Faired (rounded) fuselage sides are evident here in Sopwith photograph S.49—otherwise captioned as is number S.48, reproduced above.

the First World War quite directly with the 1930s—the Snipe eked out its presence in the RAF; and even Hawker Fury pilots were to see its presence in their printed guides.

All of which seems far removed from the first ideas of the Snipe, as they grew—generally conforming to the Air Board A.1(a) specification, then already some months old—in the summer of 1917. Six examples (B9962—B9967) were ordered for experimental and development flying, and the first of these was a very different aeroplane from the Snipe so well remembered by so many pilots. This difference was chiefly by reason of the short-span—25 ft 9 in (7.9 m)—single-bay wings; the flat-sided fuselage behind the rounded engine-fairings; and the Camel-type tail. Camel-reminiscent also was the 150 hp Bentley B.R.1 engine. The top centre section was somewhat cut back at the leading edge; was open between the spars; and was carried on near-vertical struts. Main-panel root cut-outs on the top wings were prominent, though differing in shape from those at one time schemed. In sum, this was one of the trimmest, most compact, and most aggressive-looking fighters ever built, affording a clearer view for the pilot than the Camel and—hopefully, it may be surmised (from the equal dihedral, for instance)—restoring some of the Pup's tractability.

The next version (conceivably a rebuild of the first machine) was seemingly almost identical, though having a B.R.2 engine of 230 hp and also the increased dihedral applied at some stage to the first. In any case, this first B.R.2 version (and the B.R.2 was to become well-nigh identified with the Snipe) was numbered B9963—and on 23 November, 1917, it was at the Royal Aircraft Factory, Farnborough.

Chiefly for a report on the B.R.2 engine, a single-bay Snipe (confusingly logged officially as '9965') was intended to go to Martlesham Heath, after crashing at Brooklands on 19 November, 1917, and being repaired at Kingston. It did in fact arrive at Martlesham a month later (18 December), and there it was briefly tested (before another mishap) without ammunition, attaining a speed of 119 mph (192 km/h) at 15,000 ft (4,570 m)—a height to which it climbed in 14.8 min. As the level at which the speed-run mentioned was made seems uncommonly high, and the B.R.2 was not supercharged, it may be explained that the experimental unit fitted was not run at full throttle below 5,000 ft (1,524 m).

Snipe B9965 in early form appears to have had not only single-bay wings but a wider centre section, with splayed-out struts; faired (rounded) fuselage sides; and vertical tail-surfaces of new design (small fin, with angular horn-balance for rudder). Later the machine so numbered was given two-bay wings of greater span—about 30 ft (9.1 m.)—on the uppermost whereof a Lewis gun was mounted (pivoted on the rear spar of the centre section) in addition to the two standard fuselage-mounted synchronised Vickers guns. Such alterations inevitably confuse any survey of the 'development' Snipes—a survey further blurred by the Snipe's near relationship to the Salamander; by rebuilds and structural-test specimens; and—not least—by the apparent initiation of the Snipe's construction as a private venture (under Licence No. 14, whereby the Rhino triplane bomber was also built), with retrospective numbering of individual airframes not being decided or confirmed until November 1917. Contractural convenience, or even 'security', may have been considerations here.

In any case, one major consideration in the development programme was the chosen engine—the Bentley B.R.2, as already affirmed, becoming 'well-nigh identified with the Snipe', although in November 1917 the eleven-cylinder Clerget 11E, as installed in the Hippo and Bulldog two-seaters, was still envisaged as an

Two other views in sequence: Sopwith photographs S.51 and S.52, both bearing the inscription '2nd Machine'.

alternative. (A Clerget of this type was eventually and temporarily installed in Snipe F2340, though seemingly in connection with the Salamander programme).

By February 1918, B9965, with two-bay wings and third (Lewis) gun—though in other respects deficient in military equipment and even in rigging—had been officially tested (B9966 not so, though this had been intended). Whatever criticisms may have been made of the Snipe by this time, however, the pilot's view was considered 'excellent'—which was more than could be said of the Pup or the Camel. And here one would emphasise that by this time also field of view was deemed increasingly important not for combat alone but equally for systematic search.

By March 1918 the Snipe had been chosen for adoption rather than any of its direct competitors, the Austin Osprey triplane, the Nieuport B.N.1 unconventionally braced biplane and the Boulton & Paul Bobolink, which last-named almost merited the description just applied to the B.N.1. Factors in the choice were the well-established 'acceptance' of Sopwith types by the Services and manufacturers and the use of standardised Sopwith components; though in all frankness the Snipe did not illuminate the fighter scene with dazzling brilliance. However, B9965 went off to France, where the 23-year-old officer who was to become Air Chief Marshal Sir Leslie Hollinghurst, and who was to know Snipes well between the wars, made a few criticisms but (once again) praised field of view—as well as manoeuvrability. These same characteristics were stressed by other RFC pilots, though (like Hollinghurst) they remarked on tail-heaviness and recommended removal of the Lewis gun. That the rudder was too small was a general criticism, and that manoeuvrability was not, perhaps, one of the Snipe's stronger points after all, was suggested.

In any case, on 20 March, 1918, Snipes to a total of 1,700 were ordered from Sopwith themselves, Boulton & Paul, Coventry Ordnance Works (shades of young Tom's test-flying days!), Napier (not such a new name in the airframe, as distinct from engine, world as might be supposed), Nieuport & General, Portholme Aerodrome and Ruston Proctor.

Still B9965 continued as the chief test-specimen, and in May 1918 was returned to

Two-bay wings, with a Lewis gun offset on the centre section of the uppermost of these, characterise the aircraft shown, and captioned by Sopwith 'S.162—3rd. Machine—Jan. 21/18'.

Sopwith to be given a new engine-cowling, jointly with a huge open-fronted spinner (a precursor of those on the Snapper, Cobham, post-war Schneider and Rainbow) with lower-segment holes in the cowling for the B.R.2's exhaust efflux. By the beginning of July this same Snipe was at Farnborough; but B9966 had been at Martlesham in May, with an adjustable tailplane (later standardised), larger fuel and oil capacity and a new fuel system (Badin type, with venturi tube on the starboard front centre-section strut, used to transfer petrol from the main tank under the pilot's seat to the gravity tank behind the engine). A mounting for a Lewis gun was retained, and a slightly bigger rudder fitted, though before October 1918 Snipe B9966 was at Martlesham with a new 'production-type' fin and rudder (large curved fin), a centre-section with inter-spar cut-out revised once again (there were several variations), and, most remarkable of all, an experimental tailplane/elevator assembly of puzzling complexity and near-triangular plan form, the rear part of the tailplane having variable incidence—all this together with horn-balanced ailerons (with inverse taper, like the elevators) on the top wings.

The designation Sopwith "Snipe" 7.F.1/5 which figures in the maker's captions to these—their photographs numbered S.210, S.211 and S.212—almost certainly connotes B9966, or a version thereof. The pictures are dated 'March 11/18'.

An experiment at the front, instead of at the rear—the special engine installation, with huge spinner, fitted to B9965 after its return to Sopwith in May 1918.

Though unidentified by numerals, this is probably B9965 with a less exotic engine installation.

The indubitable B9966 which had an adjustable tailplane. (Note that the left-hand part of the picture, with motor ambulance, is apparently identical with that appearing in Sopwith photograph number S.212).

True production Snipes had nevertheless been under test since mid-August 1918—and here an anomaly can be recorded, in that E7996 was with No. 43 Squadron at the Front on 13 August, whereas E7987 did not arrive at Martlesham until five days later. The term 'true production', furthermore, is open to misconstruction, for the first Snipes off the lines had plain ailerons and a small fin, though the larger fin, and rudder with shielded horn balance, as tried on B9966, together with horn-balanced ailerons on the top wings, soon became standard fitments (not, however, the 'fancy' horizontal tail surfaces!).

B9966 with the completely redesigned experimental tail unit, which incorporated a two-part tailplane.

These three views of the early-production Snipe E7989, built by Sopwith, bear the company's photograph numbers S.553, S.554 and S.555 respectively.

Even so, our earlier appraisal of the Snipe—that it 'did not illuminate the fighter scene with any brilliance'—remained valid, special engine-tuning and other recourses notwithstanding.

On 24 September, 1918, No. 43 Squadron (the first to be Snipe-equipped) could put up fourteen of the new Sopwiths, and by 31 October there were 97 Snipes in France, three squadrons being equipped by the Armistice. Of all these early-production Snipes the most famous was E8102, wherein Maj W. G. Barker won his Victoria Cross on 27 October, 1918, in an action hardly less worthy of renown than the last fight of Sir Richard Grenville's *Revenge* or the Charge of the Light Brigade. Alas, there was no Tennyson to commemorate the glory of a solitary Snipe against fifteen Fokkers, and John Masefield was sticking to his beloved ships.

Nevertheless, we now retell in essence the stirring tale of Barker's action in E8102:

Major Barker, who was on a refresher course from England with No. 201 Squadron, RAF, while on patrol in this Sopwith-built Snipe attacked an enemy two-seater at 21,000 ft over the Forêt de Mormal, and the E.A. broke up in the air. He was then fired at from below and wounded by a Fokker biplane, and fell into a spin, from which he pulled out in the middle of a formation of fifteen Fokkers, two of which he attacked indecisively. He then got on the tail of a third, which he shot down in flames from a range of ten yards. He was again wounded, and fainted, but on recovering he regained control of his Snipe and was attacked by a large formation of E.A., one of which he shot down in flames from close range. He was then hit in the left elbow, which was shattered, and again he fainted, the Snipe falling to 12,000 ft before he recovered. Another large formation then attacked him and, noticing heavy smoke coming from his machine, Barker believed it to be on fire, so he tried to ram a Fokker. He opened fire on it from close range, and the E.A. fell in flames. Finding his retreat cut off by eight of the enemy, at which he fired a few bursts and succeeded in shaking them off, Barker returned to our lines a few feet from the ground, finally crashing near our balloons.

During the latter part of this combat Barker was without the use of both legs and one arm.

Thus the analogy of the *Revenge* ('the one and the fifty-three') is surely a fitting one; and with the sea now in mind it must not go unrecorded that, but for the ending of the war, Snipes would have been in Naval service. By October 1918, E8068 had, in fact, been fitted with a hydrovane ahead of the jettisonable wheels to render ditching less risky—and, if nothing else, tests showed that the speed-reduction with the hydrovane was only 3 mph (5 km/h). Slinging gear, for salvage from the sea, was another feature that had to be developed for the 'Ship's Snipe', though arrester-hooks for deck-landing were seemingly not developed for the Snipe until 1923/24—the comparable Nieuport Nightjar then being already in service.

For Home Defence the Snipe was just beginning to enter the service stage as the war finished; one Camel squadron, indeed, would have been re-equipped in January 1919 and by March of that year five would have had the new and larger Sopwith single-seater. In this Home Defence context especially it may be noted that, having its two-bay wings of greater area, and also the lifting power of some 230 hp, the Snipe could carry a warload that might involve relatively elaborate night-flying equipment—including, for example, navigation lights, Holt electrically ignited wingtip flares, and a wind-driven generator on the front starboard landing-gear strut—in addition to two Vickers guns with a greater ammunition supply than a Camel's (with provision also, on some aircraft, for an optional Lewis gun on the top wing). Wireless, 'safety' or self-sealing petrol tanks, oxygen equipment, ply-covered leading edges for the wings and special finishes were all investigated and used in varying degrees and at various stages; and E8137 at least was adapted to take a Calthrop Guardian Angel parachute. One clearly visible item that distinguished standard production Snipes from the experimental models was the steel-tube (instead of wooden) landing gear, and in postwar years the Badin fuel system (using a venturi) was sometimes replaced by a wind-driven pump mounted under the fuselage near the rearmost pair of steel-tube struts.

The contraction 'Pro.' in the bottom right-hand corner of this superlative Snipe-study presumably connotes 'production'. The picture is numbered S.559. Note the Sopwith trademark on the side of the aircraft.

'Long distance' Snipe (Mk.Ia), characterised by increased tankage—though this view is especially valuable in showing the installation of B.R.2 engine and the two Vickers guns, with trigger motors on top and ejection chutes below.

Respecting fuel supply, however, by far the most interesting development was the increased tankage for the Mk.Ia 'long distance' Snipe (Mk.I was a retrospective designation for the standard type, and, indeed, the name Snipe had not been officially adopted until February 1918). The Mk.Ia was becoming available at the end of 1918, its chief distinction being the fitting of a special main tank—of 50 gal (227 litre) capacity instead of 32 gal (145 litres) behind and beneath the pilot's seat, the rearward movement of the centre of gravity being compensated by a very slight sweep-back of the wings, some components of which (like some fuselage members) were strengthened. Snipe E8089 was the first to be converted to something approaching the 'long distance' standard, though its Dolphin tail unit was not perpetuated. This particular Snipe was on test at Martlesham Heath in October 1918; but handling was poor, and although 'long distance' Snipes went overseas, the Ia sub-type was never truly operational. In any case, with the machine virtually non-aerobatic at heavy loadings, and 'straightforward' flying only being foreseen, one may speculate on its bomb-carrying propensities, even with the normal Camel/Snipe loads of 4 × 20 lb or 1 × 112 lb. An 'escort fighter' inhibited from fighting in defence of bombers might at least help them with the bombing.

From the purely technical, as distinct from operational, aspect the most significant development was the adaptation of Snipe B9967 (the last of the original six that had been ordered, as already recorded, for experimental and development flying) to take the A.B.C. Dragonfly nine-cylinder air-cooled radial engine of well over 300 hp. This 'Dragonfly Snipe' had been completed as early as April 1918, and though the early-pattern small fin and rudder was retained (as originally fitted on B9966) the fuselage was lengthened by 1 ft 10 in (0.6 m) to give the longer lever-arm necessitated by the greater side-area of the new radial. Though the Dragonfly was very closely cowled (the crankcase and inter-cylinder cowling being in one piece,

This 'Dragonfly Snipe'—B9967—was completed as early as April 1918 and was the precursor of the similar aeroplane that was later renamed Dragon.

and cable-secured in Sopwith style) the cylinder heads were exposed. Such was the importance (as then judged) of this new powerplant installation that in October 1918 the name Dragon was officially adopted.

Less significant (being in a tradition established by the Camel) was the trainer version of the standard Bentley-engined Snipe, a two-seater wherein there were cockpit variations (notably in width), though Sopwith's own design of late 1918 or early 1919 provided for the second cockpit to be immediately behind the standard cockpit of the single-seater. Postwar RAF Snipe fighter squadrons usually had at least one, and sometimes two, of these instructional Snipes, and the type was still serving at Flying Training Schools after the fighter Snipes had gone in 1926. Here, however, the Sopwith part of the Snipe story reaches some degree of finality, for the Hawker company was certainly converting Snipes to two-seaters (as well as reconditioning single-seaters) by early 1922.

Horn-balanced upper ailerons proclaim this as a Snipe of late production. The photograph is dated 'Nov. 25/1918', but though the print or negative number has been removed by damage, the Snipe depicted is, in fact, E8184.

The handling of the Snipe was generally good, though it always had the reputation for being heavy laterally—in which connection observe that in the following excerpts from the *Handling Notes* (which are evocative as well as educational) there is a special warning that 'lateral control is heavier than on the Avro'. Thus a few sequential fragments:

'Petrol taps must be turned on. The throttle should be fully opened and the fine adjustment opened about one-third of the quadrant. Mechanics must close the air intake pipes by holding their hands over them. After sucking in, the fine adjustment should be closed, leaving the throttle lever one-quarter open. When both switches have been switched on and the engine first fires, the fine adjustment should be opened slowly until smooth running is obtained at about 600 r.p.m. or slower for about half a minute. This should be continued until the oil can be seen moving in the pulsometer glass . . .

'The Snipe taxies steadily, but no attempt should be made to turn in a strong wind without the assistance of mechanics on the wing-tips. When the aeroplane is being taxied, the engine should not be run on the switch. The tailplane lever should be in the half-way position. A slight forward pressure on the control column is necessary to lift the tail off the ground. The Snipe will take herself off after a short run at a flying speed of 50 m.p.h.

'An average cruising speed of about 90 m.p.h. can be maintained with the engine throttled back to 1,000 r.p.m. At low altitudes, a considerably greater flying speed can be attained without strain, provided the engine does not exceed its maximum r.p.m.

'The best climbing speed is about 65 m.p.h. The bottom of the front centre-section struts are just above the horizon. The tailplane should be adjusted so that the aeroplane climbs "hands off".

'The Snipe has a good steady dive but gathers speed rather quickly. The pilot should set forward the tailplane adjustment lever when diving, and should use it when coming out of a dive.

'The lateral control is heavier than on the Avro. On a turn to port, the nose has a tendency to go up, and a lot of bottom rudder is required to keep the nose on the horizon. In a steep turn to port, the control column should be pulled back slightly; if it is pulled right back, the aeroplane will stall and spin. In a turn to starboard, the nose has a tendency to go down. Top rudder is therefore necessary, and the control column may be pulled right back. For quick turns, the tail lever should be set right back.

'When the Snipe is being glided, the tailplane adjustment lever should be set right back. The normal gliding speed is 65 m.p.h. . . .'

In no action were the Snipe's fighting qualities more apparent than in Barker's 'scrap' with the Fokkers in October 1918; but this must not obscure the fine work done on the type by Australians—an interesting reflection and connection here between this present chapter and that to follow on the Dove. Barker was a Canadian, and in Ottawa today may be seen the fuselage of E8102—the most honoured survival from 1,550 Snipes completed. Canada had a few other Snipes (as had the USA) and the French evaluated a Snipe in 1918, though whether there is any link here with a Dragonfly engine being sent to France is unsure.

Five Snipes only passed to the British Civil Register, one of these, G-EATF (ex J365) being used—apparently without great success—as a demonstrator by the Aircraft Disposal Co. The others were G-EAUU (J459), G-EAUV (J453), G-EAUW (J455) and G-EBBE (J461)—this last specimen having been delivered to the Belgian Air Force in January 1922. An all-British 'Tally-ho' was the

Rigging notes would, it may be hoped, be available to the gentlemen seen with this Snipe 'in for repair'.

participation in the 1920 Aerial Derby (the Dragonfly-engined Sopwith Rainbow also) of 'UU, 'UV and 'UW, though hardly less British and sporting—even though on behalf of US movies—was the purchase of Snipes by film-star Reginald Denny, for Denny played parts as 'true-British' as the Snipe itself.

One recurrent question concerning the Snipe, and to which an answer may be attempted, is why the type should have been used by the RAF instead of the Martinsyde Buzzard, with its much higher performance. One possible answer (apart, of course, from availability of airframes) could have involved the preferred engine for the Martinsyde—a French Hispano-Suiza; another could have been wing flutter—a phenomenon that became increasingly apparent on RAF fighters after the Armistice.

Production orders for Snipes were:

Sopwith B9962—B9967; E7987—E8286; F2333—F2532; F7001—F7030 (some at least with Dragonfly engine); H4865—H5063 (last contract not completed).

Boulton & Paul E6137—E6536; J451—J550 (most of second batch probably not delivered).

Coventry Ordnance Works E6537—E6686; F9846—F9995 (second batch possibly cancelled).

Kingsbury J6493—J6522

March, Jones & Cribb J301—J400; J681—J730.

Napier E6787—E6936.

Nieuport & General E6937—E7036.

Portholme Aerodrome E8307—E8406; H8663—H8762 (second batch probably not delivered, but some Ruston Proctor airframes were completed by Portholme Aerodrome).

Ruston Proctor E7337—E7836; H351—H650 (some of second batch probably not delivered).

N.B. Several other Snipe contracts were cancelled, or changed for Sopwith Dragons or Nieuport Nighthawks.

7F.1 Snipe (typical production)
(230 hp Bentley B.R.2)

Span (over horn-balanced ailerons) 31 ft 1 in (9.5 m); length 19 ft 2 in (5.8 m); wing area 274 sq ft (25.5 sq m). Empty weight 1,305 lb (592 kg); maximum weight 2,015 lb (914 kg). Maximum speed at 10,000 ft (3,050 m) 121 mph (195 km/h); maximum speed at 15,000 ft (4,570 m) 113 mph (182 km/h); maximum speed at 16,500 ft (5,030 m) 108.5 mph (174 km/h); climb to 10,000 ft (3,050 m) 9 min 25 sec; climb to 15,000 ft (4,570 m) 18 min 50 sec; service ceiling 20,000 ft (6,095 m); endurance 3 hr (Snipe Ia 4½ hr).

T.F.2. Salamander

The fact that this present chapter—dealing with Sopwith's second type of armoured single-seat 'trench fighter'—is quite distantly removed in the book from that on the Camel T.F.1 (so that this Camel variant could be grouped with others of the same basic type) calls for repetition of one sentence from the T.F.1 account in order to establish a time-scale. Thus:

'On 7 March, 1918, the T.F.1 was flown to France, though its visit was apparently brief—possibly because the Salamander was already far advanced.'

This information we may now expand by noting that it was only in January 1918 that Sopwith had been asked to design the highly specialised Snipe-related machine which bore the designation T.F.2—an appropriate one not only because this was a later design in the Sopwith series (the T.F.1 having been an interim or stop-gap type, possibly offering alluring prospects of ready conversions from the F.1 design) but more especially because the Snipe-related, instead of Camel-related, machine was intended to comply with British Expeditionary Force Specification No.2 (though for that matter, so had the T.F.1, handicapped though it was by lower power).

Compared with dead-rear views of Snipes in the preceding chapter, this aspect of Salamander E5429 shows similarities and dissimilarities—the latter chiefly respecting fuselage and head-fairing. The Sopwith caption is essentially as on page 252, but the photograph number is S.361 and '200 hp Bentley Rotary Engine' has been added.

The cowling of the Bentley B.R.2 engine and the unstaggered Vickers guns show well in this close-up of the first Salamander—E5429. Maker's caption: 'S.365—Sopwith Salamander Trench Fighter T.F.2—May 1/18'.

By the end of the month in which the above-mentioned request was made—January 1918—the first of three T.F.2 airframes (E5429—E5431) was being erected. On 26 April it was sent to Brooklands, and on the very next day was first flown—the T.F.1 Camel having first been airborne at Brooklands only on the preceding 15 February. By the time the T.F.2 had flown the name Salamander had been officially recognised—and never was an aeroplane named more fittingly.

The three factors most vitally affecting the specialised aspects of design and employment (apart from the Salamander's overall compatibility with the Snipe) were engine, armament and armour. Respecting the engine, an air-cooled rotary (Bentley B.R.2, though the eleven-cylinder Clerget 11E was considered as an alternative) was preferred to a water-cooled inline type, not only by reason of its reduced vulnerability to return fire from the ground but because by its very form it constituted a protective metal shield (the merits of such protection still being acclaimed for RAF radial-engined fighters between the wars).

As for armament, although it was at first intended to have one fixed synchronised forward-firing Vickers gun only (or a Lewis gun having a restricted arc of fire—as had the top-wing gun on the T.F.1 Camel) in addition to two downward-firing Lewis guns through the floor (à la T.F.1, except that they would be adjustable in depression over a range of 35–55 degrees) the ultimate, and firmly-favoured, scheme was twin fixed forward-firing synchronised Vickers guns only.

Armour protection for the pilot and petrol tanks was a pre-eminent feature of design; no longer was 'a bit of armour underneath or an armoured bucket seat' sufficient—so the entire front fuselage was made as a box composed of armour plate (sufficient plates being ordered from Firths, in fact, for six machines, though three only were initially built). The box made up of armour plate was of these thicknesses: front (forming engine back-plate) 8 mm—the engine itself giving added protection, as already noted; bottom 11 mm.; sides 6 mm.; back 10-gauge sheet with 6-gauge close behind it. The pilot's head-fairing (not fitted on the Snipe) was also armoured, the frontal armour of this portion being in continuation of the forward of the two back-plates.

As flown at Brooklands on 27 April, 1918, Salamander E5429 had Snipe mainplane panels (in essence if not in detail) though the tailplane differed from the Snipe's in having an all-duralumin front spar. This metal spar has more than a passing interest, for during 1918 it was suggested at least once that three happenings only would bring down a Salamander—a direct hit by A. A. fire, the shooting-away of two flying wires or (author's own italics now) severe damage to *a main spar*. Prominent external differences were the flat-sided fuselage; early-type vertical tail-surfaces, as at first fitted to Snipe B9966—that is, with slightly bigger rudder than formerly, but still with the horn-balanced part unshielded; differences in aileron rigging (the ailerons, in any case, being aerodynamically unbalanced, as on the early Snipes)—and unstaggered guns (staggered guns being a distinctive feature of production Salamanders).

To France for Service test went E5429 on 9 May, 1918; and on 30 June E5431 was tested at Martlesham Heath. Flight-reports were generally favourable, though aileron control was judged to be heavy. On E5431, at some stage, the guns were staggered (port gun a few inches behind starboard) the reason for this arrangement apparently being connected with the need to carry 1,000 rounds (earlier 750 rounds) for each gun. Thus, all things taken into account a fully equipped Salamander weighed about 500 lb (227 kg) more than a comparable Snipe. (A point to make in this regard is that although potential bomb load was 4 × 20 lb or 1 × 112 lb, this was not considered in any sense as primary armament—except, perhaps, for 'special' attacks, as envisaged for Naval units and as pioneered by Camels).

Although it might be supposed that, being called upon to operate at low level and high power, the Salamander would have a larger fuel capacity than the Snipe Mk.I, the reverse was the case—weight being a prime consideration, though the quantity of petrol actually carried varied. Moreover, the Salamander's petrol-delivery system and piping layout differed from the Snipe's. In addition to the Badin vacuum-feed system, the venturi for which was on the front starboard centre-section strut of the Salamander (as on contemporary Snipes) there was a Weyman hand-operated pump for stand-by use and protected by armour. The fifth instruction in *Manufacturers' Order of Erection* was, in fact: 'Fit Main Petrol Tank and connect up piping as far as V.P. cock and first rubber joints in pipe from Weyman pump at bottom of Main Tank.' A gravity tank was fitted close behind the engine.

In the dive the Salamander's speed accumulated quickly. Gliding angle was steep, and heavy landings led to landing-gear strengthening. Nevertheless, the opinions of some pilots were manifestly over-critical or ignored the aircraft's unique character, for its armour gave adequate protection against German armour-piercing bullets fired from short range. Rigging presented problems of its own, and in January 1919 it was formally notified that the machine was in rigging position when the thrust line—which was scribed on the armour plate 'immediately aft of the side cowl on the port and starboard sides'—was level longitudinally and the fuselage was 'also level transversely'.

By this time, it was also made clear, the Salamander's tailplane was braced not by wires, as initially, but by 'four adjustable streamlined steel tubes', two above and two below. By means of these tubes the tailplane was to be adjusted until level transversely and until the incidence was correct (2 deg on the centre line).

The bottom wings were attached to 'fuselage housings' on the armour plate by means of a 'joint rod' and the rear end of the fuselage was attached 'by means of the four bolts through the joint box'. Other points of interest were that in truing-up the wings it was advisable first to adjust the upper wings until the dihedral was correct, and afterwards to adjust for stagger. It would then be found that any inaccuracy would be confined to the first bay from the fuselage on the lower wings. Such inaccuracy would occur at the lower-spar housings on the armour plate, and would be due to the 'very considerable' amount of distortion inherent in these plates, caused by the hardening process.

Clearly, the Salamander was a 'difficult' aeroplane from everyone's viewpoint—designer, constructor, pilot and rigger especially; though the difficulties must in the main have been foreseen when it was ordered into production (by the early summer of 1918). As production proceeded the type acquired the 'production-Snipe' fin-and-rudder assembly as well as the Snipe's horn-balanced top ailerons; though the slab-sided fuselage, prominent headrest, staggered guns and fixed tailplane still gave it visual distinction—not to mention the camouflage that was applied in more than one scheme, was administered 'on the line' after October 1918, and was to be seen in Egypt as late as 1922.

Production continued after the Armistice (at Sopwith's own Ham works until the summer of 1919)—totalling nearer 300 than 200, and, as far as can be discovered,

Comparable views, showing the first Salamander E5429 (opposite) in Sopwith photograph S.362 and, in close-up above, a production machine before final assembly. The panel showing between the unstaggered guns of E5429 bears the word 'Petrol' and a figure for 'gallons' ending in '$\frac{3}{4}$'. Especially interesting on the production machine, with its staggered guns, is the large case chute, with small link chute above it—and the loading handle for the starboard gun peeping up above the cockpit rim just behind.

every Salamander built having the B.R.2 engine. Nevertheless, Salamanders never equipped a squadron, although many were stored, and the type was not unfamiliar at RAF stations.

One example, F6533, went to the USA. But although some of its postwar history is cloudy, the Salamander was never strong in private-owner appeal—in the post-Armistice years at least, though the same would hardly be true today!

Production orders for Salamanders were:
Sopwith E5429—E5431; F6501—F7000 (not all of main batch delivered, but F6602, with strut-braced tailplane and horn-balanced top ailerons, was photographed on 25 January, 1919, and F6660 was at Farnborough in November of the same year).

N.B. A few in the range J5892—J5991 were completed by Glendower Aircraft Co Ltd, but large Salamander orders placed with the Air Navigation Co Ltd, National Aircraft Factory No.1, Palladium Autocars Ltd and Wolseley Motors Ltd were cancelled or curtailed by reason of the Armistice.

T.F.2 Salamander

(230 hp Bentley B.R.2)

Span (over horn-balanced ailerons) 31 ft 2⅝ in (9.5 m); length 19 ft 6 in (5.9 m). Empty weight 1,844 lb (835 kg); maximum weight 2,512 lb (1,139 kg). Maximum speed at 3,000 ft (915 m) 125 mph (201 km/h); maximum speed at 6,500 ft (1,980 m) 123 mph (198 km/h); maximum speed at 10,000 ft (3,050 m) 117 mph (188 km/h); climb to 6,500 ft (1,980 m) 9 min 6 sec; climb to 10,000 ft (3,050 m) 17 min 5 sec; service ceiling 13,000 ft (3,960 m).

N.B. Having regard to the Salamander's greater weight, largely attributable to its armour, it is of special interest to note that at 10,000 ft (3,050 m) the rate of climb had fallen off to about 330 ft/min (100 m/min) only, whereas the Snipe would still be climbing at about 700 ft/min (210 m/min).

8F.1 Snail

Of all the names in the Sopwith 'menagerie' the one that heads this chapter was the most appropriate—not in the sense that the aeroplane which bore it was slow (for it was not—any more than the Camel had been a clumsy beast) but because it had a shell-type, or monocoque, fuselage. That only one version of the type did, in actual fact, embody such a fuselage is of incidental, rather than salient, interest, for, as we shall see, the monocoque form of fuselage was perpetuated in the Snark.

Perhaps of greater importance is the fact that the Snail was built specifically to use a brand-new type of engine, the 170 hp seven-cylinder A.B.C Wasp, which was deemed conducive to compactness and light weight. These virtues the engine itself exhibited in notable degree. No less significant was the armament installation, which featured 'buried' Vickers guns.

So, all in all, the tiny Snail (for it was smaller even than the Camel or the Pup) was a fighter much out of the ordinary—the more so as, into the bargain, it had in one of its forms, a slight back-stagger. This last-named feature, allied with the fact that the wing span was almost exactly that of the D.H.5, gave a lingering impression that D.H.'s fifth wartime effort had outlived, or lived-down, its own unpopularity. One would not, however, overemphasise the backward stagger of the Snail, even though this was a feature which it shared with other Sopwith types (notably, of course, the Dolphin) for it was 5 in (127 mm) only, whereas on the D.H.5 it was no less than 27 in (686 mm). At the same time, it is fitting to give the wings priority of consideration over the two very different types of fuselage that were embodied, for the cellule—though differing in rigging and centre-section design—was essentially common to both Snail C4284 (conventional fuselage) and C4288 (monocoque).

Just how deceptive first appearances can be is illustrated by these comparative views of the 'conventional' Snail C4284 (top) and the monocoque Snail C4288. Apart from fuselage differences, C4284 has backward stagger and forwardly-placed pilot, with a centre-section cut-out for his head. Positive stagger is only one identifying feature of C4288. (Sopwith captions read, respectively: 'S.301—Sopwith Snail 8.F.1 160 hp A.B.C. 1st. M/c—April 13/18' and 'S.375—Sopwith Snail 8.F.1 Monocoque. 1st. M/c—May 9/18').

These two were the only Snails completed—though six had been ordered, to comply with Air Board Specification A.1(a), the '8F.1' suffix (or prefix, according to taste) being Sopwith's own type-designation.

The first order for Snails was placed on 31 October, 1917, the specified engine, as already noted, being the A.B.C. Wasp (which had first been submitted for official tests in that same month) and the form of fuselage construction conventional. Within a month—on 23 November, 1917—Sopwith were asked to build two additional examples, these to have a monocoque fuselage of plywood—a form of construction to which official attention had already been redirected by the British Aerial Transport Co, with their Koolhoven-designed, A.B.C. Mosquito-powered F.K.22. This last-named design, like the Sopwith Dolphin and Snail, provided for the pilot to sit with his head in a centre-section cut-out.

Here, in detail, is seen the installation of the A.B.C. Wasp engine and other salient features of Snail C4284. (Sopwith caption reads, after vacant space for number, 'Sopwith Snail 8.F.1 160 hp A.B.C. 1st.M/c.').

Whatever significance may be attached to the dating of orders, the 'conventional' Snail C4284 and the 'monocoque' C4288 were built concurrently. That some fairly basic changes in design-thinking then occurred is implicit in the intended reduction in back-stagger on 'conventional' (though never completed) Snails C4285—C4287 from 5 in (127 mm) to 3¾ in (95 mm)—and vastly more so on the monocoque C4288, the wings whereof had a *positive* stagger of no less than 22 in (599 mm). This being so, the pilot was seated further to the rear, his centre-section cut-out now being transferred to, and confined to, the trailing edge. (A point of some interest here is that the rival Westland Wagtail had *both* forms of cut-out in conjunction).

The facts behind these wide discrepancies in stagger may have some connection with the obvious fact that an aeroplane having so short a fuselage as the Snail's—

measuring well under 20 ft (6.1 m)—would be extremely sensitive to weight distribution. This same shortness in fuselage length may also have been a factor in the adoption of relatively large horizontal tail-surfaces—possibly, it appears, those of a Snipe.

One other Snipe-related feature was the top-wing mounting for the free-firing Lewis gun. This gun was mounted above the starboard edge of the centre-section cut-out, the rear end picking up a special fitting. In firing position, the gun was parallel with the aircraft centre line, but a pivoted arm attached to the rear face of the front spar allowed the gun to be swung inboard for reloading.

The forward centre-section struts on C4284 were uncommonly long (though the fact was partly concealed by the rounded fuselage sides) because they were attached to the lower longerons, and not to the upper ones as usual. The reason for this arrangement was that the upper longerons carried the two synchronised Vickers guns, the entire installation of which was exceptionally neat, for the rounded-section fuselage, of ample diameter, enabled the guns to be mounted in the sides of the cockpit. The sole external evidence that the guns were, in fact, present was a glimpse of the extreme muzzle-ends projecting into short troughs, and the chutes for the spent cartridge cases and belt links in the flanks of the cockpit. The Vickers guns appear to have been actually installed only on the first, or 'conventional', Snail (C4284); but whatever the extent of armament investigations may have been, an official inspection report recommended that flash tubes (or blast tubes, as they might otherwise have been called) should be fitted to the guns, the muzzles of which nested closely beneath the main petrol tank.

Photographs of the monocoque Snail C4288 suggest that not only were the Vickers guns themselves absent but that armament had been entirely abandoned, or that some wholly different scheme was intended or enforced; for apart from panels and holes associated with the Wasp engine, the beautifully finished fuselage appears immaculate, and—more significant—the centre-section strut arrangement is quite different.

Complete with direct-reading pitot-head installation on the port interplane struts, Snail C4284 taxis at Brooklands. The maker's caption reads: 'S.297—Snail. 160 hp A.B.C.—April 1918'. Clearly, there is no risk of confusion between the pitot head and the pilot's head.

The monocoque fuselage itself was elliptical in cross-section; the structure embodied many hoops and—in the rear part—four bulkheads. There were no longitudinal stringers or formers, the plywood skin being nailed to the hoops.

That the A.B.C. Wasp engine was installed with great care for a good aerodynamic entry is clear from the close-up photographs. Alas, like the larger Dragonfly, the Wasp was troublesome, and to this fact may be attributed delay in delivery of the engine for C4284 until 18 March, 1918. However, as a Sopwith-captioned photograph shows, C4284 was taxying (if not flying) at Brooklands during the following month; and meanwhile work was going forward on C4285 and

(Top) Snail C4284, immaculate without the test instrumentation shown in the foregoing view, though the Sopwith caption is dated 'April 13/18'. (Lower) Snail C4288—showing its positive stagger and monocoque fuselage.

The number on the tail proclaims this Snail's identity (C4288). The Sopwith photograph is numbered S.379, the aircraft is described as 'Monococque 1st.M/c', and the date is 'May 9/18'. Beyond are a D.H.9 (with curious cowling) and an S.E.5.

C4286—in addition to the monocoque C4288, which was among a number of Sopwith types shown off at Brooklands on 27 April, 1918.

Official interest in Wasp-engined fighters having lapsed, components of '85 and '86 were used as spares for the only completed specimens '84 and '88. Both these machines were sent to Martlesham Heath for trials in May 1918; but though speed and climb were good, compactness and four ailerons failed to confer manoeuvrability equal to the Camel's. There was little enthusiasm, in fact, on the score of handling generally, control at low speed being downright poor. The structural interest of C4288, however, led to the despatch of this machine to Farnborough, though performance figures quoted below apply to C4284.

Snail 8F.1

(A.B.C. Wasp)

Span 25 ft 9 in (7.9 m); length 19 ft (5.8 m); wing area (C4284) 228.6 sq ft (21.2 sq m). Empty weight (C4288) 1,390 lb (630 kg); maximum weight (C4288) 1,920 lb (870 kg). Maximum speed 124.5 mph (200 km/h) at 10,000 ft (3,050 m); climb to 10,000 ft (3,050 m) 9 min 55 sec; climb to 15,000 ft (4,570 m) 19 min 15 sec.

Buffalo

To dismiss the Buffalo—the last new Sopwith type to be produced during the war—merely as a two-seat counterpart of the Salamander would be to understate its purpose and technical merits and to underrate its potential. In essence, nevertheless, there was the same inhibiting provision of extensive armour plate—this by the requirement for operation at levels easily attained by small-arms fire from the ground.

A trim, positively fighter-like appearance (for in this regard the Buffalo outdid the Bristol Fighter and rivalled even its own companion-type the Bulldog—not to name the Austin Greyhound, Bristol Badger and Westland Weasel) quite belied its poor performance; for any new warplane of 1918 having a maximum speed at low

The first Buffalo—H5892 (top) —is seen here in an almost direct comparison with the second (H5893—lower). The Sopwith caption to the upper picture reads: 'S.596—Sopwith Buffalo Trench Fighter Two Seater—1st. Experimental—Sept. 19/18'. Note the different gun-mountings and extent of armour.

A superb study of Buffalo H5892. (Sopwith caption reads: 'S.599—Sopwith Buffalo—Trench Fighter Two Seater—1st. Experimental—Sept. 19/18'). Note especially the pilot's top-wing cut-out.

level of about 110 mph (177 km/h) and a service ceiling of 9,000 ft (2,740 m) must be accounted poor indeed in this respect. And yet, when it is considered that the Buffalo's crew numbered two, each with a plentifully-fed machine-gun; that the engine was of 230 hp only; and that the weight of armour, though not precisely known, must in itself have been a formidable hindrance, then the figures seem far less distressful.

The Buffalo was, in fact, quite an object-lesson in design. In the first place, it was remarkably compact, for the two-bay wings spanned only about 3 ft (0.9 m) more than those of the single-seat Salamander or Snipe. Commensurately modest were the fuselage dimensions—a fact to which the photographs attest by emphasising the relative bulk of the Bentley B.R.2 engine and its associated fairings. The small cross-section behind the engine represented not merely economy in weight but a wide downward field of view (and of fire) for the gunner, while presenting—in modern parlance—a 'low profile' to return fire from the ground.

To Sopwith the Buffalo was known initially as a 'trench fighter'—following the precedents of the special 'T.F.' Camel and the Salamander; but officially its purpose was 'contact patrol'. This specialised and dangerous function called for very low flying over the battle zone to determine by visual observation or 'contact' the dispositions of the infantry; hence the protective armour. 'Offensive patrol' was a secondary consideration—implicit in the absence of bombing capability.

Clearly, a wide field of view for the pilot was a primary requirement, and this was met by seating him high, close behind the engine, with his head in a large, oblong centre-section cut-out. Very close behind him was the observer/gunner, whose upward view was enhanced by a trailing-edge cut-out, though whose downward vision was impeded—on the first machine especially—by the bottom wing.

Two Buffaloes—H5892 and H5893—were ordered in July 1918, the 'fighter' element in the design, already touched upon, being accentuated by the stipulation that as many Bulldog parts as possible should be embodied. This stipulation may well have been responsible in part for the speed with which H5892 was prepared for flight-trials at Brooklands, where it arrived on 18 September, 1918, and was photographed (as two pictures herewith testify) on the following day. Very soon thereafter—on 20 October—H5892 was flown to No. 1 Aeroplane Supply Depot at Marquise, and experience in France brought recommended changes, though to what degree these were incorporated in H5893 when it was delivered to Martlesham Heath on 18 November (one week exactly after the signing of the Armistice) is unsure.

Conspicuous differences between the two Buffaloes were concerned with the engine-fairing, armament, armour, and tail surfaces, and may be summarised thus:

The second example, H5893, had an extended and conical-shaped engine-fairing—to port at least—though this embodied (as on the earlier H5892) a link

The distinctive form of engine-fairing on Buffalo H5893—the second example—is seen here to advantage. (Sopwith caption reads: 'S.674—Sopwith Buffalo No. 2 Armoured Trench Fighter 2 Seater—200 hp Bentley Rotary Engine—Nov/8/18.).

chute near the feed block of the pilot's synchronised Vickers gun—which was offset to port—with a case chute lower down and an air intake even lower. For the rear Lewis gun on the second aircraft there was a Scarff ring-mounting instead of the rocking-pillar mounting which traversed in a slot at the rear of the second cockpit of H5892; further, this new gunner's emplacement had additional armour protection, because the plating (which was structural, as on the Salamander) now extended rearward for an extra bay. The downward view for the gunner (or observer/gunner) was a little improved by sizable cut-outs in the bottom-wing trailing edges. Tail design on H5893 was much in contrast with the earlier form, for the rudder was newly shaped and far deeper in chord.

On both Buffaloes a prominent feature was the ring-sight for the pilot's gun, stayed by struts to the engine cowling, though there was provision also for a central Aldis sight.

Had the war not ended when it did there is little doubt that the Buffalo would have seen service in quantity; but development was not pursued, though H5893 was used for carburetter tests, and one Buffalo (possibly the same) served in another useful capacity, for as Harald Penrose recorded: 'By the end of February [1919] the Sopwith [Atlantic] was flying. Before dismantling for freighting it was filmed on duration trials by Lieut Engholm of Jury's International Pictures, who cranked his camera from a Sopwith Buffalo flying alongside.'

Performance figures quoted below apply to Buffalo H5893 when carrying a military load of 158 lb (72 kg) and 375 lb (170 kg) of fuel and oil.

Buffalo

(Bentley B.R.2)

Span 34 ft 6 in (10.5 m); length 23 ft 3½ in (7 m); wing area 370 sq ft (34.4 sq m). Empty weight 2,178 lb (988 kg); maximum weight 3,071 lb (1,392 kg). Maximum speed at 6,500 ft (1,980 m) 105.5 mph (169 km/h); service ceiling 9,000 ft (2,740 m).

Scooter and Swallow

It is hardly to be supposed that Tom Sopwith's pre-war dictum 'If you want speed the monoplane has it' was the direct reason why the two above-named delightful little parasol machines were built, though Sopwith's French associations, stemming largely from that early love of his, the Blériot, may have had something to do with the matter. However this may be, Harry Hawker's was the name most closely linked with 'Sopwith Monoplane No. 1', as the Scooter was first known; for not only did he use it as a runabout (or 'scooter') after its completion in July 1918, but it was he who bought it in April 1921. By that time it had been registered to the Sopwith company, first as K.135, then—in mid-1919—as G-EACZ. To complete the peacetime history of the dainty little Scooter before describing its inception: after Hawker's death it was placed in storage until it was overhauled for C. Clayton of Hendon, the new C of A being dated 1 August, 1925. In August 1926 it was sold to Dudley Watt, who used it frequently until—after it had flown in the Lympne Open Handicap of 18 September, 1926—it was sold as scrap in 1927.

The Scooter's fuselage was that of a standard 130 hp Clerget-powered F.1 Camel; nor did the tail organs differ noticeably. The swept-back monoplane wing, however, was altogether new. It was mounted very close indeed to the fuselage

Hardly surpassable are these comparative views of the Scooter (top) and Swallow. The Sopwith captions read respectively 'S.534—Scooter—130 hp Clerget Engine—Monoplane—July/1918' and 'S.635—Sopwith Swallow Monoplane No. 2—110 hp Clerget—Experimental—Oct/1918'. The fin and rudder of the Swallow both bear the marking 'AMA & E'.

(though in parasol fashion, as noted) on short splayed-out struts, and with a tall pyramidal cabane above these struts to anchor four landing wires which ran from its apex to the wing upper surfaces on each side. Below the wing, running up from the bottom longerons, were corresponding flying wires. So close was the wing to the fuselage that a trailing-edge cut-out was a necessity rather than a convenience.

As aeroplanes for aerobatic demonstrations, it has long been known that parasol monoplanes possess—for no other reason, perhaps, than their 'different' appearance—a definite advantage; and this in part accounts for the tremendous public successes achieved by Marcel Doret in his Dewoitines. Thus, anyone who saw Doret perform will relish especially this *Flight* account of the Scooter's act at the 1920 RAF Pageant—immediately after the Relay Race, for Avros, Bristol Fighters and Snipes:

'A mere civilian then gave an excellent exhibition of stunt flying. Our old friend, Harry Hawker, ascended on the pretty Sopwith Swallow [*sic*: its registration proclaimed it as the Scooter, though the two names were sometimes used without discrimination]—the only monoplane on the day's active list—and executed many extraordinary evolutions. They seemed quite different from the others we had been witnessing previously—but what the exact difference was we cannot well define. For one thing, Hawker had a peculiar way of terminating each loop, roll or spin with a sort of "jerk"—the machine coming to normal attitude quite suddenly and evenly.'

Yet even this 'mere civilian' on his 'pretty' Sopwith monoplane must have conceded some applause for the next item on the programme: the 'shooting-down' of a kite-balloon by a Snipe flown by Flt Lieut Hazell DSO, MC, DFC—pilot's war record, 34 aeroplanes and 16 balloons. This was especially true because in 1920 the Snipe was still the RAF's latest standard fighter; which brings us round again to the Scooter's lineal successor the Swallow.

Now having already mentioned Sopwith's own French associations, and likewise—in a wholly different context—the aerobatic displays of Marcel Doret, one refers once again to a Dewoitine monoplane, though this time to the D.373, used in the 1930s by the French Navy as a carrier-borne fighter, in succession to another parasol-winged fighter, the Wibault 74. That these French machines were indeed the first carrier-borne monoplanes to be used as standard equipment by any of the world's navies can hardly be disputed. But that the Sopwith Swallow of 1918 was itself intended for deck-landing is, in itself, a fact of real historical significance, for it was built and flown a full five years before that other pioneer British carrier-borne fighter the Handley Page H.P.21. Over the Vickers Vireo of 1928 its lead was far greater.

One other point of interest is that the very first Hawker (post-Sopwith) production—the Duiker reconnaissance aircraft of 1923—was of parasol-

During its life of nearly ten years the Scooter was variously marked. This view was taken by Sopwith on the same occasion (July 1918) as the ¾-rear aspect already shown. The reflection of the roundel seen on the wing is itself a reflection on the quality of finish.

monoplane form, though this was not for Naval use as was its near-contemporary the Blackburn Airedale. As for the name 'Scooter', conferred on the Swallow's aviary-mate, this was already familiar when adopted, though in early postwar years small motor-cycles so named became popular, and in 1919 Granville Bradshaw, whose A.B.C. engines have prominence in this book, produced his tiny 1 hp Skootamota ('decades ahead of its time', to quote an expert view).

Although the Sopwith Swallow (completed October 1918) greatly resembled the Scooter it was, in fact, considerably different, having, in the first place, a wing of greater span and larger area—a fact that might be related to the joint demands of shipboard operation and the fitting of armament. Even so, the area was much less

figures being 162 sq ft (15 sq m) and 221 sq ft (20.5 sq m)—and this figure for the Camel was appreciably less than that for the corresponding land-based version. Stiffness was augmented by multiple chordwise external strips between the spars—to a greater extent than on the Buffalo biplane; the wing was higher set above the fuselage than on the Scooter, necessitating longer attachment struts beneath the landing-wire pylon; and—partly by reason of a new wingtip shape, but also because of their greater span—the ailerons were increased in area.

While the raised mounting of the parasol wing may in some measure have been dictated by considerations of pilot-view, this feature also allowed easier installation of, and access to, the two synchronised Vickers guns that formed the armament. These guns were more widely spaced than on the similarly armed Camel (thus further improving the field of view) and lay almost wholly exposed—certainly lacking the familiar 'hump'—with combined large ejection chutes for cases and

Another cause for reflection—this time on the Swallow, with the fuselage roundel, as it were, 'superimposed' on the huge wing-roundel, exactly as in the directly comparable view of the Scooter. The Sopwith caption reads: 'S.636—Sopwith Swallow Monoplane No.2. 110 hp Clerget—Experimental—Oct/1918'.

Two aspects of the Swallow, showing its widely spaced Vickers guns. In the $\frac{3}{4}$-front view the loading handle for the starboard gun is prominent, and a ring sight is just visible between the front struts. In the dead-front view (Sopwith No. 633—their caption as for their S.636) the line between the guns marks the meeting of the wings. There was no centre section.

links immediately below them. Ring-and-bead sights were fitted for Martlesham Heath trials.

Apart from the gun installation and cowling, the Swallow's fuselage adhered closely to standard Camel F.1 practice (not 2F.1 be it noted, though with shipboard intentions still in mind, this point may be of scant significance—for shorebased operation would hardly have been out of mind, and may, indeed, have dominated). Indeed, the Swallow—which was sometimes called 'Monoplane No. 2'—still bore the number B9276, proclaiming its Boulton & Paul origins as one of a Camel F.1 batch. Nevertheless, the name 'Swallow' was applied in neat capital letters on the line of the top longerons just behind each roundel.

Having regard to the fact that the Camel airframe was fitted with engines of upward of 150 hp, overmuch may have been made of the fact that, as officially tested at Martlesham Heath (where it was delivered on 28 October, 1918) the Swallow was powered with a 110 hp Le Rhône—or, as Sopwith photograph-captions have it, a 110 hp Clerget. Development in the direction of the 230 hp Bentley B.R.2 (which was, in fact, used in the carrier-borne Nieuport Nightjar, before the arrival of the radial-engined Parnall Plovers and Fairey Flycatchers) could well have been intended, though this did not occur.

That in round figures the Swallow proved about 5 mph slower than a similarly engined Camel is not, perhaps, as significant as it might appear (the propeller, for instance may have been a compromise between demands of speed and climb) though fuel-system trouble which delayed official trials had presumably been fully rectified. In any case, the official test report was dated May 1919; and that was no propitious time for ordering development and production of the RAF's first monoplane.

Scooter

(130 hp Clerget)

Dimensions uncertain, but wing area about 135 sq ft (12.5 sq m). Maximum weight 1,300 lb (590 kg). Maximum speed approximately 115 mph (185 km/h).

Swallow

(110 hp Le Rhône)

Span 28 ft 10 in (8.8 m); length 18 ft 9 in (5.7 m); wing area 162 sq ft (15 sq m). Empty weight 889 lb (400 kg); maximum weight 1,420 lb (645 kg). Maximum speed at 10,000 ft (3,050 m) 113.5 mph (182 km/h); climb to 15,000 ft (4,570 m) 20 min; service ceiling 18,500 ft (5,640 m).

Dragon

It has been told in the chapter on the 7F.1 Snipe how B9967, the last of the original six Snipes that had been ordered for experimental and development flying, was fitted with an A.B.C. Dragonfly nine-cylinder air-cooled radial engine, and how this 'Dragonfly Snipe' was completed as early as April 1918—though the name Dragon was not adopted for the Snipe-development so powered until October of that year. (The Armistice was signed on 11 November).

After trials at Brooklands, B9967 was sent to the Royal Aircraft Establishment, Farnborough, on 11 May, 1918—little more than a month before the Dragonfly-engined two-seat Bulldog X4 unarmed flying test-bed, which went there on 25 June. These two Dragonfly installations—in the special Snipe and unarmed Bulldog—were clearly among the first that were made; and although it initially had two Vickers guns, B9967 itself was later used as an unarmed flying test-bed.

Having regard, however, to the Dragonfly's disastrous record of failure these same installations recalled unhappily the fitting, early in 1912, of an A.B.C. engine in the so-called 'Sopwith-Wright'—formerly an American-built Burgess-Wright, acquired by Tom Sopwith, brought to England in 1911, and rebuilt. 'Unhappily' one says because with the A.B.C. engine of 1912, Harry Hawker (see 'Other Men's Aeroplanes') had set up a British duration record of 8 hr 23 min; whereas—although it returned some remarkable performance figures for military aircraft when it was not more or less destroying itself—the later 300 hp-plus radial designed for A.B.C Motors Ltd of Walton-on-Thames by Granville Bradshaw—

In the chapter on the Snipe a view was included of the 'Dragonfly Snipe' B9967. That it was indeed 'completed as early as April 1918' as then stated is borne out by these Sopwith pictures of the same machine. All three views bear the date 'April 30/18' and the aircraft is credited with having a '360 hp A.B.C. Engine'. (Front view, S.348; rear view, S.349; $\frac{3}{4}$-front view, S.351. In S.351 only is the aircraft called Snipe; in the others the name 'Dragon—1st. Exp.' is used.

could not aspire in this direction. Nevertheless, at its very end in 1920 the Sopwith company was building A.B.C. motor-cycles, and production was continued by the new Hawker company. An A.B.C. Dragonfly engine, moreover, powered the Nieuport Goshawk biplane in which Harry Hawker was killed in 1921—likewise its Sopwith Rainbow counterpart of 1920.

That B9967 was sometimes called by Sopwith 'Dragon - 1st. Experimental' as well as 'Snipe - 360 h.p. A.B.C. Engine' is indicated by photographs; and its right to a new name was surely established not by the new radial engine alone but by the fuselage of concomitantly greater length. Photographs also show how 'Dragon - Experimental No. 2', and also production Dragons, differed from the first 'Dragonfly Snipe' (B9967), the '360 h.p.' quoted for the latter's engine being both tentative and nominal.

Although its number is obscured, this aeroplane is E7990, with its new production-type tail. The Sopwith picture is captioned: 'S.731—Sopwith Dragon. 360 hp A.B.C. Engine—Experimental No.2—Jan. 1919'.

The aeroplane just styled 'Dragon - Experimental No. 2', and shown in photographs dated January 1919, was in fact E7990 which, although sometimes regarded as 'the first real Dragon' was merely another Snipe conversion. This same machine, in fact, retained the plain top ailerons and wooden centre-section struts of B9967 and other early Snipes, but differed in having the larger (and curved) fin/rudder assembly of later-production Snipes; a two-piece engine-cowling, the longitudinal joints whereof are seen in photographs; and a Badin-type petrol-delivery system, the venturi for which was on the front starboard centre-section strut. But visible details were of relatively small consequence—and Heaven knows how many modifications, to engine and airframe alike, were made to the Farnborough 'hack' B9967. The engine, of course, was the more worrisome, though even by April 1918, when that same aircraft was first completed, the Dragonfly had been ordered in great quantities, primarily—it was officially declared—for use in a fighter to replace the S.E.5 and Sopwith Dolphin, both of which had the excellent though troublesome water-cooled Hispano-Suiza. (The Snipe was uncompromisingly a Camel replacement similarly with air-cooled engine). That the converted Snipe E7990 was at one time allocated to the USA may have a special significance respecting the Dragonfly engine, rather than the airframe; but though this machine seems never to have reached America it is seen, complete with engine, in Sopwith photographs taken at Brooklands in January 1919 and showing the name Dragon on the fuselage.

'The first real Dragon' though it may be considered, however, E7990 was positively not of the first Dragonfly-engined Snipe production contract, which had been awarded to Sopwith in June 1918 and specified thirty machines, numbered F7001—F7030. This contract was founded on the recognition of the far greater speed promised by the big new radial engine, which only the lifting power of the two-bay Snipe wings made a practical proposition—the Dragonfly engine weighing over 600 lb (270 kg) whereas the B.R.2. rotary weighed less than 500 lb (225 kg). It was apparent also that with electrical heating and other services the all-up weight must rise in service.

Thus one has hoped to show that the Dragonfly radial engine was an element in the Snipe development and production programmes from early times—without prejudice, however, to this present separate treatment of the Dragon as a distinct Sopwith type.

The true production-type Dragon had the inversely-tapered horn-balanced top ailerons of the later-production Snipes; similar (rounded) vertical tail-surfaces, with inset rudder horn-balance; a tailplane adjustable over a range of 0 deg 10 min to 3 deg 50 min—which was less than on the Snipe; and the longer fuselage, which gave the impression of being longer still when bearing a four-figure J registration—so 'modern-looking', like J3704 here shown! There were alterations in cowling details, and the muzzle attachments of the two Vickers barely cleared the valve gear on the Dragonfly engine's top cylinder. Though this latter distinction applied equally to B9967 and E7990, the ammunition ejection chutes were re-arranged on true production Dragons. Near the attachment-point of the starboard rear centre-section strut the steeply-sloping fuselage decking was cut away to a greater degree than to port; and the cockpit itself was deepened by virtue of a deeper fuselage section behind the coaming. In this same area there was an oblong door, appreciably behind the cockpit on the port side.

The cowling alterations seemingly included a gap behind and below the propeller hub, partially exposing the generators for the Constantinesco gun-synchronising gear and affording accessibility for timing purposes. The guns, as on the Snipe, were of the Vickers Mk.I* (Mark One, star) pattern, with a 1.8 in Aldis tubular optical unit-magnification sight bracketed from the fore-and-aft tube that crossed the centre-section cut-out, though this sight demanded no windscreen cut-out or perforation (as on B9967) for no transparent windscreen was fitted—the fuselage decking being formed accordingly, as intimated, and coming almost to a point, located just in front of the cockpit. Additionally, there were the usual ring-and-bead sights.

Thus it appears true that in some respects at least the production Dragon followed very early Camel practice—by getting the gun-muzzles close up against the propeller and thus, theoretically at least, simplifying synchronising, and also by making the decking (scuttle, hump or point) serve in some degree instead of a conventional transparent windscreen.

One further point to make concerning armament is that although the protective pads on the rear ends of the guns—and the loading handles—are clearly visible in both the ¾-rear view of J3704 and in a now-familiar close-up of the gun installation on Dragonfly Snipe B9967, in the latter there are no trigger motors on top of the guns, as are apparently fitted on J3704 (far forward on the rear covers). The point is possibly not significant—except for the fact that the RAF standardised on C.C.

Two Sopwith photographs—S.1029, ¾ rear; S.1030, starboard side. The original captions otherwise read 'Sopwith Dragon. 360 h.p. A.B.C. Production—July 1919.

hydraulic synchronising gear (involving the trigger motors, which were really plungers acting against stiff springs) whereas the Americans and others (including some quite strong elements in Britain) favoured mechanical gear.

The particular Dragon that finally reached the USA, where it was flown at McCook Field as Project Number P-149, was J3628, which had the asymmetrical top decking forward of the cockpit and was apparently in essence a standard RAF model. It seems to have survived until January 1926.

In the United Kingdom, development work on the Dragonfly engine (and production likewise) continued after the Armistice; and though the Sopwith Dragon was certainly deemed to be a standard RAF fighter in 1921 it never equipped a squadron before being declared obsolete in April 1923. Nor does the intended unarmed two-seater version, which Sopwith had designed by April 1920, seem to have materialised, the absence of armament on a modified production-type Dragon at the Royal Aircraft Establishment (the aircraft having heater-muffs on the carburetter air-intakes jointly with drooping exhaust tail-pipes which collected from the lower group of cylinders) having no apparent connection with this variant.

Though a number of Nieuport Nighthawks (originally designed for the Dragonfly engine) were later re-engined with either the nine-cylinder Bristol Jupiter or the fourteen-cylinder Armstrong Siddeley Jaguar radial, Sopwith Dragons were not so converted. Even so, it is worth noting that the original (two-bay) form of the Hawker Woodcock (1923) was fitted with an engine of each type, and that in tail design, as well as in other respects, bore a resemblance to Sopwith's decidedly reluctant Dragon.

The number of Dragons completed is not known, but all were Sopwith-built, the orders placed with the company being: F7001—F7030 and J3617—J3916. Of neither batch is precise production known, though of the second J3809 has been recorded. Some airframe parts may have been used for Snipes or Salamanders, and orders for many Dragons from other companies were cancelled.

Dragon

(A.B.C. Dragonfly)

Span 31 ft 1 in (9.5 m); length 21 ft 9 in (6.6 m); wing area 274 sq ft (25.5 sq m). Empty weight 1,405 lb (637 kg); maximum weight 2,132 lb (967 kg). Maximum speed below 10,000 ft (3,050 m) 150 mph (240 km/h); maximum speed at 15,000 ft (4,570 m) 141 mph (227 km/h); climb to 10,000 ft (3,050 m) 7 min 30 sec; climb to 15,000 ft (4,570 m) 13 min; service ceiling 25,000 ft (7,600 m).

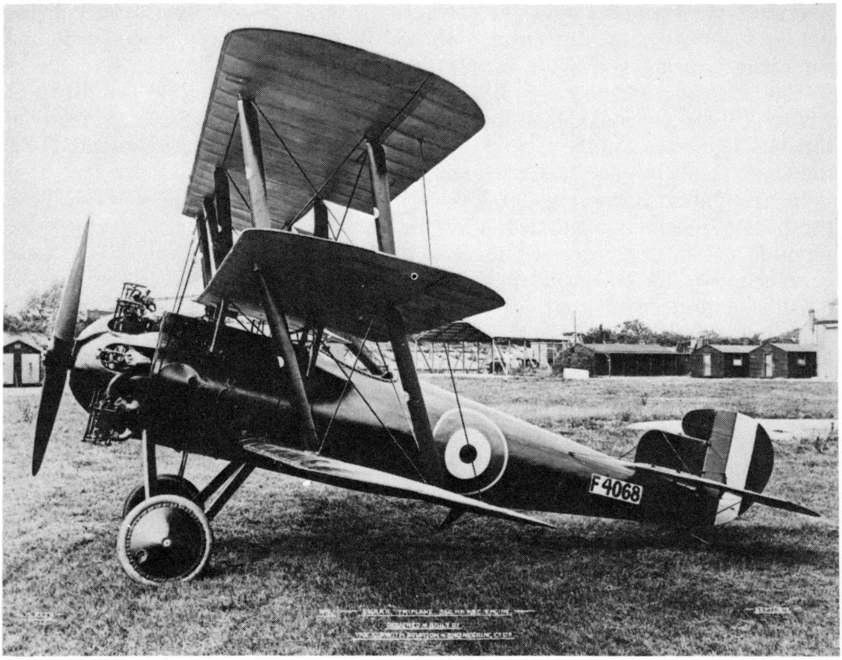

The first Snark, F4068, complete—except perhaps for the Lewis guns, attachments for which are nevertheless visible. The maker's photograph number is S.1079, the machine is identified as 'No.1 Snark Triplane 360 hp A.B.C. Engine' and the date is September 1919.

Snark

In respect of airframe and armament the Snark was one of Sopwith's most remarkable creations, and only in fairly recent times has its full significance been recognised.

Reversion to the triplane formula was interesting from several aspects, notably in that the Snark was designed to meet the same general requirements—those of the RAF Type I specification—as the similarly powered Snapper biplane, described in the following chapter. The generous wing area afforded by this formula—322 sq ft (29.9 sq m) against the 292 sq ft (27.1 sq m) of the Snapper—was considered to be beneficial both to performance and handling at altitude (especially when two Vickers guns were the sole armament) or, alternatively, to the carrying of a greatly increased armament, though obviously, in this case, with some sacrifice in performance generally. The maximum speed was, in fact, at least 10 mph less than the Snapper's.

The wing arrangement was far more complex than that of the famous rotary-engined Triplane single-seater of 1916 (the design of the Snark dated from early 1918, and Herbert Smith signed drawings on 20 April of that year), for not only were the interplane struts of 'normal'—that is, double—form, but stagger was sharply unequal (about 21 in on the upper bays and 12 in on the lower ones). As formerly, there were ailerons on all six main wing panels.

Of even greater interest perhaps (though it had a precedent in the Snail C4288) was the monocoque fuselage, with the basic armament of two Vickers guns almost literally built-in to it, slightly below the centre line and almost at the level of the rudder bar. The rounded cross-section of the fuselage blended very happily with the installation of the A.B.C. Dragonfly radial engine, whether a spinner was fitted (as on the third machine) or not (as on the first).

Three Snarks—F4068—F4070—were ordered in April 1918, and by 26 June following the still-uncovered wings of F4068 had been attached to the fuselage (itself completed, as were the landing gear and tail). The Vickers guns had been installed (prudently, if not necessarily, as will later become apparent) during May. Nevertheless, it was decided in October that only this first example of the Snark should be finished and delivered—a decision which (as in the case of the Snapper also) was not, in the event, acted upon. By October the airframe of F4068 had already been cleared for flight (this clearance having in fact been given in September) though an acceptable engine was not delivered until December. Peacetime-pace and magneto trouble apparently shared the blame for the installation not being completed until April 1919. In July there was an engine-change, and trials did not take place (at Brooklands) until September. By the end of 1919, Snarks F4069 and F4070 had been completed, and on 12 November (the war having then been over for a year) F4068 had been delivered to Martlesham Heath, though it was soon returned to Sopwith for another engine-change before resuming its official trials at Martlesham in March 1920.

Concerning the second Snark, F4069, little is known, though F4070 ultimately flew early in 1921. As tested at Martlesham Heath F4070 had a very large spinner (not open-fronted as on one of the Snappers, the Cobham I and the Rainbow), a modified engine cowling and very prominent air intakes for the carburetters, these intakes extending downwards from behind the engine almost half-way to the axle. The vertical tail-surfaces—similar in form to those of the early production-type Snipe—appear to have been identical with those of F4068; thus the side area of the new engine installation could have been of little consequence in this connection, and 'the hunting of the Snark' may not have become a serious phenomenon.

Having begun by remarking that one advantage of the Snark's generous wing area was considered to be that of benefiting performance and handling at height, and that another was the feasibility of augmented armament, it is fitting to conclude with two quotations bearing on both considerations. First we have the words of Oliver Stewart who, as a Martlesham Heath test pilot was qualified to affirm, in comparing the type with its Clerget-powered precursor: 'The Snark, with radial engine, although it possessed a better performance and was a satisfying aeroplane to fly, did not achieve the supreme handling excellence of the earlier model.'

As for armament, the present writer remarked some years ago: 'The Snark was the most heavily armed single-seat fighter of the 1914—18 war; but by the time the Gloster S.S.19 appeared in 1932 with a similar armament—two synchronised Vickers guns and four wing-mounted Lewis guns—its existence seems to have been entirely forgotten. That only one of the three specimens built appears to have been thus armed (the basic war load being the two Vickers guns) does not detract from the type's significance. The two Vickers guns were entirely 'buried' in the fuselage, and the Lewis guns were in two close-set pairs, one pair under each bottom wing, in which were provided a pair of staggered access panels. There were brackets for an

Aldis sight forward of the windscreen, offset to starboard, and a ring-and-bead sight a little further to starboard, the pedestals being canted outwards by the curved monocoque fuselage.' (The monocoque construction, of course, probably accounted for the early fitting of the Vickers guns during construction of F4068, as previously mentioned).

Yet even now, one fact of truly historic significance has yet to go on record: The Snark was the last RAF fighter, experimental or otherwise, to have a wooden monocoque (or something akin thereto) fuselage—until the arrival of the de Havilland Mosquito W4052, which made its first flight on 15 May, 1941. The only possible exception, one believes, might have been one of the B.A.T. machines—but the point would be an academic one indeed. (Should the Avro Avenger of 1926 be advanced as a claimant, then one would submit that this was a private-venture type which was never to bear Service markings).

A final footnote: The Snark's only known public appearance was on the occasion of a demonstration of the Wallaby, when it 'chucked stunts', seemed 'uncommonly fast', landed, and disgorged Harry Hawker, who was flying coatless, though 'everybody else was cold enough though well wrapped up.'

Snark

(A.B.C. Dragonfly)

Span 26 ft 6 in (8.1 m); length 20 ft 6 in (6.2 m); height 10 ft 1 in (3 m); wing area 322 sq ft (29.9 sq m). Maximum weight 2,283 lb (1,035 kg). Maximum speed 130 mph (209 km/h) at 3,000 ft (914 m); range 300 miles (480 km).

Snapper

'The machine is, we believe, known as the Sopwith Snapper', coyly ventured *Flight* in commenting on the single-seat biplane, wearing racing number 17, that was to have been flown by Harry Hawker in the 1919 Aerial Derby (the 'Victory Aerial Derby' as this fourth of the series was promoted). Apart from one or two scintillating snippets of intelligence—such as the aeroplane concerned having 'one pair of struts on each side'—it was further disclosed that 'the authorities' had refused to give permission for the machine to take part in the race, for the reason (it was believed) that the engine was Government property.

Although it was otherwise declared that the ban had been imposed because the engine was still on 'the Secret List' it can now, at least, be confidently asserted that this hapless aeroplane was indeed a Sopwith Snapper; that the engine was an A.B.C. Dragonfly I radial of 320 hp; that although the aircraft bore the registration K-149 on the fuselage side-panels it was later allotted the letters G-EAFJ; that this particular Snapper was seemingly one of three that had been designed (as the Snark had been) to the RAF Type I specification; and that all three of these were at the Royal Aircraft Establishment, Farnborough, as late as June 1920.

Aerodynamically and structurally the type had inter-related features of special interest, notably that, although having a single-bay wing cellule, this structure was uncommon in embodying a broad-span top centre section that was strut-braced well inboard of the attachment points for the outer panels, and also in having the single set of interplane struts placed far outboard. The result was to emphasise that the Snapper was no mere biplane version of the Snark (though there were common

A particularly informative aspect of Snapper F7031. The padded rear-ends of the two Vickers guns are in evidence, and this study shows in addition the large ejection chute for the cartridge cases and belt links from the port gun.

points in geometry), for the Snark's top centre-section struts were splayed out to the main attachment points. Simply stated, it looked as though the Snapper was asking to become a 1½ Strutter once again, so that *Flight's* seemingly naive remark about 'one pair of struts on each side' may have been less superficial than it seemed.

The wings were relatively broad in chord and the moderate aspect ratio gave a lower service ceiling than was attainable by the Dragon or the Mk.II Dolphin. Nevertheless, this very feature of broad chord accentuated the Snapper's trim appearance, though this was somewhat marred because the two staggered Vickers guns were largely exposed (even though they were emplaced in troughs) by reason of the small cross-section of the fuselage. Had a Snark-type monocoque fuselage been used, as was at first intended (hence, perhaps, the 'M' in one recorded designation—R.M.1) the guns might have been enclosed, with advantage to appearance and performance. As things turned out, the Snapper bore a striking resemblance to the Pup—and so (allusion having already been made to the 1½ Strutter) Sopwith fighter design appeared in the Snapper to have turned almost full circle.

Although three examples, numbered F7031—F7033, were ordered early in 1918, and by May/June work on the first was well advanced (the monocoque scheme having by then been abandoned) it was at one stage intended to reduce the order to one, with an ordinary wire-braced wooden fabric-covered fuselage. In the event, all three Snappers were completed (though well after the Armistice) the first of these, F7031, appearing at Brooklands in April 1919, apparently in the form shown in photographs reproduced—that is, with the Dragonfly engine having a large rounded crankcase-cowling but no spinner. Quite shortly afterwards—in June 1919—the civil-registered K-149, referred to at the outset as a would-be participant in the Aerial Derby, was briefly and prematurely in the public eye—as will have been gathered from the story of the 'secret engine'. The RAF identity of this machine—if any—is indeterminate, and the fact that K-149 was unarmed, and faired accordingly, has scant significance. It could well have been F7031 in a new guise—as indeed could the Snapper that was tested (with Service markings, and

The first Snapper, F7031, showing that the smooth-lined crankcase cowling for the A.B.C. Dragonfly engine still left the greater part of each cylinder, with its prominent valve-gear, exposed.

with armament installed) at Martlesham Heath in September 1919. The most obvious modification on this last-mentioned version, however, was a much-revised installation of the Dragonfly engine. In this instance the nose fairing was of such proportions that it could no longer be termed a crankcase-cowling, leaving, as it did, much less of each cylinder exposed to cooling air. It was fronted, moreover, by a very large blunt-nosed open-centred spinner, which left the front flange of the propeller hub exposed and conformed in all essentials with that used on the Rainbow racer, and shown in close-up in the rightful context.

That a considerable measure of official interest in the Snapper (of a technical nature, perhaps, rather than military) was sustained until well after the Armistice is suggested by work on F7033 that was still in hand as 1919 ended and by the presence of all three specimens at the RAE in June of the following year. General superiority was, nevertheless, conceded to the Nieuport Nighthawk, for—its two-bay wings notwithstanding—it was as fast as, if not faster than, the Snapper and its service ceiling was higher. Both these fighters carried two Vickers guns, synchronised by C.C. hydraulic gear, but the guns of the Nighthawk were internally mounted; both types carried 40 gal (182 litres) of petrol and 4 gal (18 litres) of oil.

Characteristic Sopwith features perpetuated in the Snapper were the staggered guns, as on the Salamander, and the form of tail that was first seen on the Snipe—with an almost rectangular fin partly overhung by the rudder horn-balance.

Snapper

(A.B.C Dragonfly)

Span 28 ft (8.5 m); length 20 ft 7 in (6.2 m); wing area 292 sq ft (27.1 sq m). Empty weight 1,462 lb (663 kg); maximum weight 2,190 lb (993 kg). Maximum speed at 3,000 ft (910 m) 140 mph (225 km/h); maximum speed at 17,000 ft (5,180 m) 126 mph (203 km/h); climb to 16,800 ft (5,120 m) 17 min 10 sec; service ceiling 23,100 ft (7,040 m).

Cobham

Insofar as it has achieved any distinction whatsoever in aeronautical history, the Cobham three-seat bomber has attained that status firstly by reason of the fact that it was the only multi-engined Sopwith to be built; second because it was the largest of the company's triplanes. Thus, in the absence of further aeronautical distinction, one may proffer a topographical note, for this aeroplane was named after the Surrey town not far removed either from Kingston-on-Thames, where it was built, nor from Brooklands, where it was flown. Town-names, indeed, distinguished its generally comparable near-contemporaries the Nieuport London, Avro Manchester, Boulton & Paul Bourges and de Havilland Oxford—not forgetting that far bigger triplane bomber the Bristol Braemar. (The Siddeley Sinaia one might be forgiven for forgetting, though like the Cobham Mk.II this had advanced-type Siddeley engines).

From the foregoing it must not be assumed that the Cobham was devoid of technical interest, and although that interest is largely concerned with the powerplant, the airframe and armament also invite some comment.

The triplane arrangement—though it distinguished also the London and Braemar—was obviously a Sopwith speciality, and one that continued to be favoured, by Herbert Smith for instance, for the Mitsubishi torpedo-dropper that

One of the finest studies of any Sopwith aeroplane (and likewise A.B.C. Dragonfly engines installed) is this close-up of the Cobham Mk.I. This was the second Cobham to fly, and the full maker's caption reads: 'S.1259—Cobham No. 2. 360 hp A.B.C. Twin Engines. Designed and built by The Sopwith Aviation & Engineering Co Ltd—March 4/1920'.

was completed in Japan in 1922. Inasmuch as the degree of stagger differed between the upper and the lower wings—and the interplane struts were accordingly 'cranked' in side elevation—the Cobham resembled both the Snark fighter and the Japanese machine just mentioned. However, whereas the Snark had positive stagger (though of different degrees) between the three sets of wings, the Cobham, in its Mk.II form at least, had noticeably back-staggered top wings—and even in the Mk.I form the centre wings were so set back from the others as to result in positive stagger for the 'upper storey' and neutral or negative stagger for the lower one.

The Cobham Mk.I had A.B.C. Dragonfly engines, was designed in the early summer of 1918, and was inspected in mock-up form in August of that year. The intended primary role was bombing over short and medium ranges, though an armed-reconnaissance capability was a secondary requirement. Internal bomb-stowage was to Sopwith no novelty; and on the Cobham this provision was made between the roots of the bottom-wing main spars. One quoted load of 750 lb seems reasonable; but in any case external storage of heavy-calibre bombs under the lower longerons would appear to have been compromised by the inverted-pyramid tubular structure which anchored the wide-track four-wheel landing gear to the fuselage (one pair of wheels beneath each engine). A fuselage access panel, beneath the root of the port centre wing, probably signifies vertical stowage—especially so as, at about the time when the Cobham was being designed, it was officially notified that: 'When a machine has to carry bombs vertically stowed—and this is by far the best method of stowage and can be employed for bombs of 50 lb., 65 lb., 100 lb., 112 lb., 230 lb. and 250 lb.—overhead girders must be fitted above the bomb cells to take the weight of the bombs.'

The study of a close-up photograph discloses another point of interest in the general area mentioned—this point concerning the roots of the bottom wings. These wings are apparently attached below the bottom longerons, and not directly to them; and this particular feature may have resulted from a decision—taken in October 1918 or thereabouts—to make the wings readily detachable. Some complications may thus have resulted, the cable-connected ailerons, for instance, being carried on all six wing panels.

Three Cobhams, H671—H673, were ordered, but in July 1919 work on the third was stopped. Although the intended engine for all three may well have been the A.B.C. Dragonfly—and thus powered the bomber was formally designated Cobham Mk.I—the chronic troubles that beset the Dragonfly caused the first machine, H671, to be fitted with high-compression Siddeley Pumas, and thus to be designated Cobham Mk.II. Though the Pumas had arrived at Kingston in November 1918, even at the end of March 1919 the Puma-powered Mk.II was incomplete, while H672 still awaited the Dragonfly engines shown in fine detail in the close-up photograph already mentioned (and dated almost exactly one year later—4 March, 1920). Even in July 1919, Cobham H672 was still incomplete, though it is clear that a firm decision to fit Pumas in at least one airframe had been taken appreciably before the end of 1918, whereas the Cobham itself (as already remarked) was designed only in the early summer of that year. Therefore—now noting that the Dragonflies in H672 were ultimately installed with quite exceptional neatness, the nacelles being small in diameter for maximum propeller efficiency and the propellers themselves having large open-fronted spinners as, for instance, on one version of the Snapper and the Rainbow racer—close attention to the Siddeley powerplant is clearly warranted.

It will have been noted that the Puma engines as initially mentioned in this account were described as being of high-compression type, and it is therefore

Comparative views of the A.B.C. Dragonfly-engined Cobham Mk.I (H672)—uppermost, although it was the second of the two Cobhams to fly—and the Siddeley Puma-powered Cobham Mk.II. Apart from the engines, the rudders are distinctive.

important to differentiate between this form of the Puma and the standard pattern. The latter—over 625 examples of which had been delivered by the Armistice—was officially described thus: 'This engine is of the stationary, water cooled type, with six vertical cylinders 145 mm. by 190 mm. It is rated at 240 h.p., but is capable of developing some 250 b.h.p. at the maximum permissible speed of 1500 r.p.m.' The weight was given as approximately 625 lb, and though no compression ratio was quoted, the figure was in fact 4.9 to 1. (As many unkind things have been said about the Puma, it seems only fair to add that one man at least who knew it well declared that it was a great improvement on anything else of its size and capacity, and that it was very reliable—as long as it was 'carefully handled and looked after'.).

During the latter part of 1918 a high-compression version of the Puma was under development, having a maximum power of 290 hp at 1,700 rpm, largely by virtue of the compression ratio now being raised from 4.9 to 1 to 5.4 to 1. For air-tests of this new version a D.H.9 was used in October 1918. In Sopwith Cobham H671—the Mk.II, which was sent to Martlesham Heath in the summer of 1919 and was at Brooklands at that year's end—two high-compression Pumas were installed in heavily louvred deep and slim nacelles mounted on the bottom wings and having frontal radiators. A Scarff ring-mounting for a Lewis gun was emplaced in the sharply rearward-sloping nose of the narrow fuselage, and there was a second mounting of the same type just abaft the wings.

That the fitting of high-compression Pumas in Cobham H671 is of more than passing interest is sure (even though similar engines were installed in the Avro Manchester, the Bristol Fighter and the D.H.9); and that it was no mere stop-gap measure to get a Cobham airborne in default of the Dragonfly is possible, for there was evidently a serious proposal to install in the Cobham (as in other British 'twins' and two-seaters) an engine called the Armstrong Whitworth Ricardo Patent R.H.A. And here it must be emphasised that 'Whitworth' is correct—not 'Siddeley' as might be supposed. The 'R.H.A.' signified 'Ricardo-Halford-Armstrong', and the engine concerned was of twelve-cylinder vee form, fitted (to quote a contemporary statement) 'with a special supercharging device, by which means the engine is capable of maintaining 300 b.h.p. at 10,000 feet.' On another occasion it was claimed that, when full use was made of the supercharger *at ground level*, the output was 360 hp.

Before the earlier-quoted statement was made (correctly emphasising performance at height) an officially-issued publication was listed as having the title *Supercharging as Applied to Aero Engines (Ricardo System)*; and though a 'Siddeley Puma R.A.' engine was one designation current, and though the Ricardo-Halford 'Inverted Supercharger' engine was evidently tested at Farnborough in a D.H.4, there is still good reason to place the preceding facts on record in connection with Sopwith's one-and-only 'twin'. This belief is supported because two R.H.A. engines had been completed when the Armistice came, and four more were almost ready. Of the first pair, one was claimed to have been sent to Farnborough before the Armistice and to have been tested in a de Havilland airframe, though development was soon abandoned.

The Dragonfly-engined Cobham Mk.I (H672) was eventually to be air-tested, though not until the spring of 1920. This aircraft differed in tail design, the rudder (now having a lower horn balance, as well as the upper one as on the Mk.II machine H671) being extended well below the fuselage and necessitating a tall and elaborate tailskid. More basic were the changes in the setting of the triplane wings, as noted in the third paragraph of this present chapter.

Cobham Mk.I

(Two A.B.C. Dragonfly)

Span 54 ft (16.5 m); length 38 ft (11.6 m); height 13 ft (4 m). Maximum weight 6,300 lb (2,858 kg).

Cobham Mk.II

(Two Siddeley high-compression Puma)

Data essentially as for Mk.I.

Atlantic

The early chapter in this book which dealt with 'Circuit Seaplanes' recorded how, early in 1913, the *Daily Mail* newspaper had offered £10,000 for the first flight across the Atlantic Ocean. Though hopeful preparations had been put in hand—notably by Martin & Handasyde—the war came both as an interruption and as a technical stimulant; so that in July 1918, shortly before the Armistice, T. O. M. Sopwith was able to declare: 'The Transatlantic flight could be made and the prize won this month . . . Crossing from America to England by air is not the problem it was a few years ago . . . A dozen machines of today could do it. They could do it at once if aeroplane makers and pilots were not all busy with war demands.'

No equivocation here; and, in fact, the number of machines that *was* eventually assembled in 1919 for the great adventure did approach Sopwith's 'dozen' (actually eight). Eminent among these were—as hoped so long before—a Martinsyde (*The Raymor*, manned by F. P. Raynham and Capt C. W. F. Morgan); the Sopwith Atlantic, now to be studied, and crewed by H. G. Hawker, with Lieut-Cdr K. Mackenzie-Grieve as navigator; and a certain Vickers Vimy manned by Alcock and Brown, and concerning which it would be impertinent to write more here. We may note, nevertheless, that all three of the equipages mentioned were closely tied with Brooklands (and carried 'Tabloid' first-aid kits!).

Although the Sopwith Atlantic—as the inevitable Harry Hawker's challenger was properly called, though Rolls-Royce was often linked with the name, or with some variation, and Sopwith also used the appellation 'Transport'—was based on the design of the B.1 Bomber, and on accumulating experience with the Cuckoo, and was built in a mere six weeks (it was flying by the end of February 1919) it represented no patched-up operation either in design or equipment. The man chiefly responsible for the metamorphosis was W. G. (George) Carter.

The chief innovations, apart from the extra-large fuel-tankage (between the cockpits and the engine) were the jettisonable landing gear, saving drag, weight and petrol—beneficial to cruising speed and seemingly attributable to Hawker's own insistence; wooden skids to reinforce the bottom longerons and serve as landing gear; an upturned lifeboat which formed the aft decking of the deepened rear fuselage; the fitting of special wireless (initially directional, but changed, after tests, to a T.55A transmitter, loaned by the Air Ministry); a retractable wind-driven generator to provide power for the wireless; and staggered seats for the two occupants—for easy communication and (as one account put it) to facilitate changing watches during the long journey.

The most common criticism of the aeroplane was that it should have one engine only, though another concerned the releasable landing gear, this being considered (though not by Hawker himself) to put the crew at risk in alighting and deny them the hope of taking-off again.

The all-important engine was the very best that Britain could produce—a Rolls-Royce Eagle VIII water-cooled Vee-12 unit of 360/375 hp. The radiator was frontally mounted, with adjustable cooling-shutters, and there were long exhaust tailpipes extending straight aft to a point behind the cockpits. Early in its operational career, the Eagle engine (designed by Henry Royce in collaboration with his young chief assistant A. G. Elliott, and first run in October 1915) had been known as the '250 h.p. Rolls-Royce'; but an official 'R-R' instruction book dated December 1917 ascribed the following outputs (at normal rpm.) to successive Marks: Eagle I 225 bhp; II 266; III 284; IV 284; V 322; VI 322; VIII 350. Certainly,

So revealing is this view of the Sopwith Atlantic that, studied with other pictures reproduced—especially those taken at Brooklands on the same occasion and bearing maker's reference numbers accordingly—it obviates a lengthier textual description. The Sopwith caption reads: 'S.804—Sopwith Transport 375 hp Rolls Royce. Trans-Atlantic M/c—Feb. 21/19'.

by the time of the Atlantic attempts the Eagle VIII was already well-proven, notably in Handley Page bombers. Its cubic capacity was 20.32 litres, and an epicyclic gear gave the left-hand tractor propeller a reduction ratio of 0.6 to 1. (*N.B.*: The A. G. Elliot just mentioned must not be confused with A. F. Elliot, who helped to design the A.B.C. Wasp, Dragonfly and Gnat).

Although a four-blade propeller had been used on early tests of the Sopwith Atlantic that were made in England, the pattern selected for the transoceanic flight—with take-off performance especially in mind—had two blades only, giving a diameter of 9 ft 6 in (2.9 m).

The tense adventure-story of the attempted crossing from Newfoundland has been written at length by so many and various pens that a brief profile only is called for. Thus:

Hawker and Grieve airborne 5.42 GMT, 18 May, 1919. Almost immediately ran into fog off the Grand Banks. Pressed on through fog. After fog cleared, Grieve estimated that aircraft had flown 400 miles. Massive clouds now loomed, but

The curious staggered cockpit arrangement is particularly evident in this dead-rear aspect (Sopwith number S.803—otherwise captioned as side and $\frac{3}{4}$-front views).

Hawker climbed above them. After nearly six hours' flying engine-water temperature rose. Radiator shutters failed to work, so, with engine throttled, Hawker dived for 3,000 ft to cool it and dislodge any obstruction. Situation improved—but only temporarily, and on renewed climb the centre of the top wing iced-up from freezing steam. Early on the 19th (a Monday) cooling trouble returned and worsened. More bad weather; and Hawker headed south for shipping lanes, coaxing engine. Came the dawn, and, through the mist, the ss *Mary*, of Denmark. Signals (Very lights) from the Sopwith; Hawker steered aircraft about a mile ahead of ship. Text-book ditching (see Hawker's own remarks later) and the Sopwith's lifeboat launched as aircraft was sinking. Boat arrived from *Mary*. Everyone safely aboard—but *Mary* had no wireless, and the world was left in an agony of waiting.

An account of the Atlantic attempt, ascribed to Hawker himself, was factual and vivid, and in its very first sentence it proclaimed the Brooklands camaraderie by recording that, before take-off, Hawker and Grieve sent their respects—'and hopes of seeing him at Brooklands'—to Fred Raynham, whose rival Martinsyde nevertheless crashed on setting forth from Newfoundland.

Hawker described his own take-off as 'just a bit ticklish', the 'bit' being measurable by the number of inches by which he missed a drainage ditch. Then—true to form as an observant and technically minded test pilot—'As soon as the coast had been passed, I pulled the undercarriage release trigger and away it [*sic*] went into the water. Simultaneously the finger of the air speed indicator went over to another seven miles an hour.' Nor was this Hawker's only observation concerning the jettisonable wheels, for of the period immediately preceding the historic splash-down near the good ship *Mary* (which occurrence, Hawker thought,

In this study (Sopwith S.801—and otherwise captioned as the side and dead-rear views) the original four-blade propeller and the radiator shutters are prominent.

went 'quite nicely', with the aircraft riding clear of the water by reason of the partly empty petrol tanks) he recalled: 'There was no lack of rain squalls, the wind was getting stronger and gustier and bumpier every minute, and the sea rougher. I was very glad to be without the undercarriage, and would rather have been as I was than have floats under me, for the waves looked too heavy for any ordinary seaplane to stand.'

Clearly, the wartime lessons learnt by the Navy and the RAF in the ditching of Pups, 1½ Strutters and Camels had been put to good account—though Grieve, as a Naval officer, observed the traditions of the Silent Service while Hawker shouted for joy.

A moment, this, to have set the hearts of all mankind athrob; though for days the world was left, not to wonder, but to wait and worry. As Hawker said: 'We had hoped to fall in with a ship equipped with wireless so that we could communicate with our people in England, and of this Captain Duhn, who spoke excellent English, thought we had a good chance, but later on the storm got considerably worse and he had to heave to, only making very little way in a northerly direction and so going further away from the busier shipping route. So it was not until we were off the Butt of Lewis that we could communicate with home and the world that at one time had seemed so distant.'

Then the old Brooklands comradeship shone through once more, as Hawker's story concluded: 'The men who should have the reception are Raynham and Morgan, for what they did was a magnificent act of pluck. The east-north-east wind was not by any means a bad one for our getting off, for it suited our aerodrome pretty much as well as any other and better than most, but it was almost the worst possible wind for the Martinsyde aerodrome. But knowing this Raynham and Morgan never hesitated to attempt the flight . . . They were visited with cruel hard luck indeed.'

And just what kind of luck, it might be asked, had Hawker himself encountered? But in conveying these personal glimpses of fine men one is seeking merely to establish sturdy links not only between the Sopwith and Martinsyde companies, but likewise between Sopwith and the succeeding H. G. Hawker Engineering Co Ltd, established late in 1920.

As already noted, it was a Sopwith draughtsman, W. G. Carter, who had the task of remodelling the basic B.1 Bomber airframe into a virtually new aeroplane for the Atlantic attempt—just as, in later years, Carter redesigned the Hawker Woodcock fighter into something acceptable by the RAF and designed from scratch the Horsley bomber, which was itself later specially arranged (as the Sopwith Atlantic had been in its day) for extreme long-distance work. In even later years Sir Sydney Camm—to whom the design of the Horsley is sometimes incorrectly ascribed—recalled for the present writer how he had joined the newly formed Hawker company from Martinsyde not long after Harry Hawker's death in 1921, and at about the same time as his friend and Martinsyde colleague Raynham had taken a similar step. That Hawker's high opinion of his fellow test pilot was shared by Camm was clear from Camm's recollection of flying with Raynham (I think in the Hedgehog) and of his unwonted emotion in mentioning 'Old Raynham'. (To Sydney Camm all men and things he held in high esteem were 'old').

Thus it was that the Hawker company had in the 1920s a personal structure buttressing the business side; and in the planning and manning of the Atlantic—one of the last and finest of the Sopwith line—this was already becoming manifest.

Links between companies and individuals being now in mind in the context of the Sopwith Atlantic, with its Rolls-Royce Eagle VIII engine, it is particularly

This view of the Atlantic (obviously taken on the same occasion as those for which numbers are quoted) shows the retractable wind-driven generator; and, such is the lighting, the cowling louvres are more than usually obvious.

appropriate to quote from a letter addressed to the press by Claude Johnson, Rolls-Royce's managing director, following the first nonstop Atlantic flight—by Alcock and Brown in a Vickers Vimy, though (as may be judged from the wording) the letter may well have been based on a draft press release prepared in anticipation of success by the Sopwith. Thus the letter read in part:

'It has been calculated that if the Atlantic crossing were completed in 20 hours by one Rolls-Royce engine running at an average of 1,800 revolutions per minute, its performances would be as follows (in a two-engined 'plane such as the Vickers Vimy the operations would naturally be doubled):—Each engine will make 2,160,000 revolutions, and each piston will travel up and down the interior of its cylinder a total distance of 440 miles. As the Rolls-Royce engine has twelve pistons, they will in all travel 5,280 miles. The valves will be operated some 25,920,000 times in each engine . . .'

Although, as already noted, a Sopwith association may be implicit here (the Martinsyde contender had a Rolls-Royce Falcon III engine, of lower power and shorter stroke, so the figures for piston-travel would not have applied) a similar link is not instantly apparent from an ensuing paragraph concerning what Johnson termed 'recent history of the Rolls-Royce aero engines.' This paragraph—the full significance of which will later be made clear—ran as follows:

'On 11th November, 1918, when armistice was declared, there were in possession of the Royal Air Force Rolls-Royce aero engines of a total horse-power of over 1,000,000, which far exceeded that of any other make of aero engine in use. Previous to armistice day there had been constructed 122 Handley Page bombers, of which no less than 113 were fitted with Rolls-Royce engines. Prior to the same date 1,524 complete Bristol Fighters had been delivered, of which number 1,364 were also equipped with Rolls-Royce engines. Rolls-Royce engines were used exclusively in the planes of the London—Paris Government courier service for the conveyance of Ministers, officials and dispatches to and from the Peace Conference . . .'

Paradoxically, the promised clarification of the foregoing lies in the fact that although allusion is made to Handley Page, Bristol, Vickers (and by 'Peace Conference' implication) de Havilland, there is no reference to the great Sopwith organisation which is our present concern. That no operational Sopwith military type was, in fact, ever fitted with a Rolls-Royce engine (and only the highly experimental L.R.T.Tr. carried the 'R.R.' trademark—and that, perhaps, by reason of Government insistence) will be apparent from the contents of this present book; but that the names of Rolls-Royce and Hawker were later to make a resounding impact on aeronautical history is made clear by another volume in this series (*Hawker Aircraft since 1920*, by Francis K. Mason). Thus the Sopwith Atlantic represented not only a glorious failure—and a no less glorious approach to the ending of the Sopwith line—but equally the beginning of one of the greatest families of aircraft ever known, broadly delineated by Horsley (Condor), Hart and Fury variants (Kestrel), Hurricane (Merlin), Sea Hawk (Nene) and Hunter (Avon). Truly the pistons had 'travelled up and down' incalculable miles since Fred Sigrist, in 1912, was writing helpful hints on how to tune the Gnome.

That even the Eagle VIII did not bear out the present Sopwith's name Atlantic (though Hawker assured 'T. O. M.' that it ran perfectly even when all the cooling-water had boiled away) the failure was by no means a catastrophe; for among the significant technical aspects was the taking of "X-ray" photographs (the term was still commonly printed in quotation marks, for Röntgen had not announced his discovery until 20-odd years earlier) and the publication of one of these photographs in 1919 with the following notations:

'An interesting experiment in connection with the engine of the salved aeroplane in which Mr. Harry Hawker was forced to descend in the Atlantic has produced a radiograph of the lower part of the engine's radiator. Owing to the science of radiography, it is now possible to take "X-ray" photographs (or radiographs) through 2 in. of solid steel. At the request of Mr. T. O. M. Sopwith, the designer [*sic*] of Mr. Hawker's machine and Mr. Gordon Selfridge (on whose premises the wrecked machine was exhibited), Harry W. Cox and Co., Ltd., "X-ray" specialists, of Wigmore Street, London, W.1, took a radiograph of the lower part of the radiator, in order to ascertain the presence, or otherwise, of any foreign matter, such as might have caused the engine failure.'

It was then explained how one radiograph clearly showed the flanges and all details of the soldered joints; how small white patches on the flanges denoted entire absence of solder; that the lightest grey patches indicated a small quantity of solder only, while darker blotches represented large masses of solder; and the report concluded:

'Owing to the aeroplane having been submerged in the Atlantic for some time, any foreign matter that may have lodged in the radiator must have been washed out, as a completely free path through the tube can be seen in the radiograph.

'The importance of the use of an apparatus of this kind is palpable. Faults in castings, for instance, which it would be impossible to detect by any external examination, or—as in this instance—three distinct grades of soldering, are accurately revealed. In the aircraft industry, where so much depends upon the absolute perfection of all parts of the machine, an examination of the essential portions, by radiography, is invaluable.'

Invaluable in many other ways were the lessons learnt from the Sopwith Atlantic. Knowing that he could not carry on indefinitely with the Eagle's cooling-water boiling away—owing, he thought, to a blockage with refuse such as solder—Hawker had 'played for safety', as he put it.

Unlike 'absolute perfection'—the term used by the "X-ray" commentator—'safety' is relative.

Atlantic

(Rolls-Royce Eagle VIII)

Span 46 ft 6 in (14.1 m); length 32 ft (9.7 m); wing area 547 sq ft (50.8 sq m). Empty weight 3,000 lb (1,360 kg). Maximum speed 118 mph (190 km/h); cruising speed 105 mph (169 km/h); ceiling 13,000 ft (3,960 m).

Wallaby

To identify this relatively obscure Sopwith 'one-off' civil aircraft—originally registered G-EAKS—these facts may be set out:

(1) It was closely related to the earlier Atlantic, and sometimes shared with that machine the denomination 'Transport'—the general form of aircraft represented by this pair being a prospective passenger-carrier or freighter.

(2) The Wallaby resembled the Atlantic not only in appearance and dimensions, but also in having a Rolls-Royce Eagle VIII engine.

(3) It differed from the Atlantic in having three-bay, instead of two-bay, wings, in its passenger accommodation and structure, and in dispensing with the extra-heavy fuel provision and special emergency facilities (notably the lifeboat and the jettisonable landing gear).

(4) As its name suggested, the Wallaby had very strong Australian associations, for it was built at Kingston-on-Thames under the personal supervision of an

Variously publicised, for instance as 'a special version of the Sopwith Transport, built for the Australia Flight', the Wallaby is studied here with a particular degree of intimacy (notably respecting the curious crew-accommodation and the installation of the Rolls-Royce Eagle VIII engine. The Sopwith caption reads: 'S.1168—Wallaby. 375 hp Rolls-Royce—Oct.3/1919'.

That the Wallaby differed from the Atlantic notably in having three-bay wings is emphasised here.

Australian (Harry Hawker) to compete for the £10,000 prize which had been offered by the Australian Government in March 1919 to the first Australian who would fly back to his homeland within the space of thirty days, before the year's end, in an aircraft of British or Commonwealth manufacture. Not only was the Wallaby's nominated crew Australian (as detailed later), but a Sopwith-associated company—the Larkin-Sopwith Aeroplane Company (of Australia)—had been established with offices in Melbourne in anticipation of competitive and commercial success.

(5) Even beyond these personal and commercial ties, there was a strong nationalistic feeling towards the undertaking, expressed with special warmth in this message from the Australian Prime Minister to Capt G. C. Matthews (former Camel-pilot, and first pilot and commander on the imminent venture): 'Wish you and Serg. Kay every success in your great adventure. While every one of your fellow-citizens hopes that an Australian aviator may be the first to fly from Europe to Australia, and so achieve what will be easily the world's record in aerial navigation, I want you to take no unnecessary risks. Plug on day after day doing your best, but do nothing foolhardy. If you cannot make Australia in thirty days never mind. The main thing is that an Australian should get here first. If you do that you need not worry. Good luck. (*Signed*) Hughes, Prime Minister.'

As first pilot, Matthews (having been a master mariner) was credited with 'twelve years of practical navigation', while the aforementioned Sergeant Kay—though primarily a mechanic—was competent to 'take turns in flying the machine'.

Much had happened since Harry Hawker had taken the Tabloid to show the folk down-under in 1914.

The following contemporary account of the Wallaby has a particular interest and value, the interest being apparent at the very outset, where the earlier Sopwith type-name 1½ Strutter is explained with unusual clarity as connoting, in effect '1½ bays'. Thus the Wallaby:

'It is a three-strutter machine, with a slight dihedral, of somewhere about the same size as the Transatlantic machine, to which, of course, it bears much resemblance. It is, however, a good deal more lightly loaded than its predecessor, as it only carries 200 gals, of petrol instead of 350. [N.B. The true fuel *capacity* of the Atlantic was closer to 400 gal (1,818 litres.)] The actual machine is slightly heavier and stronger in construction. The arrangement of the cockpit has several features of special interest. The pilot's seat can be raised so that he looks out over the top of the fuselage [Author's note: A preview of the Heinkel He 111P?] or lowered and a lid pulled down over his head so that the occupants are entirely enclosed. There are two rudder bars at different heights. The passenger's seat can similarly be moved, and there is a complete set of dual controls, the joy sticks being removable. The whole place is quite roomy, and has windows of triplex glass. Captain Matthews finds he can see perfectly well from inside the fuselage, which has a window below as well as at the sides. There is an air intake to bring fresh air to the occupants instead of air tainted with engine oil, and windows at the side can be opened. It is, of course, well fitted out with instruments; besides the usual engine one, the compasses, and the air-speed meter, there is a turn-meter, which by recording the difference of air pressure on the two wing tips, tells the pilot if he is keeping on a straight course when he is in a mist; there is a flow meter, recording the rate of consumption of petrol, which works out at about 15 gals. an hour [this seems astonishingly low—Author]; a spirit-level for sideways motion, and an inclinometer for measuring the angle fore and aft; and an azimuth mirror for checking the compass by readings from the heavenly bodies on a system patented by Captain Matthews himself. The window below the pilot too is marked in degrees so that he can observe the direction of drift. There is a wheel at the side for altering the angle of the empennage in flight. The modern pilot, especially if he is also the navigator, has plenty to attend to.'

Clearly, this last remark was true—especially on such an arduous expedition; and after setting out from Hounslow on 21 October, 1919, Matthews and Kay clearly deserved the success that was nevertheless denied them by bad weather, one arrest, and damage in Persia. Eventually they crashed when landing in Bali, in the Netherlands East Indies, on 17 April, 1920.

This chapter of accidents was not the Wallaby's end, however, for it was shipped on to Australia, rebuilt as an eight-seater, and its registration changed to G-AUDU—on behalf of Australian Aerial Services Ltd.

Whatever truth there may have been in the story that the terminal letters of the new registration signified 'down-under at the end', the Wallaby must be assessed as a true advance in the development of the modern airliner, if only by reason of the special attention given to crew comfort and navigational aids.

Thus, this particular Sopwith might best be dismissed with these thoughts: that it was actually called a '3-strutter', as distinct from a '$1\frac{1}{2}$'; that it had a special sighting panel in the floor; and—a fact not hitherto recorded—that smoke-bombs were actually dropped from the machine while in England, with the peaceful intention of 'observing the direction of drift'. Kinship with the B.1 was closer than might have been supposed . . .

While cogitating on 'swords into ploughshares' it must finally be added that the true Sopwith 'Transport' was intended to have carried either of these loads: five passengers and one pilot (four passengers inside and one in the cockpit with the pilot); or 1,500 lb (680 kg) of cargo—still with 'pilot and passenger'. Cruising at 90 mph (145 km/h) the five-passenger machine would have a six-hour endurance, while the corresponding figure for the freighter was quoted as eight hours.

Wallaby

(Rolls-Royce Eagle VIII)

Span 46 ft 6 in (14.1 m); length 31 ft 6 in (9.6 m); wing area 547 sq ft (50.8 sq m). Empty weight 2,780 lb (1,260 kg); maximum weight 5,200 lb (2,359 kg). Maximum speed 115 mph (185 km/h); cruising speed 107 mph (172 km/h).

N.B. Reference having been made in the text to the greater strength of the Wallaby as compared with the Atlantic, it may be noted that a factor of safety of 6.5 was quoted. The 'slightly heavier' construction is not reflected in the respective figures for empty weight; but for this there could be several explanations.

Though a direct development, or adaptation, of the Pup, the Dove had swept-back wings, as this rare picture illustrates.

Dove

When, as earlier recorded, Sir Thomas Sopwith mentioned a stalling speed of about 50 mph (80 km/h) for the Churchill, one noted that this figure seemed to be 'a shade high', the reason being that one carried in one's mind the speed range that the makers later quoted for the roughly comparable Dove, a little civil aeroplane of many distinctions. The quoted figures were, in fact, 35–100 mph (56–160 km/h)—with a Le Rhône engine of 80 hp only and when seating two occupants.

One might be forgiven, in this regard, for sensing a margin of give or take at either end of the scale; yet even if some such allowance is made (and likewise for the first flush of post-Armistice salesmanship) the figures represent an impressive achievement—especially so as the Dove was not, in the fullest sense, a new type, being little more than an adaptation of the 1916 Pup.

'Sporting two-seater' was the description by which this tiny tourer/trainer of 1919 was publicised—the 'trainer' element, attributed by the present writer, being implicit in the claim that, when fitted with dual control, the Dove was 'particularly suitable for instructing pilots, who have passed their preliminary tests, before placing them on high-powered single-seaters'. Powerful single-seaters being in 1919, as now, mainly of a military persuasion, the 'sporting' aspect must be viewed accordingly, and the very low stalling speed quoted must not be construed as a 'pussyfoot' approach to the postwar market. This Dove of peace had an ancestry of war-hawks, and its 'stunting' capability was one of its hoped-for selling points, in which connection a factor of safety of 6 was claimed.

Physical aspects of design-appeal were summarised as follows when the Dove was shown at Olympia in 1920 (though built, as mentioned, in 1919): 'Petrol is fed by gravity, thus eliminating all pressure troubles. Both pilot and passenger are comfortably situated, possessing very good visibility. The low landing speed, together with the extra strong landing chassis, enables it to alight on or get away from very small fields. Fuel is carried for a period of 2½ hours at a cruising speed of 85 m.p.h., and an adjustable tail plane allows of the machine being flown with the maximum comfort.'

Not without reason was the Dove to be assessed at a later period as 'the D.H. Moth five or six years ahead of its time'—though few private-owners from the mid-1920s forward would have elected to operate a rotary engine in preference to a stationary Cirrus, Gipsy or Genet. That the rotary had quickly received the 'thumbs down' as a prospective powerplant for the private or public operator was implicit, one may judge, in two early comments by *Flight*, the first relating to the Dove, now under review, and the second to the Gnu (see next chapter). Of the Pup—the Dove's progenitor—the journal declared: 'In its time it was a very efficient machine, and even to this day *except for a rather extravagant rotary engine* [present writer's italics] does not make a bad sporting machine.' As for the Gnu, when G-EAGP, later named as 'perhaps the best-known' example of that aeroplane, participated in the 1922 King's Cup Race, a quite harrowing account was rendered of how the rotary engine refused to start, culminating with this comfortless assurance: 'However, Longton [the pilot] keeps smiling, remarking that it is a long race, and that much may happen before the finish. Capt. Cockerell volunteers for a spell of prop. swinging, and the engine begins to fire, spitting flames out of its exhaust ports and setting some petrol on the ground on fire. This necessitates wheeling the Gnu back a few paces to avoid fire on board.'

Unfortunately for the rotary-engined Dove and Gnu, aviation was not—even in the lean postwar years—trying to go 'back a few paces'.

The 'ancestry of war-hawks' earlier ascribed to the Dove is most firmly established by naming men before aeroplanes, and in particular Maj W. G. Barker VC, exponent of the Snipe; for it was Barker who, on 10 May, 1919 (with Harry Hawker away in Newfoundland) ensured a truly royal debut for Sopwith's latest product. Barker, though still convalescent from his mighty battle of October 1918,

Perhaps the most famous event in the Dove's career was the trip made in the example shown by HRH the Prince of Wales—with Maj Barker, VC, of Snipe associations, as his still-convalescent pilot. On that occasion T. O. M. Sopwith was present, though as he wore a bowler hat he is not identifiable here.

not only flew the Dove to Hounslow in the following May, but carried as a passenger on an aerobatic session the still-youthful Prince of Wales (later Duke of Windsor)—much to the displeasure of the Prince's father 'KG5'. Thus, whatever peace the Dove came bearing on its little back-swept wings, that peace was not borne on to Buckingham Palace.

The wings of the Dove were, in fact, the salient visual difference between it and the Pup, and some of the old hands were quick to make comparisons, one of these observing: 'The Dove is a two-seater version of the famous Pup, one of the most delightful of Service machines as far as ease of flying is concerned. Generally speaking, all dimensions are practically the same. In the Dove the span is cut down by 1½ ft. and the overall length is increased by a few inches.' More explicitly, this gentleman might have remarked, the Dove was a little longer than the Pup because it had two seats (in tandem) instead of one, and its span was less because the wings were swept back to get the centre of lift further aft accordingly. That a single-seat version—still with swept-back wings, in one form at least—nevertheless existed is proved by a photograph; and that this version differed otherwise from standard is apparent from what appears to be a wind-driven pump or generator (as on the contemporary Gnu) visible between the V-struts of the landing gear, and possibly attached to the port bottom longeron. Other seeming peculiarities—in the engine installation, for example—may be more apparent than real, though one clear possibility is the fitting of extra petrol tankage instead of the second seat. No markings, apart from the makers' name and address on the fin, appear, and there is no reason to suppose that this machine was G-EBKY, later in life to be converted to a semblance of a Pup and famous to this day as the 'Shuttleworth Pup'. G-EBKY was, in fact, the last of ten Doves to appear on the British Civil Register, and the intimate details of its history are related at length by Harald Penrose in *British Aviation: The Adventuring Years.*

Self-explanatory is the uncommonly brief Sopwith caption to this rare and lively study of a Dove: 'S.944—"Dove" 1 Seater'. Its Pup ancestry notwithstanding, the specimen shown could almost be mistaken—from this particular aspect—for a Sopwith Camel.

Concerning the 'extra tankage' possibility just mentioned it is well worth recording another Penrose anecdote, thus (in writing about Bert Hinkler): 'Italy was well known to Bert, for in 1917 he had flown as Peter Legh's observer, before becoming a Camel pilot there, and the story was told that his skill as an engine tuner, coupled with his very light weight, enabled him to keep formation with pilots using 130 h.p. Clergets though his own had merely a 110 h.p. Le Rhône. His great ambition was to fly to Australia, for which Tom Sopwith loaned a Dove, but the start was so long delayed by lengthy arguments of the International Air Convention that by the time permissions were received the Dove was rescinded for an export order.'

To the foregoing, the present writer would only add that Bert Hinkler (full Christian names Herbert John Louis) had worked for Sopwith before he joined the RNAS—as a mechanic, the capacity in which he had served back at the old Kingston Skating Rink.

The Dove used by Barker to fly the Prince of Wales at Hounslow was G-EACM (ex K-122), though for the Royal jaunt its fuselage carried only the name 'Dove', in small capital letters, following a then-established Sopwith practice of distinguishing the flow of new types introduced during 1918/19. This same machine appears to have gone to Bishop-Barker Aeroplanes Ltd, of Toronto, in 1920, as G-CAAY. Then came G-EACU (ex K-133), registered to Maj Olof Enderlein of the Royal Swedish Air Force in 1923 as S-AFAA; G-EAFI (ex K-148) sold in Norway during 1921; G-EAGA (ex K-157) sold somewhere overseas in September 1919—as was G-EAHP (ex K-168); G-EAJI, likewise sold abroad, apparently as G-AUDN in Australia; G-EAJJ, which went to Larkin Supply Co, Melbourne, in March 1920 as G-AUJJ and was scrapped in June 1925; G-EAKH, likewise to Larkin, though as G-AUKH and in existence until June 1928; G-EAKT, apparently crashed before re-registration as G-AUDP (being destined for Tasmania); and—making a total of ten Doves in all—G-EBKY—already named as the 'Shuttleworth Pup', though equally of interest in having been fitted during 1925 with a horn-balanced rudder; in having once belonged to D. L. Hollis Williams, one of Britain's more eminent aircraft designers—and then to C. H. Lowe-Wylde, a man who might well have achieved a similar eminence in the light-aircraft movement had he not been killed in 1933, having founded the British Aircraft Company in 1930.

Yet—such were the early post-Armistice years, and the destinies that ruled them—that the ten dainty Doves fluttered vainly out from the Sopwith 'ark', which the famous 'zoo' had now become.

Dove

(80 hp Le Rhône)

Span 24 ft 9½ in (7.5 m); length 19 ft 6 in (5.8 m). Weights—see footnote. Maximum speed 100 mph (160 km/h); climb to 5,000 ft (1,525 m) 7.5 min; range 250 miles (400 km).

N.B. In 1919 the maximum weight was quoted as 1,430 lb (650 kg), the 'normal load' then being given as 365 lb (166 kg) and the 'maximum safe load' as 665 lb (302 kg). The maximum weight later certificated, however, was not 1,430 lb (650 kg) but 1,350 lb (612 kg).

Perhaps the finest and most detailed study of a Gnu extant, this starboard aspect of the first machine was clearly recorded by the company camera on the same occasion as the rear view also reproduced. The Sopwith caption to this picture reads: 'S.868—Sopwith Gnu—3 Seater. 200 hp B.R.2—March 31/1919'.

Gnu

In several respects the Gnu was the most interesting and promising of the three civilian offerings that were introduced by Sopwith to the public around mid-1919 (the other two being the Dove and the 'Transport'—the last-named type or class being represented by the Atlantic and the Wallaby). As early as 29 May, 1919, the first Gnu (K-101) was flown by Harry Hawker—now safely back from his transatlantic venture—from Brooklands to Hendon, the occasion being the subsequent reception accorded to the US Navy-Curtiss flying-boat crewmen (notably their leader, Lieut Cdr A. C. Read) who had just completed the first Atlantic crossing by air, by way of the Azores and Lisbon. This first Gnu was, in fact, the second British-registered civil aeroplane, the first having been K-100, a D.H.6 owned by Airco though still bearing RAF markings.

Sopwith's re-entry into the civil passenger-carrying business was marked with due ceremony and a sense of public relations, no less than sixty guineas being paid by Miss Daisy King of Leeds, not so much (it may be supposed) for the privilege of trying the new aeroplane as being flown by the heroic Hawker, though Harry had brought his wife with him in the Gnu from Brooklands. Among those present at the reception for the Americans was T. O. M. Sopwith, and the ticket for the flight in the Gnu was auctioned by his old rival Claude Grahame-White. It was just like old times; for not only had 'G-W' competed with 'Tom' in early sporting events but was an aircraft constructor in his own right, having built his first machine in 1910.

Moreover, just as he had championed the aeroplane for war, so, now, was he seeking to promote it as a public-transport vehicle. (All this, of course, quite apart from the fact that for so many people Grahame-White *was* 'Hendon'—in its best-known and best-loved sense).

As for the Sopwith Gnu that was central to these postwar civil promotions, this can best be introduced by turning back the pages of history—as represented by those of this present book—to reconsider the Three-seater of 1913, and to reiterate a view thus expressed: '. . . the present writer would go so far as to proclaim the early-1913 Sopwith "land tractor" (as it was sometimes called) as the true begetter of a line of British transport biplanes built with particular success to a particular formula: that is, with the passengers seated in a forward, fenestrated, compartment, with the pilot behind them, and having no more engine power than was strictly necessary to perform (jointly with generous wing area) a rigorous operation with exemplary economy.' In this regard one explained that the line of aeroplanes one had in mind came to its full fruition 'in the well-nigh incomparable de Havilland series of the inter-war years, culminating in the little Fox Moth.'

Yet instantly, now, there seems to be a basic contradiction to this view, for a prominent design-feature of the Sopwith Gnu was the seating of the pilot not behind his passengers but ahead of them. This seeming contradiction, however, is quickly explained—first by remarking that de Havilland, in some of their civil designs, adopted this same (Gnu) formula; second by observing that the seating arrangement was of psychological, rather than technical, importance, prime considerations being the pilot's view of where he was flying and (in those times) the passengers' view of the pilot. In any case, the Gnu's passengers numbered two only, as in the 1913 tractor already mentioned.

As for economy, expressed in terms of engine power, Gnu K-101 was accompanied to Hendon by another specimen, K-136, the second of twelve that were built in all, and having a 110 hp Le Rhône engine, whereas K-101 had a 200 hp Bentley B.R.2. (In this last regard, it may be noted, the quoted figure of 200 hp instead of the familiar military rating, or nominal output, of 230 hp, may have signified a civil rating, though Sopwith sometimes gave even the Snipe '200 hp').

That many modifications were made to Gnus is clear, though the Australian installation of a Wright Whirlwind engine is touched on only at the very end of this chapter. Another Australian 'mod' was the fitting of an 18-gal (82 litre) petrol tank under the centre section, and thus above the pilot's seat. With a Gnu so fitted, from a little clearing on a bend of the River Murray, F. S. Briggs essayed a take-off in this 'heavily loaded machine of small horse-power' (as he described it). Briggs climbed as steeply as he dared, but one cylinder of the 110 hp Le Rhône cut out—and a 60-ft fall into a gum tree Briggs considered 'fortunate'. Even so, as the Gnu stalled, he put his right hand over the side to switch off the engine—only to get his hand trapped, and a finger fractured, by the three-ply side of the fuselage. Into the bargain, he got an involuntary bath in 18 gallons of petrol. No fire, thank Heaven; but the petrol stung 'like Hades'.

Hopefully, the Gnu 'limousine' (as it was sometimes styled) was orginally launched much along the lines set out in the following contemporary description, based on one prepared by Sopwith themselves:

'The "Gnu" has been designed to meet the requirements of a light, high-speed machine for passengers or cargo. It can be equipped either with the 200 Bentley rotary or with the 110 h.p. Le Rhône—both engines having proved extremely reliable upon active service. Accommodation is provided for two passengers, or the equivalent in cargo, who are totally enclosed in a roofed and windowed cabin. The pilot is placed well forward in front of the cabin, and has very good visibility, being well protected from the "slip stream", enabling him to fly long distances without suffering discomfort. The "Gnu" possesses a speed variation of 100 per cent. [*Sic*—presumably meaning that the speed range was about 2 to 1] pulling up when landing and taking off [curious, though comprehensible, phrasing]—owing to its light weight—very quickly. With the 200 h.p. Bentley rotary, fuel is provided for a range of 250 miles, whilst in the case of the Le Rhône engined machine, this distance becomes 300 miles. The engine unit is extremely accessible, and in the event of necessity can be changed by two mechanics in five hours. An adjustable tail plane is fitted, enabling the pilot to trim the machine to suit the particular load that is being carried at the moment. The construction is on perfectly normal lines.'

The Sopwith number on this rear view of the first Gnu is S.869, and the date (31 March, 1919) suggests that the machine had just arrived at Brooklands.

To this account it may be added that the two passengers were seated side by side under a glazed roof which hinged outwards from the centre in two sections, though from K-140 onwards most of the Gnus dispensed with this elaboration, and the passenger compartment was open. That the hinged roof was in any case a concession to postwar 'refinement' rather than actual demand may have been implicit in *The Aeroplane's* remark that the Gnu should appeal to those who desired 'to travel by air in comfort, relative silence, and the absence of wind' quite as strongly as would the Dove to those 'who rather prefer to experience even the minor discomforts of flying rather than forgo any of its sensations.' As in the case of the Dove, incidentally, a factor of safety of 6 was quoted—so, the span and length corresponding roughly with the dimensions of the Hippo fighter—many of the Gnu's passengers may actively have sought the 'sensations' of flying. Certainly, the 'joyriders' of those times (who constituted a large proportion of the people carried by air) could have been impressed—according to temperament—either by cribbed, cabined confinement, or by the breezy environment wherein, perhaps, their friends or relations had lately fought in Sopwith aeroplanes. In this connection it may be recorded that, after its Hendon debut, K-101 was flown by C. D. Barnard to give pleasure flights at Southport, Lancs, where it was quickly crashed—on 10 June, 1919. (Barnard had served with the RFC, was a Sopwith test pilot during 1919, was shortly thereafter with de Havilland, became famous for long-distance flights, operated an air circus, and was eventually engaged in experimental flying with Flight Refuelling Ltd. In associating his name with the Sopwith Gnu and what has just been said concerning it, one may further emphasise the closeness of the links remaining after 1918 between Service and civilian flying by noting that in 1921 a former RNAS/RAF pilot, Hubert Broad, joined de Havilland to serve with Barnard and others—arriving in his own Sopwith Camel G-EAWN!).

As for the remaining Gnus, the outline of their story was as follows: K-136 (Hawker's escort to Hendon in May 1919) became G-EADB, was later variously owned, and crashed in March 1926; K-140 became G-EAEP, remained unsold, and was dismantled at Brooklands in July 1920; K-156 (G-EAFR) unsold, withdrawn from use October 1920; K-163 (G-EAGP) various owners, notably pioneer Lieut Col Sir Francis McClean—whose magnanimity had permitted the early start of British Naval aviation—crashed May 1926; K-164 (G-EAGQ) unsold, dismantled August 1920; K-169 (G-EAHQ) to Larkin-Sopwith Aviation Co, Melbourne, as G-AUBX, registration cancelled (after use by Fulham Air Transport, Melbourne) March 1922; G-EAIL to Larkin-Sopwith as G-AUBY, privately used, destroyed in freak storm April 1946; G-EAIM to Larkin-Sopwith (believed as spares); G-EAME, 'MF, 'MG and 'MH unsold, registrations cancelled September 1921.

Perhaps the best-known of the Gnus was G-EAGP, for apart from being owned by the pioneer already named (McClean, who was also Chairman of the Royal Aero Club) it won the first race for the Grosvenor Challenge Cup—competed for annually, under Royal Aero Club conditions—on 23 June, 1923. The pilot was Flt Lieut W. H. Longton, the engine a 110 hp Le Rhône (the rules admitting only British aircraft, with engines not exceeding 150 hp), and by flying the 404-mile (650 km) course at an average speed of 87.6 mph (141 km/h) Longton showed what these early British transports could accomplish on low power.

Too low, perhaps, the power. Too late, assuredly, the date.

Gnu

(200 hp Bentley B.R.2 or 110 hp Le Rhône)

Span 38 ft 1 in (11.6 m); length 25 ft 10 in (7.8 m); wing area 354 sq ft (32.9 sq m). Maximum weight 2,400/2,500 lb (1,090/1,134 kg). Maximum speed (Bentley) 110 mph (177 km/h); maximum speed (Le Rhône) 93 mph (150 km/h); landing speed 40 mph (64 km/h); climb to 5,000 ft (1,525 m) (Bentley) 5.5 min, (Le Rhône) 7.75 min; range (Bentley) 250 miles (402 km), (Le Rhône) 300 miles (483 km).

N.B. Although a maximum weight of 3,350 lb (1,520 kg) has sometimes been ascribed, this is greatly at variance with the figures quoted above, and is certainly suspect, except as a special case—even when allowance is made for the fact that in 1919 a 'normal load' of 845 lb (383 kg) was quoted jointly with a 'maximum safe load' of 1,202 lb (544 kg). Nor, apparently, can the higher weight of 3,350 lb be linked with the fitting of a Wright Whirlwind J-5 radial engine to G-AUBY in Australia—even if the equally suspect power of 300 hp is credited to that particular J-5, as it sometimes is. (A more typical output for an engine of this series would be 220 hp, while the engine weight, at about 500 lb (227 kg), would likewise approximate to that of a standard 230 hp Bentley B.R.2).

The Antelope with its original ailerons, but with its four-wheeled landing gear. The cabin door was a notable feature, and its handle shows clearly.

Antelope

Such was the general nature of the Antelope—with two passengers in a cabin behind the pilot, and an engine of about 200 hp—that the question might well be asked: 'If Sopwith had the Gnu, why should they build this too?'

The simple answer is that the Antelope was a near-relation of the Wallaby, designed to use a war-surplus water-cooled engine (instead of the air-cooled rotary round which the Gnu design took shape)—partly, at least, with a view to obtaining lower petrol consumption, allied with lower maintenance costs. Like the Wallaby (which it ultimately joined in Australia) the Antelope was strictly a 'one-off', and though lighter and less powerful than the Wallaby, it latterly had a heavier look about it because of its four-wheeled landing gear—an impression that was little mitigated by slight stagger on the wings, for these were very much greater in area than those of the comparable Gnu.

These seemingly prosaic facts notwithstanding, it may be judged from what follows that the Antelope was quite a notable aeroplane in the development of civil flying—as, indeed, it had to be to gain the second prize in the Air Ministry Small Commercial Aeroplane Competition in 1920.

The one-and-only Antelope was built by Sopwith in 1919, with a normal V-strut landing gear, and a maker's description of it in the summer of 1920 was in these terms; 'The "Antelope" is intended to serve the purpose of a utility machine, characterised by the highest possible performance compatible with great structural strength and having a wide speed range—38 to 100 mph [61–161 km/h]. Accommodation is provided for pilot and two passengers, the former being located high up between the main planes, whilst the latter are enclosed in a comfortable cabin of 50 cubic ft [1.4 cu m] capacity, aft of the planes. A door in the side of the cabin enables the passengers to enter straight from the ground. Triplex windows in the cabin provide a good field of view, whilst one of the passenger's seats is adjustable so that, on sliding open a door in the roof, the passenger may sit in the open if desired. The engine, a 180 h.p. Hispano-Suiza "Viper", is enclosed by a quickly detachable cowling, giving extreme accessibility, and is fitted with a Black and Manson self-starter, operated from the pilot's cockpit. A fireproof bulkhead is interposed between the fuel tanks and engine. There are no welded joints in the machine.' It was otherwise explained that, although in one form the Antelope had the simple V-type landing gear already mentioned, lugs were provided for attaching, if desired, 'a pair of front wheels which will protect the propeller and prevent the machine from nosing over on landing.' And finally: 'A steel tube steerable tail skid is provided, and the opening in the floor of the body through which the skid passes has a flexible cover of oilcloth which prevents dirt thrown up by the skid from getting inside the fuselage.'

Trivial or obvious as some of these points may seem today, some, nevertheless, should have a particular significance for students of airliner design, especially respecting passenger-convenience; for to be able to enter a (reasonably) comfortable cabin, having *vis-a-vis* seating as the car enthusiasts acclaimed it, 'straight from the ground' was assuredly a boon to ladies in hobble skirts. The adjustable seat mentioned was the rear (forward-facing) one, and the occupant was raised by sitting on the hinged back-rest which could be folded forwards on to the arm-rests of the wicker seat. The passenger's head then protruded through an apparently conventional cockpit opening, with vee-shaped windscreen, the so-called 'door' in the roof being, more descriptively, a panel, shaped to the turtle-decking of the fuselage (*i.e.*, the cabin roof) and sliding in side-channels running forward of the opening.

Especially interesting is this preoccupation with enabling one of the passengers at least to 'sit in the open if desired', the primitive joys whereof were a legacy from war. What passenger today could expect such individual attention? In any case, there existed, at about the time of the Antelope's design, a lack of unanimity about

Although in this view the Antelope has the revised (inwardly tapering) ailerons, the struts betokening a four-wheeled landing gear are absent. Clearly seen is the sliding roof-hatch for the second passenger.

enclosing passengers, just as there was in later years about enclosing fighter pilots—and even the big Vickers Vimy was offered with row after row of open passenger-cockpits, in a bomber-style fuselage, as an alternative to the bulbous, enclosed, and finally accepted, 'Commercial'.

As for the Antelope's engine installation, the 'quickly detachable cowling', mentioned as another special feature of the design, was, in reality, even more interesting than it sounded, accessibility being exceptional by virtue of hingeing the two heavily louvred aluminium side-panels along the bottom longerons, allowing them to be folded down after a few bonnet-fasteners had been undone (thus following automobile practice, as designers of those times were much inclined to do).

Certainly the Antelope made a good impression when shown at Olympia in July 1920, and a little later in that year its technical merits were more openly manifest when, with a new landing gear, and ailerons reduced in chord by being tapered inwards (though still horn-balanced as formerly) it participated in the Air Ministry Small Commercial Aeroplane Competition conducted at Martlesham Heath. The engine was then declared as a 200 hp Wolseley Hispano Viper, giving an 'actual output' of 210 hp at 2,100 rpm; and the aircraft and the occasion were thus discussed by *Flight*:

'The machine entered by the Sopwith Aviation and Engineering Co. [note the correct company-name then used—though the familiar 'Sopwith Aviation Co.' was still in common currency] is the "Antelope" exhibited at Olympia. A few minor alterations have, we understand, been made to various parts, but the machine is essentially as shown at Olympia. One difference will be noted, however, in the undercarriage. This is of the four-wheeled type, an extra pair of wheels having been fitted since the show. It may be remembered that one of the tests to be made at Martlesham consists in landing over obstacles 50 ft. above the ground and coming to rest inside a circle marked out on the ground. As side-slip landings are not permitted, and the machine must be brought to rest after the shortest possible run, special arrangements have been made on several of the machines entered for pulling-up quickly, and the extra pair of wheels on the "Antelope" may be expected to form part of such a scheme. What the nature of the arrangement is on this particular machine cannot be stated at the moment.'

The 'nature of the arrangement' concerning which such reticence was observed was the fitting of brakes—not quite so remarkable as might be supposed, for four-wheeled Voisins had brakes in earlier years, and the sensational Sopwith-built A.B.C. 400 cc motor-cycle (engined by Granville Bradshaw) had internal-expanding drum brakes on both wheels. During the course of the Martlesham competitions Hawker came gliding in with brakes already applied, thus bursting both main tyres and one of the smaller forward ones. (An alternative explanation was that the brakes were applied too quickly after a heavy landing). In any case, this occurrence stopped the Antelope within the stipulated distance, though this unmatched achievement failed to secure official recognition as the aeroplane was not intact. Nevertheless, the Antelope was awarded the second prize of £3,000. The winner was the Westland Limousine.

Yet this was not the Antelope's finish, for with F. P. Raynham as pilot it won the Surrey Open Handicap Race at Croydon in June 1922, and—re-engined (apparently temporarily) with a Siddeley Puma and re-registered as G-AUSS*—was transferred by the makers to the Larkin-Sopwith Aviation Company in

* The British registration was G-EASS

Australia during April 1923. This company later became the Larkin Aircraft Supply Co, Ltd, and went into liquidation in 1934. Two entries in the 1923 diary of the Australian pioneer F. S. Briggs have a particular interest respecting the Antelope and its more powerful near-relation, thus: '*September 1st.* Arrived Hay today, ferried a Sopwith Antelope 160 h.p. Wolseley Viper [*sic*]. *Adelaide September 7th.* Arrived here today, ferried Sopwith Wallaby from Hay so that it will be in position to fly first air-mail out of Adelaide.'

For good measure, Briggs' entry for 19 May read: 'Ferried a Sopwith Gnu . . .'—and also in his log book was Dove G-AUJJ!

The powerplant change mentioned—from a Viper to a Puma—may have been less drastic than might be supposed, having a parallel, for instance, in the Bristol Fighter.

Antelope

(200 hp Wolseley Viper)

Span 46 ft 6 in (14.1 m); length 31 ft (9.4 m); wing area 531 sq ft (49.3 sq m). Empty weight 2,387 lb (1,083 kg); maximum weight 3,450 lb (1,565 kg). Maximum speed 110.5 mph (177 km/h); cruising speed 84 mph (135 km/h); climb to 5,000 ft (1,525 m) 7.5 min; approximate range 450 miles (724 km).

Schneider and Rainbow

Because the Schneider Trophy had been won in 1914 by Great Britain (with a Sopwith Tabloid on floats, as recorded in the 'Tabloid' chapter) the contest of 1919 was to be flown in the British Isles. Here indeed—at Bournemouth, Hants, in September of that year—was an opportunity to show the world how British aircraft-designers and constructors had progressed in the years between, and, in particular, how advantage could be taken of the latest aero-engines (specifically, the water-cooled Napier Lion and the air-cooled Cosmos Jupiter). Though France and Italy were both well represented, it will suffice here to record that the British entries were a short-span Fairey III (Lion), Supermarine Sea Lion I (Lion), Avro 539A (Puma)—and the superb Sopwith creation which now concerns us. For this new floatplane the name Schneider had, understandably, been adopted (or re-adopted, having regard to the earlier chapter headed 'Schneider and Baby') though the name was now commonly used with contextual reference to its Cosmos Jupiter engine.

By way of preface to this present sporting excursion nothing could surpass the following brief excerpts from Peter Lewis' *British Racing and Record-Breaking Aircraft*:

'Misfortune struck the British team when the Sea Lion, taking-off again after having alighted for safety in the fog, was damaged on striking an obstruction, and the Sopwith Schneider retired as the fog enveloped the course. Subsequently, the contest was annulled because of the confusion caused by the fog . . .

'With a maximum speed of 160 mph Hawker's Sopwith Schneider was the fastest of the 1919 entries, and its Jupiter was faired carefully into the fuselage. The single-bay 24 ft span wings were rigged with slight negative stagger, and the fuselage terminated at the rear to form the lower part of the rudder and to act also as the tail float. The pair of main floats were designed with flat bottoms, and the general design of the Sopwith Schneider was drawn up under the direction of W. G. Carter. R. J. Mitchell was responsible for the Supermarine design . . .'

A singularly fine study of the Jupiter-engined Schneider—especially valuable because the engine installation may be directly compared with that of the A.B.C. Dragonfly in the Rainbow, shown in close-up on a later page. The Sopwith caption to this picture reads: 'S.1055—1919 Schneider Cup Seaplane. 450 hp Cosmos Jupiter engine—Aug. 29/1919'.

To expatiate on the significance of the names Carter and Mitchell would affront the eyes of any present reader—except, perhaps, in one regard, and that by way of stressing that George Carter (as he was always known to his intimates) was in later years to design Britain's first jet-propelled aircraft, the Gloster E.28/39, and that the Sopwith racer now surveyed was his first really fast aircraft. When Harry Hawker first tried it on 10 September, 1919, however (after its assembly at Hythe near Southampton) the machine had to be beached because the floats were too far aft. The Sopwith Experimental Shop rectified matters in two days. Shades of Monaco and 1914 . . .

This dramatic re-entry of the name Sopwith Schneider into the aeronautical vocabulary (the rising, in fact, of a new star in the Sopwith firmament—a metaphor warranted by the form of engine, if by nothing else) drew from technical observers laudations, recollections and prognostications in bewildering profusion, and one in particular that may strike the reader as the acme in pontifical pronouncements, though all may be forgiven when the perpetrator is later identified. This pronouncement—which, the purple patch aside, gives a useful appraisal—was as follows (present author's italics):

'The Sopwith Schneider Cup racer shows a number of interesting and more or less typically Sopwith characteristics. The very small span, relatively large chord, and the small gap are very reminiscent of the original Sopwith Schneider Cup machine of 1914. *If memory serves one right*, the chord and gap, 5 ft. 1½ in. and 4 ft. 6 in. respectively, were the same in the 1914 model as in that of today. The span of the new machine is some 3 ft less, and there is a small backward stagger.

'The floats and chassis are not unlike those of the original, though the section of the front part of the floats, with an outwardly flaring bottom—a sort of "Vee" bottom halved—is new to this firm.

'The main floats have also been extended aft and faired off considerably and a tail float dispensed with. The big 450 h.p. radial engine has naturally called for a big bull-nosed body—even more bull-nosed than the original—and the body has been faired off to a rounded section instead of having been left square. The tail fin and tail plane are likewise faired into the body with large radii, and altogether the machine has been carefully studied with a view to producing the minimum of head resistance.

'With a power loading of slightly under 5 lb per h.p. and with the Sopwith firm's unrivalled experience in the production of small fast machines, this machine should prove to be one of the fastest aeroplanes yet built, and Messrs. Sopwith should stand an excellent chance of winning the Cup for the second time.'

Now to identify the perpetrator of the seemingly immemorial 'if memory serves one right, the chord and gap were 5 ft. 1½ in. and 4 ft. 6 in. respectively' (and not forgetting that ½ in); for here one must admit that, of all men, this particular one was so positioned and equipped as to throw off with sincerity this classic of minutiae. He was, in fact, W. H. Sayers, who was eminent in the Experimental Construction Department of the RNAS, designer of (for instance) the Grain Kitten light ship's fighter, and technical editor of *The Aeroplane* not only in 1914, but from 1919 to 1928 as well. Thus, of all men, Sayers should have an intimate knowledge of Sopwith seaplanes old and new.

Yet, in common with other observers, even Sayers failed to stress the most significant fact of all—that here was an aeroplane which (whatever resemblance it may have borne to Pixton's tiny 100-hp Monaco mount of 1914) was designed to use a brand-new British radial engine more than four times as powerful (450 hp nominal), the successors to which were to establish the name 'Bristol Jupiter' in

Taken on the same occasion as the preceding view, this dead-rear study of the 1919 Schneider shows in particular the thick hollow bottom portion of the rudder, which displaced the tail float. Except that the Sopwith number is S.1060, the maker's caption is identical with that for the ¾-front view.

almost every country of the world—to set the pace indeed even for Wright and Pratt & Whitney in the USA.

Very clearly—even if certain dimensions did recall old practice—the new Schneider racer represented not only a substantial aerodynamic advance (for example, in the smooth, hollow portion of the rudder displacing the earlier tail float) but a major step also in the stresswork that was implicit in the adoption of a new and very powerful engine. The installation of that engine, moreover, presented problems of its own; and the manner in which that newly risen 'star' (or newly blossomed 'sunflower') the Cosmos Jupiter radial, with its relatively large diameter of about 52 in (1,320 mm) by reason of its nine big air-cooled cylinders, was blended with the airframe is attested by superbly detailed photographs.

That the world-renowned name 'Jupiter' was in 1919 still prefixed by the company name 'Cosmos' and not by the 'Bristol' of later years (though indisputably the names of the engine's designers were Roy Fedden and L. F. G. Butler) is explained by the fact that the Cosmos Engineering Company did not go into liquidation until February 1920. Technically, it is important to note that the Jupiter's cubic capacity was roughly 25 per cent greater than that of the A.B.C Dragonfly, though behind each projecting cylinder was a cone-shaped fairing. The petrol tanks, four in number, were in the top and side fairings of the fuselage, and the circular cross-section occasioned by the engine and the tanks was carried gradually into the flat sides of the rear fuselage by stringers. Though of the usual wooden cross-braced construction, the fuselage was covered at the front with aluminium and at the rear with fabric.

Bracing of the stubby single-bay wings was by streamline wires, and there were ailerons on all four panels, following standard Sopwith practice; so in respect of airframe design interest was centred on the unusual tail assembly. Still italicising 'aileron' and 'fuselage' (though 1919 was already in its autumn) *Flight* had this to say: 'The extreme front portion of the fin is built integral with the *fuselage*. The rest of the fin, although curving gradually into the top of the body, is a separate structure, bolted on after the tail plane is in place. The rudder is unusual, inasmuch as its lower portion is very thick, forming a continuation of the *fuselage*. The latter does not come to a sharp edge at the rear, but is some 6 to 8 ins. wide at the stern post. The leading edge of the rudder is made of the same width, and is covered with plywood. There is thus no external rudder crank lever, while the single pair of levers for the elevators is housed inside the fin. The hollow lower portion of the rudder serves as a tail float.'

In this new aeroplane, beyond a doubt, Sopwith had produced a high-speed aeroplane that was worthy of their name—a name, however, that was now inscribed (as distinct from the fuselage 'hoarding' of Pixton's machine) only on the fin. Registered G-EAKI, the seaplane was otherwise bedizened as the pictures show (for the era of Civil Registration was now upon us).

At this point there is brought to the reader's attention the Sopwith Rainbow, a re-engined, renamed and reduced landplane version of G-EAKI— 'reduced' connoting here that the span was shorter by 3 ft (0.9 m). The word had a double significance, however, by reason of the re-engining, for in place of the Jupiter (which was itself in the teething phase) was one of the dreaded Dragonflies—adopted, it has sometimes been said, because the Jupiter was no longer available, though all the facts may not be known. That the Rainbow was sometimes called the Sopwith A.B.C is sure.

Harry Hawker's racing number in the 1920 Aerial Derby—for which the Rainbow, with its simple V landing gear and neatly cowled Dragonfly, had been

Far from appearing displeased with the A.B.C. Dragonfly in the Rainbow as prepared for the 1920 Aerial Derby, Harry Hawker looks positively proud. Certainly, the cowling was a splendid piece of work.

prepared—was 13; and early in its race-report *Flight* declared: '. . . another machine approached the aerodrome, and we recognised Hawker's Sopwith. To everyone's surprise, instead of making the required half-circuit of the aerodrome, Hawker flew straight in [*sic*] to the aerodrome past No. 1 pylon, but not across the line. It was obvious by his time of arrival that he would have found a place in the handicap, so there was much disappointment at his not finishing correctly. On landing he said he thought the finish was the same as last year's, hence the mistake. Well, Hawker, do not have No. 13 next time!'

One bright, if minor, sidelight on this 1920 race was the presence on the Committee of Management of Grp Capt C. R. Samson, CMG, DSO—and what this one-and-only 'Sammy' didn't know about Sopwiths old and new could hardly have filled a page of his log-book.

To *Flight*'s 'hard luck' story just recounted may be appended that journal's comments on the Rainbow itself (the usual figure of 320 hp being quoted as the Dragonfly's output). Thus: 'Owing to the impossibility of obtaining a Bristol "Jupiter", which was the power plant originally installed in the Sopwith "Schneider" machine entered for that race at Bournemouth last year, a smaller engine had to be used. The one which could be fitted with least trouble was the A.B.C. "Dragonfly", and consequently this was chosen. The lighter weight of this engine permitted of reducing the area while retaining the same landing speed, and consequently slightly smaller wings were fitted. The result was that the speed was still quite good—somewhere in the neighbourhood of 150 m.p.h. (241 km/h).' In fact, the speed was considerably higher than *Flight*'s approximation.

The Rainbow, *Flight* went on to declare, had 'vertical wings (no stagger)', whereas it may be recalled that Sayers, in *The Aeroplane*'s description of the Jupiter-engined Schneider seaplane, had correctly noted that this earlier version had 'a small backward stagger'. However, that at one stage at least, the Rainbow itself had backward stagger, seems certain—though the point is a fine one indeed, for the amount of back-stagger quoted by Sopwith for the Schneider had been a mere 2½ in (65 mm).

Such trivial considerations, in any case, seemed all the more so because they were completely overshadowed by the Sopwith company's liquidation on 11 September, 1920, and by the withdrawal of the Rainbow (which had been entered as having a Jupiter engine) from the 1920 Gordon Bennett Aviation Cup race in France. (Late in September 1920, after a lapse of seven years, the contest named had been revived. The proposed British entries—apart from those by France and the USA, the latter including the astonishing Dayton-Wright cantilever monoplane with retractable landing gear, variable-camber wing etc, etc—were the Martinsyde Semiquaver (pilot, Raynham), the Nieuport Goshawk (Tait-Cox) and the Sopwith/Hawker Rainbow (H. G. Hawker). The Goshawk and Rainbow having withdrawn, only Raynham was left for Britain—and even he was compelled to retire in the first lap, leaving Sadi Lecointe to win the cup outright, for France had gained it in three successive races).

How were the (prospective) mighty fallen in this Gordon Bennett affair of September 1920!—though this was not the end of the Rainbow, even if no crock of gold awaited it. With a Jupiter installed once again, with the airframe refurbished and the legend 'Sopwith Hawker' painted on the fin so as to follow its elegant contour, G-EAKI not only appeared at Croydon for the 1923 Aerial Derby (the last of the series) held on August Bank Holiday, but finished second—to Larry Carter in the Gloster I. The pilot was Walter Longton, and his speed was 164.02 mph (264 km/h). Although it had been planned to refit floats and to try once again for the Schneider Trophy (the race was at Cowes, Isle of Wight, in late September 1923, and was won by Lieut D. Rittenhouse, US Navy) the spinner of the Rainbow came adrift while Longton was making a final test of the aircraft as a landplane, and the forced-landing near Brooklands was such that Longton was lucky to emerge unharmed—through the side of the fuselage.

The Hawker company was to build no more out-and-out racers, though on 7 September, 1953 (almost exactly thirty years after the Rainbow's end) a special Hunter set up a new World's Absolute Speed Record of 727.6 mph (1,171 km/h).

Schneider

(Cosmos Jupiter)

Span 24 ft (7.3 m); length 21 ft 6 in (6.5 m); wing area 222 sq ft (20.6 sq m). Empty weight 1,600 lb (726 kg); maximum weight 2,200 lb (998 kg). Maximum speed 160 mph (257 km/h).

Rainbow

(A.B.C. Dragonfly)

Span 21 ft (6.4 m); length 18 ft (5.5 m). Maximum speed 160+ mph (257+ km/h).

Grasshopper

If the Grasshopper—as first appearances suggested was the case—was Sopwith's answer to the Avro 504K, then (as C. G. Grey might have put it) Sopwith were simply saying 'Yes'. If—on the other hand—the machine is passed off as a modernised landplane development of the similarly powered prewar Anzani Tractor Seaplane then the answer must be 'No'. But in any case, neither answer would suffice, for the story (though concerning only a single aeroplane—G-EAIN) may well be more complex than the superficial facts suggest.

By Sopwith themselves this aeroplane was little publicised; and though its first C of A was not issued until 22 March, 1920 (the airframe having been built during 1919) the Grasshopper was clearly not regarded—as was the Dove—as a type to be 'pushed' in the postwar civil market. Yet, like so many other Cinderellas and Ugly Ducklings, it had its points of interest. Very apparent among these was its extremely low full-load landing speed of 35 mph (56 km/h); indeed, its suitability for small-field operation, equally with its physical appearance, may well have suggested its whimsical name—likewise the description 'tourer' which the makers themselves applied.

That for several years the Grasshopper enjoyed a continuing, if solitary, popularity is attested by its successive registrations, thus: Sold December 1922 to L. C. G. M. Le Champion; May 1923 to E. A. D. Eldridge; February 1925 to John Cobb; August 1925 to Dudley Watt ('Dangerous Dan', who rebuilt it); and February 1928 to Miss C. R. Leathart. On expiry at the end of May 1929, the last C of A was not renewed, though by that time the green-painted Grasshopper had afforded many people much joy—not least when flown round Brooklands motor-racing track (by J. King in 1926) with vastly more clearance between its top wings and the Byfleet bridge than between its wheels and the concrete.

Clearly, this aeroplane was a gentle, if inelegant, creation; and its character having stemmed largely from the fitting of a radial engine (100 hp ten-cylinder air-cooled Anzani) as distinct from the rotaries which generally powered the comparable Avro 504K family—and, for that matter, Sopwiths' own delightful Dove—it seems worth considering the implications of this radial engine in the post-war private-owner market.

First must come the historical background—for the pre-1914 Anzanis were already truly historic (Blériot's 1909 cross-Channel effort being only one example of their use) and were technically very advanced. Rotaries, true, were the engines which, from 1908 onwards, enabled aviation to make the strides it did; and in the 1914—18 war especially their virtues of simplicity, lightness and compactness (not to mention availability) advanced their massive prestige. Yet once the war was over, prospective aircraft-owners—many of whom had been wartime pilots—much desired to leave behind them the concomitant associations of high consumption figures (both in petrol and oil), the inevitable gyroscopic effect produced by an entire engine that spun together with the propeller—and an inseparable grimy, clammy feeling of slung oil.

That the fixed-cylinder radial air-cooled engine was established in British service before the 1914 war is a fact too little recognised; and the Anzani was here pre-eminent. Companies which tried this engine in small aeroplanes included Blackburn, Handley Page, Vickers and Wight; and even in Felixstowe flying-boats and the towering Supermarine Nighthawk Anzanis were seen. Additionally there were the curious French Caudrons, which, with Anzani engines, served Britain well, especially for training (and which, to some extent, the Sopwith Grasshopper may originally have been intended to succeed).

Probably the finest picture of the Grasshopper ever taken—an uncaptioned Sopwith study evidently made on the same occasion as the dead-rear view that follows, with provision on the aircraft for civil markings not as yet applied. Note the staggered cylinders of the Anzani engine.

The Sopwith caption to this flattering study of the Grasshopper reads: 'S.1043—Grasshopper No.1. 100 hp Anzani—July 31/1919'. The 'No. 1' appears to have been superfluous, as only one example is known.

On the debit side for the 100 hp Anzani were its ten cylinders—double the number eventually adopted for radial rivals to the inter-war Cirrus and Gipsy ranges. Heeded, too, must be Maj F. M. Green, that leading authority on engines in general and radials in particular, who said of 'post-Blériot' engines: 'The engine [Blériot's famous 'fan'-type] was very soon followed by a large number of true radial engines made by the same firm, and others designed by Viale, who had formerly been connected with the Anzani firm. These engines were not, on the whole, successful, partly, I think, on account of a number of special difficulties connected with radial engines which had not then been solved, and also on account of the comparatively poor design of the cylinders.'

Nevertheless, in the postwar light-aircraft field the Sopwith Grasshopper helped to stake a very firm claim indeed for the radial engine—jointly, it may be added, with an Anzani-powered London and Provincial biplane, five examples of which came on to the British Civil Register.

As for the Grasshopper's airframe, there is little to add to what the pictures and the table tell, except to affirm that its structure was of wood.

Grasshopper

(100 hp Anzani)

Span 33 ft 1 in (10 m); length 23 ft 1 in (7 m). Maximum weight 1,670 lb (757 kg). Maximum speed 90 mph (145 km/h).

Apocrypha

This will be a strictly 'take it or leave it' sort of chapter—an appendage to the book rather than an appendix, comprising merely a tenuous trailer and not a tangible tract for clinical scrutiny. Some old and tender spots it can hardly fail to touch, though also possibly new areas inviting exploration.

In any case, there may be something here to afford stimulation, if not satisfaction. For which the author offers no apology, just as he will advance no claims, adduce no formal evidence—and expect no sympathy whatever should his dubious candidates for Sopwith's 'zoo' (or museum of curiosities) prove to be nothing more than chimera, and best ignored in the hope that they will go away.

Yet even such diffidence is surely more excusable than rank indifference; and a little sideshow may yet afford some entertainment, if nothing more, at this back end of the book. Even so, one must give each of these shadowy creatures a label of some kind; and thus we have:

Possible Twin-engined Gun Bus

In the chapter on 'Pushers and Gun Buses' mention is made of a seaplane known in the summer of 1913 as 'the 80-ft. span machine', and in this regard these points are stressed: (1) That Sopwith appear to have built the largest British aeroplane of its time, and (2) the company's early use of increasingly powerful engines ('in the 200 h.p.+ bracket').

By reason of this second fact—the single engine concerned being a Salmson (Canton-Unné)—no allusion is made to the apparent thought that was given to the fitting in this airframe of not one, but *two* engines, these being Austro-Daimlers ('Austrian-Daimlers' as they were termed) of 120 h.p.

But whatever the facts of this matter, it is worth remarking that at about the same time (1913) the Vickers company was busy with designs for a pusher aircraft (and actually made and tested floats for it) from which it was proposed to develop a twin-engined gun carrying machine. Had it materialised, this proposed machine would have probably been only about 10 ft less in span than the '80-ft.' Sopwith.

Whatever claim Sopwith may have had to large dimensions in early aircraft, however, a supplementary one (for twin engines) might now be cautiously advanced, though the description given of the Cobham triplane bomber in its own chapter—'the only multi-engined Sopwith to be built'—appears to stand inviolate.

The 'Possible Twin-engined Gun Bus' (as this present note has been prudently headed) was intended to carry wireless, in addition to one or more guns; and the prudence exercised with regard to its powerplant, and to the remote possibility of the aircraft itself having actually existed, is warranted by the fact that the gun-carrying seaplane No.93 appears to have had a *single* 120 hp Austro-Daimler, as noted under 'Pushers and Gun Buses'.

Further, the (nominally) 200 hp Salmson (Canton-Unné), which might conceivably have delivered 240 hp, was sometimes regarded before the 1914—18 War as being, in effect, 'two engines in one', as witness this comment of 1914: 'The fourteen-cylindered [*sic*] 200 b.h.p. design consists of two groups of seven cylinders acting upon a *one-throw* crank, the junction of the crankshaft being made in the common crank-pin between two big-end pin cages. This fourteen-cylindered engine is noteworthy as it acts upon a *six*-stroke and not upon a four-stroke cycle, the valve cams being in consequence specially arranged, and the cam-shaft driven at *one-third*, and not one-half, of the crankshaft speed. This is done in order to obtain working impulses at equal angular intervals of crankshaft revolution with two seven-cylindered groups operating all on one extended crank-pin . . .'

Small wonder that such a fascinating—as well as a powerful—piece of mechanism should appear to have captivated the Admiralty, and that official interest should have been transmitted to their new aircraft-builders, Sopwith. Therein may conceivably lie the background to this present note, though the Navy had long since had an introduction to twin-engined aeroplanes, for Cdr C. R. Samson had flown in the Short Triple Twin during 1911; thus an installation of two Austro-Daimlers in 1913 would have presented no drastic technical departure.

The 'Kitten' mentioned under the heading 'The 'Sopwith Kitten''.

The 'Sopwith Kitten'

Although Commander Harry Busteed was at one time in charge of the RNAS Experimental Flight at Eastchurch, and—in the course of his work—also had a close relationship with the Sopwith company (being, in any case, one of the three original 'Aussie Harrys' together with Hawker and Kauper) there is scant reason to suppose that the name 'Sopwith Kitten' was justly applicable to the ship's ultra-light armed scout which was otherwise called the Port Victoria P.V.8 Eastchurch Kitten. The ascription might, nevertheless, have been fostered by structural features, the most obvious being the plank-type interplane struts. (The true designer was, it seems, Capt Gilbert Henry Millar, who, after a period as a prisoner of war, escaped. Millar had yachting experience, had joined the RNVR and was transferred successively to the RNAS and the RAF. Although he became a pilot himself, he served also as an observer with the Fleet).

The photograph herewith of the charming little aeroplane just mentioned is reproduced for two reasons—apart from the discredited name 'Sopwith Kitten' and from the fact that this particular picture appears not to have been previously published: (1) The photograph was given to the present writer by Sir Frank Spriggs, who was with Sopwith from 1913 to 1920, and later became managing director of

Hawker Siddeley. Though received together with views of indubitable Sopwith types, this Kitten picture had no accompanying text. (2) This same photograph shows a tailplane of seemingly unfamiliar form, though one of another shape was apparently fitted after the aircraft had been initially flown with no horizontal fixed tail-surfaces whatever.

The 'Sopwith Mouse'

The above name was conferred by John ('Jack'—later Sir John) Alcock himself upon the single-seat 'fighting scout' built at his instigation in mid-1917. Alcock had flown the Triplane and the Camel, and in his own little two-bay machine used major Sopwith components. The front fuselage and bottom wings, for instance, were adapted from the Triplane, while the top wings were in essence those of the Pup (though with longer ailerons, and these on the top wings only—the bottom ones, of course, being much smaller). From the Camel came the horizontal tail-surfaces.

Alcock's delightful little creation had a 110 hp Clerget engine. Apparently it performed well, and was flown at Mudros after Alcock himself had been taken prisoner, the recognised designations, apart from 'Sopwith Mouse', being 'Alcock Scout' or A.1.

Later (June 1919) Capt John Alcock and Lieut Arthur Whitten Brown, in a Vickers Vimy, made the first nonstop air crossing of the Atlantic. Earlier—before the 1914 war—Alcock had helped in developing the 150 hp Sunbeam engine, using a Farman pusher biplane, and thus presaging the installation in the Robey-built Sopwith Gun Bus. He had, in fact, been engaged by Louis Coatalen, the Sunbeam engine-designer himself.

Possible 'Convert-Convert' Monoplane

A picture herewith shows part of an unidentified—and, one must affirm, probably misrepresented, monoplane (misrepresented, one remarks, because the picture is evidently a retouched photograph, with important features possibly removed or 'modified'). All that can reasonably be said of this picture now follows:

It seems to have originated in the Mediterranean area soon after the First World War, and the aircraft which it shows appears to have been adapted from one of the Sopwith Baby series—possibly a Fairey Hamble Baby or Parnall Hamble Baby Convert, aircraft of the former type at least certainly having served in the Mediterranean area. Any comparable view of a Hamble Baby Convert will show in particular the form of engine cowling—as used, it must be added, on Sopwith-built and Blackburn-built aircraft of the Clerget-engined Baby family. The metal-sheathed propeller-tips on the 'mystery monoplane' are of course, suggestive of a seaplane association, though possibly more important is that some Fairey connection might well imply the utilisation of that company's Patent Camber Gear, thus reconciling a relatively low landing speed (perhaps for aircraft-carrier operation) with a fairly high wing-loading.

Remarkably enough, the 'carrier-borne' suggestion appears to have a degree of support, however flimsy the evidence may seem, for it is simply this:

Early in 1922 that greatly respected Naval artist/historian Dr Oscar Parkes made a drawing showing an aeroplane taking-off from the aircraft-carrier HMS *Eagle* (a converted battleship, commissioned during 1922)—and the aeroplane shown most certainly resembled the monoplane depicted here in every major respect.

Artist's licence? A vision of the future? Quite possibly; but let these further points not be altogether overlooked. Dr Parkes, although a surgeon by training and by Naval experience, had also served in Naval Intelligence, and one of his hobbies was

The (apparent) monoplane which is associated in the text under the heading 'Possible "Convert-Convert" Monoplane' with the Sopwith Baby series.

the collecting of Naval photographs. (From 1919 to 1920 indeed, he was Director of the Naval Photographic Section of the Imperial War Museum). Certainly he was blessed with Admiralty guidance or inspiration in the making of many of his drawings; and thus his 'carrier-borne monoplane' commands additional attention. (For further observations on such monoplanes the reader is referred to the chapter on 'Scooter and Swallow').

There is one point yet to add. In their book *Shackleton*, published in 1957, Marjorie and James Fisher alluded to (Sir) Roderick Carr, a member of the Antarctic expedition of 1921, as having been able to work with 'Hawker's chief designer' on a 'modified single-seater Fairey-type monoplane'—though this allusion was seemingly mingled (and understandably so) with A. V. Roe & Co Ltd and their special 'Antarctic Baby'.

Yet even now one is impelled to add that Sopwith had a 'Baby' too—also that, as early as 1915, a shoulder-wing Deperdussin monoplane had taken-off from a trackway on the light cruiser HMS *Aurora*, a 30-knotter of the *Arethusa* class.

INDEX